Pharmaceutical Crystallography

A Guide to Structure and Analysis

"There is no such thing as [...] applied science; there is science and its applications, which are related to one another as the fruit is related to the tree that has borne it."

—Louis Pasteur

Few scientific disciplines know this better than the pharmaceutical sciences that are held together by the object of their studies, rather than by their methodology. Due to the overwhelming importance of the solid state in drug and dosage form research and development, crystallography is a branch of science that is of the utmost importance for students and researchers in the pharmaceutical sciences alike, who need to have a thorough understanding of this discipline rather than just to "use the instruments". The current book provides this knowledge and understanding in an exemplary way. The book is expertly written, in an easily accessible language, without compromising on the required scientific breadth and depth. This, together with the frequent use of pharmaceutical examples, make it the perfect guide to crystallography for pharmaceutical scientists and students, to arrive at truly "fruitful" results in their research on pharmaceutical crystals.

Thomas Rades
Research Chair in Pharmaceutical Design and Drug Delivery, University of Copenhagen, Denmark

Pharmaceutical Crystallography: A Guide to Structure and Analysis is a very well written textbook that provides an end-to-end overview of the process of structural analysis. It will be a fantastic reference text for many starting out in crystallography but also for those who need to reacquaint themselves with the fundamentals. This is made easy through the overall structure of the book and, in particular, the structure of the chapters. Each chapter possesses a helpful 'Key Points' and 'Case Studies' section at the end to help solidify concepts. The Case Studies are a very good addition with different examples in every chapter except for the chapters on symmetry where the author, rightfully, sticks to the same example to help cement ideas. Bond does not shy away from complex molecules as exemplars and they cover many of the different types of molecules/complexes one would observe during studies of pharmaceutical materials. The figures in the book are excellent and give clear visual aid to the topic of discussion. Overall, I think that this book will find its place on the shelves of many scientists that use crystallography no matter what stage or career path they have chosen.

Iain Oswald
Senior Lecturer, University of Strathclyde, UK

3   Symmetry in Crystals   28

3.1   Introduction   28
3.2   Point Symmetry Operations   29
3.3   Point Symmetry in Crystals   38
3.4   Space Symmetry Operations   40
3.5   Summary of Key Points   43
3.6   Case Study: Formoterol Fumarate Dihydrate   44
      Reference   47

4   Space Groups   48

4.1   Introduction   48
4.2   General Concepts of Space Groups   49
4.3   Crystal Systems   54
4.4   Centred Unit Cells   56
4.5   Space Group Symbols   61
4.6   Other Classifications of Space Groups   62
4.7   Summary of Key Points   63
4.8   Case Study: The Monoclinic Form of Paracetamol
      (Acetaminophen)   64
      References   67

5   Planes and Crystal Morphology   68

5.1   Introduction   68
5.2   Defining Planes in Crystals   69
5.3   Crystal Planes and Crystal Faces   74
5.4   Summary of Key Points   78
5.5   Case Study: BFDH Morphology of Famotidine Form A   78
      Reference   81

6   Crystal Structures and Diffraction Patterns   82

6.1   Introduction   82
6.2   X-ray Diffraction   83

Contents

1 Pharmaceutical Solids   **1**

1.1 Introduction   1
1.2 Classifying Pharmaceutical Solids: An Overview   2
1.3 Structural Classification: Crystalline and Amorphous Solids   4
1.4 Chemical Classification: Single- and Multi-component Solids   8
1.5 Classifying Pharmaceutical Solids: More Details   11
1.6 Summary of Key Points   15
References   15

2 What Defines a Crystal?   **16**

2.1 Introduction   16
2.2 Lattices   17
2.3 Unit Cells   20
2.4 Fractional Coordinates   23
2.5 Metric Properties   24
2.6 Summary of Key Points   24
2.7 Case Study: Mefenamic Acid   25
Note   27
References   27

Pharmaceutical Crystallography: A Guide to Structure and Analysis
By Andrew Bond
© Andrew Bond 2019
Published by the Royal Society of Chemistry, www.rsc.org

crystallographic content. There may be some discussion of powder X-ray diffraction, but there probably will not be anything specific about the single-crystal technique. This skews understanding because powder diffraction is really best understood as an extension of single-crystal diffraction.

My aim with this book is to "demystify" the techniques of X-ray crystallography for pharmaceutical scientists, with specific reference to relevant pharmaceutical compounds. There are some omissions, and no doubt some over-simplifications, but I have tried to consider from scratch the concepts that are actually needed to operate modern X-ray instruments and to understand crystallographic results. I have not avoided reciprocal space, partly because it enables better understanding of crystal morphology and habit, and principally because it is so important during practical analysis with a modern instrument. Hopefully the description is accessible. The intended overall message is that there is nothing complicated about crystallography, and it is absolutely possible for anyone to make their own crystallographic measurements. Of course, accessibility brings responsibility to make reliable measurements and sensible scientific interpretations. Understanding the material in this book should enable more effective and informed use of modern X-ray instruments, and hopefully will help to provide a realistic interpretation of crystallographic results in the pharmaceutical literature.

I am grateful to the teachers and co-workers that have helped me to develop my interest for crystal structures and X-ray crystallography. Principally John Davies, who taught me most of what I know about working with single-crystal instruments and has given some helpful comments on the structure and content of the book. I thank ex-colleagues at the University of Copenhagen, particularly Jukka Rantanen, who helped to convince me that this book might be helpful in the pharmaceutical sciences, and the commissioning editor at Royal Society of Chemistry, Michelle Carey, whose patience with this project surpassed all reasonable expectations.

Andrew Bond

Preface

This book is aimed at students and researchers in the pharmaceutical sciences. It is concerned with pharmaceutical crystals, which are vitally important in the industry in the context of solid-form selection and control and are widely studied in academia within the broad research fields of pharmaceutical materials science and crystal engineering. The term "pharmaceutical crystallography" could (and probably should) encompass macromolecules, protein–ligand complexes, and other such things, but the emphasis here is on small-molecule active pharmaceutical ingredients (APIs). The title was chosen to highlight this emphasis. The assertion of the book is that students and researchers in the pharmaceutical sciences should confidently understand and be able to contribute to structure-based discussions in the pharmaceutical literature. This comes to the fore in the "Case Studies" at the end of each chapter, which are chosen from the literature to illustrate key concepts, good practice and potential pitfalls.

Concepts from the traditional field of crystallography are obviously central, but the book is not really a formal crystallographic text. The focus is on understanding and describing molecular crystal structures, and on the practical aspects of using modern X-ray diffraction instruments. It is clear that the material is relevant to *any* molecular crystal, and I hope that the book might also be read by chemists working with molecular materials. The desire to focus on the pharmaceutical sciences reflects to a large extent my own experience as a researcher and teacher in university chemistry and pharmacy departments. Compared to chemistry courses, undergraduate courses in the pharmaceutical sciences are (even) less likely to include formal

Pharmaceutical Crystallography: A Guide to Structure and Analysis
By Andrew Bond
© Andrew Bond 2019
Published by the Royal Society of Chemistry, www.rsc.org

Print ISBN: 978-1-78262-966-5

EPUB ISBN: 978-1-78801-851-7

A catalogue record for this book is available from the British Library

The Royal Society of Chemistry is a charity, registered in England and Wales, Number 207890, and a company incorporated in England by Royal Charter (Registered No. RC000524), registered office: Burlington House, Piccadilly, London W1J 0BA, UK, Telephone: +44 (0) 20 7437 8656.

Visit our website at www.rsc.org/books

Printed in the United Kingdom by CPI Group (UK) Ltd, Croydon, CR0 4YY, UK

Pharmaceutical Crystallography

A Guide to Structure and Analysis

Andrew Bond

University of Cambridge, UK
Email: adb29@cam.ac.uk

ROYAL SOCIETY
OF CHEMISTRY

6.3 The Geometry of Diffraction from a Single Crystal   83
6.4 The Intensities of Diffraction from a Single Crystal   93
6.5 A Review   101
6.6 Summary of Key Points   102
6.7 Case Study: Ropivacaine Hydrochloride Form 2   103
     Reference   106

7 Symmetry in Diffraction Patterns   107

7.1 Introduction   107
7.2 Symmetry in Diffraction Patterns   108
7.3 Centrosymmetric and Non-centrosymmetric
     Structures   116
7.4 Summary of Key Points   122
7.5 Case Study: Ropivacaine Hydrochloride Form 2
     (Continued)   123

8 Single-crystal X-ray Diffraction (Part 1)   125

8.1 Introduction   125
8.2 Overview of a Single-crystal X-ray Diffractometer   126
8.3 Setting Up the Analysis   131
8.4 Measuring the Geometry of Diffraction   135
8.5 Measuring Crystal Morphology   143
8.6 Summary of Key Points   144
8.7 Case Study: Aspirin Form I   145
     References   148

9 Single-crystal X-ray Diffraction (Part 2)   149

9.1 Introduction   149
9.2 Measuring the Diffracted Intensities   150
9.3 Measure First, Analyse Later   160
9.4 What's the Outcome of a Single-crystal Data
     Collection?   161
9.5 Summary of Key Points   163
9.6 Case Study: Aspirin Form I (Continued)   165

10   Powder X-ray Diffraction   168

10.1 Introduction   168
10.2 Relating Powder Diffraction to Single-crystal
      Diffraction   169
10.3 Powder X-ray Diffraction Measurements   175
10.4 The Sample   184
10.5 Extracting Information from a PXRD Pattern   185
10.6 Summary of Key Points   189
10.7 Case Study: Celecoxib   190
      References   193

11   Solving X-ray Crystal Structures   194

11.1 Introduction   194
11.2 Building the Electron Density From the
      Structure Factors   195
11.3 The Phase Problem   202
11.4 Structure Solution in Practice   206
11.5 Developing the Structure Model   207
11.6 Summary of Key Points   209
11.7 Case Study: Sulfathiazole   211
      References   213

12   Refining X-ray Crystal Structures   214

12.1 Introduction   214
12.2 Structure Refinement in Theory   215
12.3 Structure Refinement in Practice   220
12.4 Absolute Structure   230
12.5 Summary of Key Points   231
12.6 Case Study: Sulfathiazole (Continued)   232
      References   235

13   Disorder and Twinning   236

13.1 Introduction   236
13.2 Disorder   237

13.3 Solvent Masking 240
13.4 Twinning 244
13.5 Summary of Key Points 252
13.6 Case Study 1: Griseofulvin–Nitroethane (1 : 1) 252
13.7 Case Study 2: L-Aspartic Acid 255
References 259

14 Crystallographic Results 260
14.1 Introduction 260
14.2 Uncertainties 261
14.3 Interpreting Crystallographic Results 266
14.4 Reporting Crystallographic Results 278
14.5 Summary of Key Points 282
14.6 Case Study: Perindoprilat Monohydrate 283
References 285

Subject Index 286

1 Pharmaceutical Solids

Summary

This chapter describes common classifications for active pharmaceutical ingredients (APIs) in the solid state and places them within a scheme based on independent chemical and structural categories. An overarching chemical classification is that pharmaceutical solids are either single-component (containing only one type of API molecule) or multi-component (containing API molecules and some other type(s) of molecule(s)). For multi-component solids, specific chemical sub-classes are co-crystals, salts and solvates/hydrates. Structural classification distinguishes crystalline and amorphous solids based on their degree of order. The aim is to provide a scientific overview, without dwelling too much on legal/regulatory definitions.

1.1 Introduction

This book is about pharmaceutical crystals. The focus is on small-molecule active pharmaceutical ingredients (APIs) rather than polymers or macromolecules such as proteins. Typical small-molecule APIs are paracetamol, carbamazepine and piroxicam (Figure 1.1). An effective definition of a crystal is that it exhibits long-range order in three dimensions. This means that the spatial arrangement of the molecules repeats regularly in three dimensions on a scale that is much larger than the basic repeating unit of the structure. Amorphous solids, which do not show long-range order, are also important for pharmaceuticals but it is much harder to discuss their structure in any general way.

Pharmaceutical Crystallography: A Guide to Structure and Analysis
By Andrew Bond
© Andrew Bond 2019
Published by the Royal Society of Chemistry, www.rsc.org

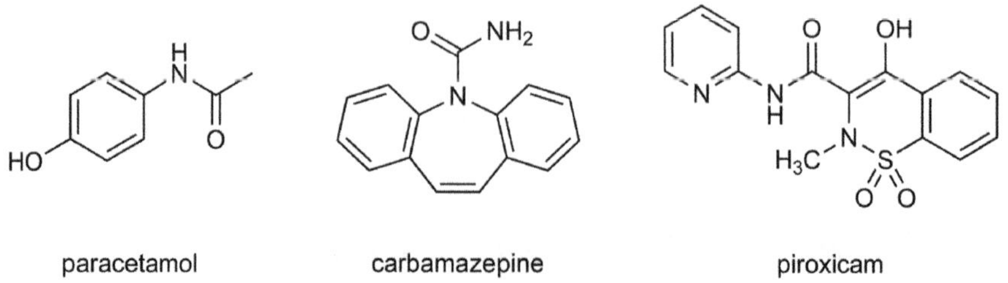

paracetamol   carbamazepine   piroxicam

Figure 1.1 Typical active pharmaceutical ingredients (APIs) considered in this book.

This chapter sets the scene by providing some definitions and classifications. The need to make such classifications is largely practical. We need to know what it means when someone talks about a polymorph or a pharmaceutical co-crystal, and we need to have some appreciation of the implied structure if we are to understand how it is likely to influence the solid-state properties. Some of these definitions may have differing viewpoints expressed in regulatory documents compared to the academic literature. The aim in this chapter is to give a scientific overview, without intending to be definitive and without dwelling too much on legal/regulatory definitions.

1.2 Classifying Pharmaceutical Solids: An Overview

A systematic scheme to classify pharmaceutical solids is suggested in Figure 1.2. Two principal classifications are defined, indicated by the two axes of the diagram. The horizontal axis is associated with *structural identity* while the vertical axis is associated with *chemical identity*. The chemical identity refers to the types of molecules that are present. Assuming it is possible to define molecules unambiguously (which it usually is), it is straightforward to say whether a given solid contains one type of molecule or more than one type of molecule. This provides a clear chemical classification: a pharmaceutical solid is either *single-component* (containing one type of molecule), or *multi-component* (containing more than one type of molecule). Single-component solids are pure chemical compounds which do not need any further chemical classification. Multi-component solids, however, are commonly divided into numerous chemical sub-classes. Three classes commonly defined for pharmaceutical solids are salts, co-crystals and solvates/hydrates. These classifications will be discussed in more detail in Section 1.4, but for now we will just accept that they exist and describe what they mean.

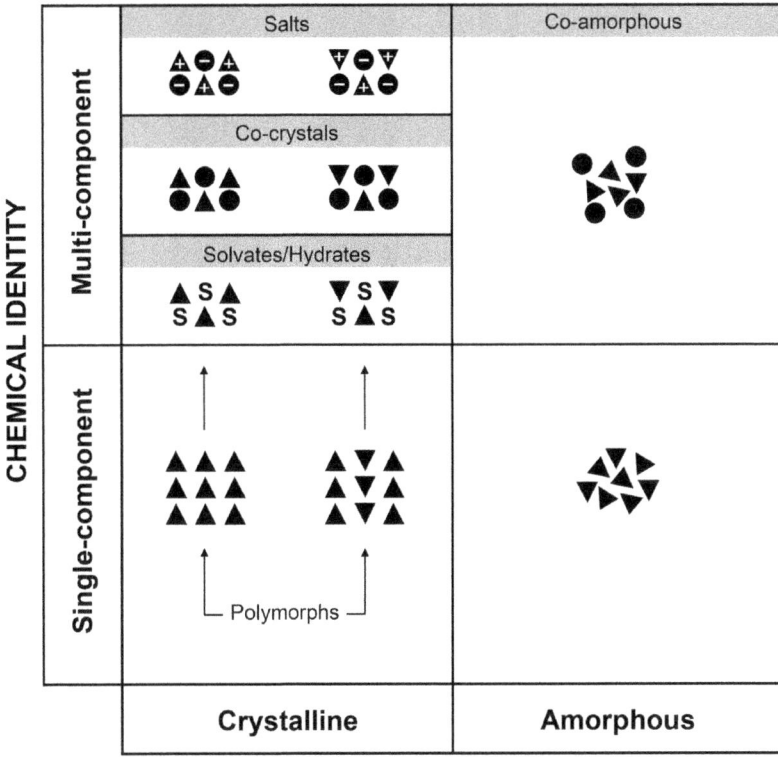

Figure 1.2 An outline classification of pharmaceutical solids. Key: ▲ = API molecule; ● = co-crystal/salt former molecule; S = solvent molecule.

On the horizontal axis of Figure 1.2, the structural identity refers to whether the solid is crystalline or amorphous. A crystalline solid exhibits long-range order in three dimensions, while an amorphous solid does not. The definitions of "crystalline" and "amorphous" will be discussed in Section 1.3; for now, we state that a solid must be either crystalline (being perfectly ordered in three dimensions) or amorphous, and draw a hard boundary between them.

The two-dimensional nature of Figure 1.2 highlights that the different structural classifications can be applied to each of the chemical categories. So, any given chemical composition might potentially exist in both crystalline and amorphous forms. Different crystalline forms with the same chemical composition are called *polymorphs*. Both single- and multi-component solids could be polymorphic. A given API (such as those in Figure 1.1) might exhibit numerous crystalline polymorphs and an amorphous form; it might be combined with a wide variety of other molecules to give a range of solvates, co-crystals and salts; each multi-component solid might be polymorphic or amorphous, and so on. The wide variety of solid forms that might be encountered for a particular API is sometimes referred to as its *solid-form landscape*.

1.3 Structural Classification: Crystalline and Amorphous Solids

> *A crystalline solid exhibits long-range order. An amorphous solid does not exhibit long-range order.*

According to the definitions above, the key feature that distinguishes crystalline and amorphous solids is *long-range order*. An ordered arrangement is one where the positions and orientations of the molecules exhibit a regular repeating pattern, more precisely a *periodic* pattern, through the entire solid. Thus, the arrangement of the molecules in one small part of the solid is reproduced regularly throughout the entire solid. This does not necessarily mean that all molecules in the solid have exactly the same environment, but usually there will be only a small number of different molecular environments. Chapters 2–4 relate crystalline order specifically to the concept of symmetry. At this stage, it is sufficient to note that Figure 1.3(a) shows an ordered (crystalline) arrangement of paracetamol molecules, while Figure 1.3(b) shows a less ordered (amorphous) arrangement. It should be immediately clear from these diagrams that an absolute distinction cannot be made between crystalline and amorphous. It is not the case that a solid must be either fully ordered or not at all ordered. Instead, we should consider the *degree of order*, and we should view crystalline and amorphous solids as the end points of a structural continuum.

1.3.1 Polymorphism

> *Polymorphs are crystalline solids with the same chemical composition but different crystal structures.*

This definition states explicitly that polymorphism refers to the crystalline state, and that polymorphs must have the same chemical

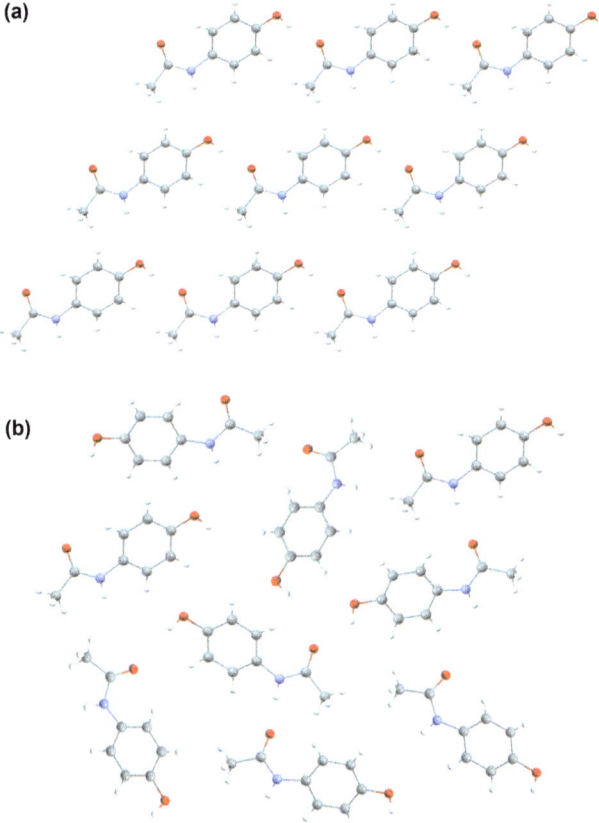

Figure 1.3 Schematic solid-state arrangements of paracetamol molecules: (a) an ordered (crystalline) arrangement; (b) a less ordered (amorphous) arrangement.

composition. The specification of a crystalline solid distinguishes polymorphs from amorphous forms, while the chemical specification means that a multi-component solid, for example an API hydrate, is *not* a polymorph of the corresponding API. In principle, the word polymorph should only be applied to a crystalline solid if it is established that different crystal structures exist for the same chemical composition. In other words, polymorphs are *sets* of crystal structures. Typically, polymorphs are labelled using Roman numerals (form I, form II, *etc.*) or Greek letters (form α, form β, *etc.*). However, the notation of polymorphs in the pharmaceutical literature is highly inconsistent and there is no standard practice. Indeed, it is quite common for the same polymorph of a given API to be given different names by different authors.

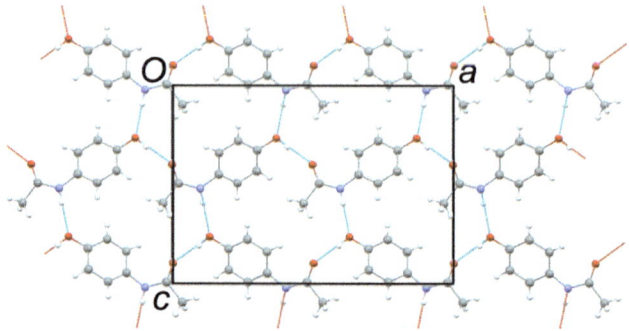

Figure 1.4   Hydrogen-bonded layers within paracetamol form I.

All three APIs in Figure 1.1 are known to be polymorphic. For paracetamol, crystal structures have been determined for three polymorphs to date: two by single-crystal X-ray diffraction and one from powder X-ray diffraction data. The molecular conformations are closely comparable and the primary intermolecular interactions are essentially the same in all three polymorphs: the NH group of the amide forms a hydrogen bond to the O atom of the hydroxyl group, which in turn forms a hydrogen bond to the O atom of a neighbouring amide (Figure 1.4). In two polymorphs (forms II and III), these interactions link the molecules into identical layers, with the distinction between the polymorphs arising mainly from the way in which the layers are stacked on top of each other. By contrast, the molecules in form I adopt layers with a more "corrugated" arrangement (Figure 1.5). This structural difference has a significant influence on the compression properties of the polymorphs, with form I being less susceptible to plastic deformation.

1.3.2   Polyamorphism

The concept of "polyamorphism" is encountered in the pharmaceutical literature, but it is much less clear than its crystalline counterpart, polymorphism. Principally, this is because polymorphism is a structural concept, and the structures of amorphous solids are inherently less well defined than those of crystalline solids. For crystalline solids, it is possible to define the structure of the entire solid clearly and unambiguously, as we will see in Chapters 2–4, and therefore to specify whether two crystal structures are the same or different. For amorphous solids, it is questionable whether we can define the structure at all. Since the defining feature of an amorphous

(a)

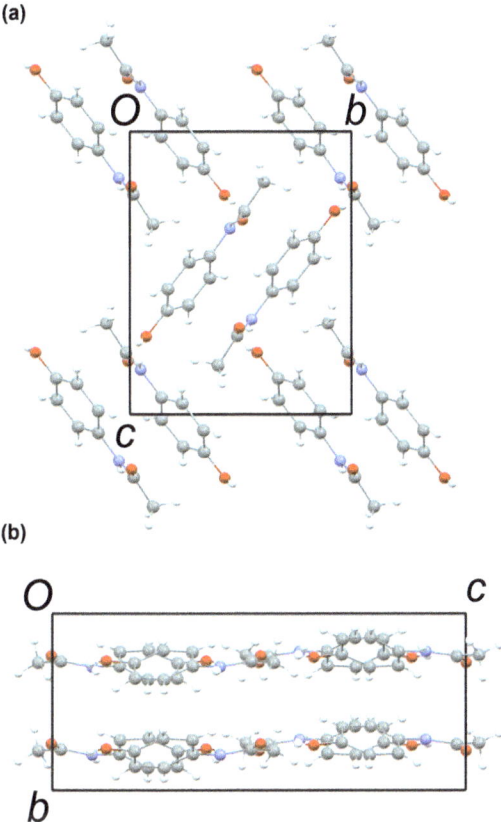

(b)

Figure 1.5 Polymorphs of paracetamol: (a) form I, containing corrugated hydrogen-bonded layers; (b) form II, containing flat layers. Form I is less susceptible to plastic deformation.

solid is that it lacks long-range order, the structure in one part of the solid cannot be representative of the whole solid. One feature of an amorphous solid that *might* be representative is its local structure. This means that there could be some consistency in the way that neighbouring molecules are arranged relative to one another. For example, molecules might be consistently arranged into pairs through some specific intermolecular interaction but the molecular pairs are arranged in a disordered (amorphous) fashion. Alternatively, the molecules could arrange themselves into higher-dimensional motifs such as chains or sheets, but these motifs could extend over only a limited range and/or be aligned in some irregular way. Assuming that local motifs could be defined clearly, polyamorphism might describe amorphous structures with different types or degrees of local order. Again, however, this is problematic because we would have to define what is meant by "local", and it is in any case practically much more difficult to determine structural information for an amorphous solid. Thus, polyamorphism is a poorly-defined and largely conceptual counterpart to the well-defined polymorphism of crystalline solids. This book focuses on crystalline materials.

1.4 Chemical Classification: Single- and Multi-component Solids

Single-component pharmaceutical solids contain only molecules of one particular API. Multi-component pharmaceutical solids contain molecules of an API plus some other chemical component(s). The additional components may be molecules or single-atom ions such as Na^+ or Cl^-. Commonly, multi-component pharmaceutical solids are grouped into several sub-classes, based on their chemical composition. The following sections provide some working definitions for three chemical classes that are commonly defined: salts, co-crystals and solvates/hydrates. As stated in Section 1.1, these classifications arise largely from practical needs, and they may become ambiguous at the borderlines. A flexible and coherent outlook is therefore encouraged.

1.4.1 Salts

> *A salt is a multi-component crystalline solid containing charged molecules and/or ions in a stoichiometric ratio.*

The "stoichiometric ratio" specified in this definition is a physical necessity within the solid because any positive charge must be balanced by an equal negative charge. As such, there may be no need for it to be stated explicitly. However, including it provides neat consistency with the definition of a co-crystal given in Section 1.4.2, which is helpful to encourage a coherent view of a salt/co-crystal continuum (Section 1.5). A pharmaceutical salt contains charged API molecules (either positively or negatively charged) and some counterion that is sometimes referred to as the *salt former*. For practical pharmaceutical applications, the salt former should obviously be safe for human ingestion. A high-profile example of a marketed pharmaceutical salt is sildenafil citrate (Figure 1.6), more commonly known as *Viagra*.

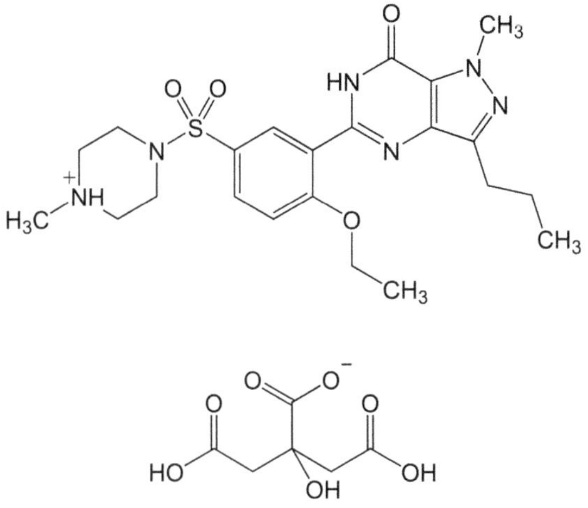

Figure 1.6 Molecular structure of the pharmaceutical salt sildenafil citrate (*Viagra*).

1.4.2 Co-crystals

> *A co-crystal is a multi-component crystalline solid containing uncharged molecules in a stoichiometric ratio.*

For a pharmaceutical co-crystal, at least one of the molecular components should be an API, and the other molecular component(s) is commonly referred to as the *co-crystal former* (or just *co-former*). For practical pharmaceutical applications, the co-crystal former should be a molecule that is known to be safe for human ingestion. Common examples include mono- or dicarboxylic acids, such as acetic or oxalic acid, tartaric acid, *etc.* In this definition, the specification of un-charged molecules distinguishes co-crystals from salts. The specification of a stoichiometric ratio is again intended to facilitate the view of a salt/co-crystal continuum (Section 1.5). It may also provide some practical distinction between co-crystals and solvates/hydrates (Section 1.4.3) because the latter are frequently found to exhibit non-stoichiometric ratios between the molecular components. One example of a pharmaceutical co-crystal from the literature, which includes carbamazepine and saccharin in a 1:1 ratio, is shown in Figure 1.7.

Referring to Figure 1.2, we should consider the case where a multi-component solid may have an amorphous structure. A logical term for this is *co-amorphous*. All of the comments regarding amorphous solids in Section 1.3.2 apply equally to co-amorphous solids, which means

(a)

(b)

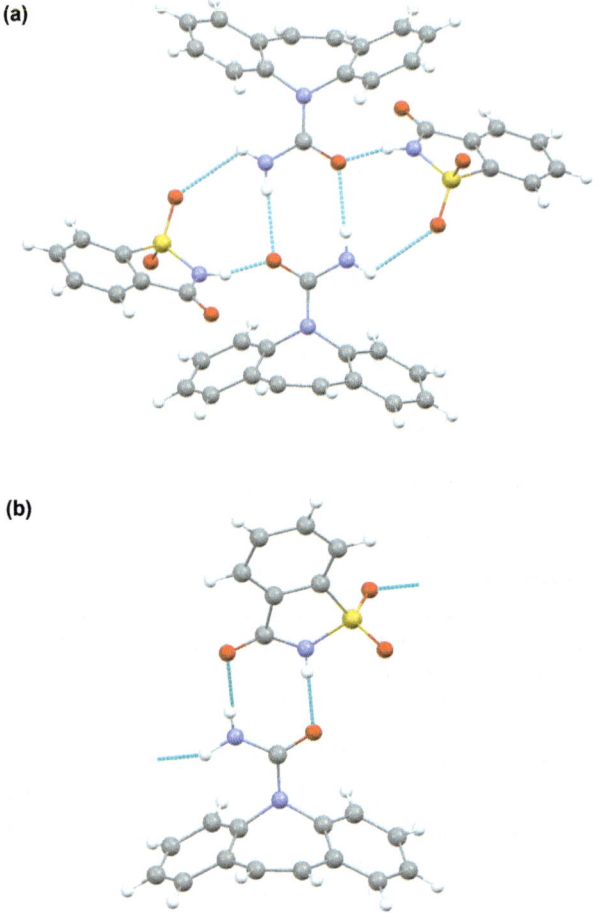

Figure 1.7   An example of a pharmaceutical co-crystal: different hydrogen-bonded motifs in two polymorphs of carbamazepine-saccharin (1 : 1).

that it may be difficult to determine how the molecules interact at a local level, whether they are homogeneously or inhomogeneously mixed, *etc.* We will simply state that the term co-amorphous exists in the pharmaceutical literature, and note its place in Figure 1.2.

1.4.3   Solvates and Hydrates

A solvate is a crystalline solid that includes solvent molecules in either a stoichiometric or non-stoichiometric ratio. If the solvent is water, the solvate is more specifically called a hydrate.

According to this definition, and the one given in Section 1.4.2, the distinction between a solvate/hydrate and a co-crystal hinges on identification of a molecule as a "solvent". We will consider the extent to which this can be justified in Section 1.5. Examples of common solvents for typical APIs include methanol, ethanol or acetone. The special distinction made for water reflects its ubiquitous presence in the environment. When referring to hydrates, it is common also to define an *anhydrate*, which refers to the corresponding API without inclusion of water. Obviously, the term has little meaning on its own, since it just means the pure API. The fact that solvates/hydrates are commonly non-stoichiometric reflects the fact that the interactions between the solvent/water molecules and the API are commonly non-specific and/or weak, and that the crystalline structure can often be retained even if some fraction of the solvent/water molecules is removed.

In the more recent literature, the word *pseudopolymorph* has been introduced as a generalisation of solvate/hydrate. The intention is to stress that solvates/hydrates are alternative crystalline forms of a given API in the same way that polymorphs are, but their different chemical composition means that they must be "pseudo" rather than genuine polymorphs. Another term that appears to serve the same purpose is *solvatomorph*. We will not use either term in this book.

1.5 Classifying Pharmaceutical Solids: More Details

As stated in the introduction to this chapter, the need to classify pharmaceutical solids is largely practical. However, it has significant legal/regulatory implications because lawyers/regulators must make clear and absolute distinctions in matters of chemical identity and intellectual property. Scientifically, there is no real need to make any distinction. But if a distinction is to be made, it should be logical and rigorous. One distinction that is generally rigorous is the chemical identity of the solid, in particular whether it contains one type of molecule or more than one. In some cases (for example, where a molecule might exist as different tautomers) it could be debatable whether molecules are the same or different, but the chemical identity of a molecule is clear enough in the majority of cases, and measurable by a variety of techniques, so the distinction between single- and multi-component solids is generally robust. Further chemical classification of multi-component solids, however, is fraught with difficulties.

Salt or co-crystal? By the definitions specified in Sections 1.4.1 and 1.4.2, the distinction between a salt and a co-crystal depends on whether the molecules are charged or not. In pharmaceutical solids, charged molecules are usually the result of proton transfer from an acidic group to a basic group. Often the groups involved in the proton transfer make a hydrogen bond in the solid. Like most things in science, we should not ask whether the acidic proton has been transferred or not, but rather *to what extent* has the proton been transferred? In very strong hydrogen bonds, the equilibrium position for the proton might be exactly half-way between the donor and acceptor atoms. Is this a salt or a co-crystal? Worse still, the extent of proton transfer can depend on temperature, so the solid might be formally classified as a salt at one temperature but a co-crystal at another. A sensible and flexible approach is to consider salts and co-crystals as the conceptual end points of a *salt/co-crystal continuum*. This description allows for a continuous view of proton (or charge) transfer, which can also accommodate the concept of temperature-dependent proton (or charge) transfer. Often it *is* straightforward to say whether a particular multi-component solid is a salt or co-crystal at room temperature and it might be argued that temperatures outside of the common ambient range have no practical relevance to pharmaceutical solids. However, the concept of the salt/co-crystal continuum seems fundamentally correct, and it enables flexibility in ambiguous cases.

Solvate or co-crystal? The distinction between *stoichiometric* solvates and co-crystals certainly has a weak scientific basis. Whether or not one of the molecular components acted as a solvent during the crystallisation process is subjective—one person's solvent might be another person's co-crystal former! More fundamentally, the identity of a solid should not depend on how it was prepared. For example, a multi-component solid prepared by evaporation crystallisation from a solvent might be called a solvate, but the *same* solid might also be prepared by mechanical grinding of a stoichiometric mixture of the two components. Thus, a stoichiometric solvate is exactly the same as a co-crystal. Attempts to limit solvates to solids containing molecular components that are liquid under a specified standard condition could conceivably define a hard boundary, but it would be an arbitrary one. The scheme given in Figure 1.2 requires us only to define whether a solid is single-component or multi-component.

Things become more problematic with non-stoichiometry, because the common perception of a co-crystal is not really consistent with non-stoichiometry. In particular, co-crystals are usually considered to be the result of robust intermolecular interactions between complementary functional groups, while non-stoichiometry is usually associated with weak, non-specific interactions between molecules. Specifying a stoichiometric ratio of molecules for a co-crystal is convenient for the concept of the salt/co-crystal continuum and that is probably the best reason for it to be included in the definition specified in Section 1.4.2. Solids that are usually considered to be solvates can more frequently be non-stoichiometric because solvent molecules are typically volatile and often form weak non-specific intermolecular interactions within crystals. This does not mean that we should define a co-crystal as something that is never non-stoichiometric and a solvate as something that can be non-stoichiometric: does removing one molecule from a co-crystal turn it into a solvate? In Figure 1.2, variable stoichiometry could be accommodated by considering that the chemical composition axis (vertical) is in principle continuous. Many pharmaceutical solids do take on specific stoichiometries, and therefore they are represented as discrete points along the vertical axis. However, others might span a range along the vertical axis without changing their structure.

Hydrates. Classifying hydrates is largely unproblematic—either a solid contains water or it doesn't. From a scientific viewpoint, a molecule of water is no different from any other molecule, so hydrates are just another type of multi-component solid and hydrates could just as well be classified as co-crystals. However, pharmaceutical hydrates have a practical importance that warrants their status as a specific sub-class, especially in the context of practical discussion. Every pharmaceutical scientist has heard about hydrates and knows what a hydrate is. In that sense, the classification is robust. Frequently, hydrates can be stoichiometric or non-stoichiometric and a distinction is sometimes made between the two.

In addition to the word anhydrate, the literature contains descriptions such as *dehydrated hydrate* (or *desolvated solvate*). Such terms violate the principle that the identity of a solid should not depend on its preparation route, and they are usually based on the *properties* of the solid rather than any known structural difference. For example, it is commonly observed that anhydrates formed by dehydration of a hydrate can behave differently from anhydrates prepared by direct crystallisation. These differences must be attributed to factors other than the idealised crystal structure, such as internal defects or different surface properties. Thus, they are applied in a slightly different context compared to the structure- and composition-based descriptors considered here.

1.5.1   Inconsistencies Between the Literature and Regulatory Guidelines

At the time of writing, the United States Food and Drug Agency (FDA) refer to all solid forms as polymorphs. Specifically, FDA guidelines from 2018[1] refer to polymorphs as *"different crystalline forms of the same API. This may include solvation or hydration products (also known as pseudopolymorphs) and amorphous forms."* There is possibly some justification for such a universal view because the literal meaning of polymorph is "many forms", which could just as well encompass forms of different chemical composition. In the academic literature, however, the term polymorph is consistently applied to mean different crystalline forms with the same chemical composition. The European Medicines Agency (EMA) agrees with the academic view.[2] It is certainly difficult to understand the inclusion of amorphous forms at the end of the FDA definition when the opening statement specifies that polymorphs are crystalline. These structural terms are very well established as antonyms, as described in Section 1.3.

The term *co-crystal* has been subject to much debate, especially as this type of material has grown in practical relevance for the pharmaceutical industry. The distinction between a co-crystal (uncharged molecules) and a salt (charged molecules) is widely accepted, and the concept of the salt/co-crystal continuum (Section 1.5) seems to be gaining acceptance, at least in the academic literature. The distinction between co-crystals and solvates, however, remains problematic. A statement that has persisted is that a co-crystal is comprised of molecular components that are solid in their pure forms while a solvate/hydrate includes (at least) one molecular component that is liquid in its pure form (at room temperature). The scientific basis for this classification is questionable, and the EMA have stated that is has *"little scientific value"*.[2] It also appears to have been partially abandoned by the FDA. The 2018 guidelines[1] define pharmaceutical co-crystals as: *"crystalline materials composed of two or more different molecules, one of which is the API, in a defined stoichiometric ratio within the same crystal lattice that are associated by nonionic and noncovalent bonds"*. This seems broadly in line with the definitions given in this Chapter. However, the same document goes on to define a co-former as *"a component that interacts nonionically with the API in the crystal lattice, that is not a solvent (including water), and is typically non-volatile"*, thereby retaining the subjectivity in relation to what constitutes a solvent.

1.6   Summary of Key Points

- There are numerous motivations to classify pharmaceutical solids, which may have different aims or requirements. A common language to enable coherent day-to-day communication can be more flexible than regulatory/legal definitions.
- Classifications can generally be broken down into structural or chemical categories.
- A structural classification distinguishes crystalline (ordered) and amorphous (disordered) solids. The degree of order in a solid is potentially continuous. Crystalline solids represent one well-defined and common extreme, while amorphous solids are far less well-defined.
- An overarching chemical classification is that pharmaceutical solids are either single-component (containing only one type of API molecule) or multi-component (containing API molecules, and some other type(s) of molecule(s)). This classification is generally unambiguous.
- Common specific chemical sub-classes for pharmaceutical solids are co-crystals, salts, and solvates/hydrates. These definitions are broadly understood in day-to-day communication, but formal classification can be problematic and should be applied with a flexible scientific outlook.
- Differences may exist between perceptions in the academic literature and legal/regulatory definitions. A flexible and coherent outlook is encouraged.

References

1. *Guidance for Industry: Regulatory Classification of Pharmaceutical Co-Crystals*, US Food and Drug Administration, Feb 2018. https://www.fda.gov/regulatory-information/search-fda-guidance-documents/regulatory-classification-pharmaceutical-co-crystals (accessed February 2019).
2. *Reflection Paper on the Use of Co-crystals of Active Substances in Medicinal Products*, European Medicines Agency, May 2015. http://www.ema.europa.eu/docs/en_GB/document_library/Scientific_guideline/2015/07/WC500189927.pdf (accessed February 2019).

2   What Defines a Crystal?

Summary

The effective defining feature of the crystalline state is translational symmetry, which can be exploited to develop an efficient and practical description of a molecular crystal structure. A crystal structure contains a common building block that is translated in three dimensions over a length scale much greater than the dimensions of the building block. This leads to the simplifying concepts of lattices and unit cells, which provide a general framework to describe crystal structures. The positions of the atoms within a crystal are expressed as coordinates relative to the edges of the unit cell and the periodic nature of the structure means that a description of the unit cell is in effect a description of the entire crystal structure. The definitions of the lattice, unit cell and fractional coordinates are independent of the metric properties of the crystal, but the metric properties provide the link between the fractional coordinates and the actual distances and angles between atoms.

2.1   Introduction

A typical API crystal that might be appropriate for single-crystal X-ray diffraction has dimensions around $0.1 \times 0.1 \times 0.1$ mm. Thus, its volume is *ca.* 1×10^{-6} cm^3. Taking paracetamol as an example, the density of the crystal is *ca.* 1.3 g cm^{-3} and the molecular mass is *ca.* 150 g mol^{-1}. This gives a mass of *ca.* 1.3×10^{-6} g for the crystal, which corresponds to *ca.* 8.7×10^{-9} moles of paracetamol molecules. Since 1 mole $= 6.022 \times 10^{23}$ (Avogadro's number), the crystal contains somewhere around 5.2×10^{15} paracetamol molecules. Each molecule

Pharmaceutical Crystallography: A Guide to Structure and Analysis
By Andrew Bond
© Andrew Bond 2019
Published by the Royal Society of Chemistry, www.rsc.org

contains 20 atoms, so to describe completely the structure of this paracetamol crystal we must describe the positions of *ca.* 1×10^{17} atoms! Fortunately, we are aided by the fact that crystals are highly symmetrical objects. Indeed, the statement that crystals exhibit long-range order is an alternative way to say that crystals exhibit *symmetry*. It is symmetry that allows crystal structures to be described in an efficient and practical way. Effectively, *translational symmetry* defines the crystalline state. Thus, there is one common building block within the crystal that can be translated in three dimensions to define the entire crystal structure. This leads to the simplifying concepts of *lattices* and *unit cells*, which are applicable to all crystals and which we will discuss in this chapter. Other types of symmetry, such as inversion, rotation or mirror symmetry, are also commonly present in crystals but they are not the defining feature. Those aspects of crystal symmetry will be considered in Chapter 3.

2.2 Lattices

Figure 2.1 shows a schematic 2-dimensional arrangement of paracetamol molecules which should be considered to extend infinitely in the plane of the page. The dots in the diagram indicate equivalent points, which in this context refer to points in the structure having exactly the same surroundings in exactly the same orientation. The equivalent points are related to each other by translational symmetry. If we take the structure and translate it (without turning) so that any

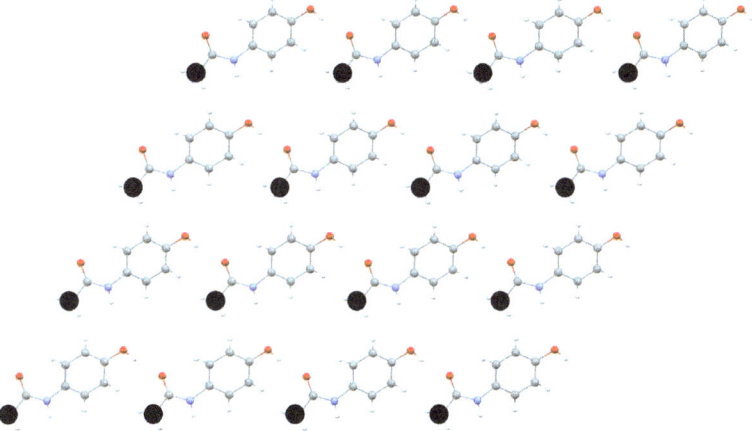

Figure 2.1 Schematic 2-D periodic structure of paracetamol molecules with the black dots indicating points that are equivalent by translation.

equivalent point is moved to the position of any other equivalent point, the structure will appear to be exactly the same. This is a practical description of a symmetry operation and hence translation is a type of symmetry operation. Symmetry operations that can be applied to individual objects, such as proper and improper rotations, are probably already familiar. When we come to describe crystals, translational symmetry is just another type of symmetry operation that should be viewed in the same way as all the others.

The idea of equivalent points in a crystal structure related by translation leads to the concept of the *lattice*. If we take the picture in Figure 2.1 and delete the molecules, we are left with a regular pattern of equivalent points (Figure 2.2)—this is the lattice. The mathematical definition of a lattice in three dimensions is that it is constructed from three vectors (called **a**, **b** and **c**, with bold text to indicate vectors), which should not all be parallel or lie in a common plane. It is usual in crystallographic descriptions to refer to **a**, **b** and **c** as the *lattice vectors*. Any lattice point is related to any other by adding integral multiples of the lattice vectors. Often, the concept of the "crystal lattice" can be confused because it is used to refer to a crystal in some physical sense (for example, in the FDA guidelines at the end of Section 1.5.1). We will avoid this ambiguity and limit use of the word "lattice" to the geometrical construction that arises from the existence of translational symmetry within the crystalline state.

Defining the lattice simplifies the task of describing a crystal structure because it means we only need to describe the lattice vectors and the part of the structure that is repeated at each lattice point. The presence of translational symmetry thereby allows a crystal containing 5.2×10^{15} paracetamol molecules to be described by just three lattice vectors and the positions of a few representative molecules relative to each lattice point. In the schematic paracetamol structure in Figure 2.1, there is only one molecule associated with each lattice point.

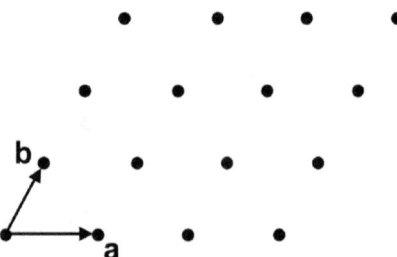

Figure 2.2   The lattice associated with the 2-D periodic structure of paracetamol molecules in Figure 2.1. Suitable lattice vectors are indicated.

In Figure 2.1, the lattice points are drawn at the position of the C atom of the methyl group in each paracetamol molecule. What would happen if we specify instead that the lattice points correspond to the position of the carbonyl O atom in each molecule? Of course, we would end up with exactly the same lattice because the carbonyl O atoms in each molecule are related by translation in exactly the same way as the methyl groups. As long as the lattice points are defined at points in the structure that are truly equivalent by translational symmetry, the result will always be the same. Thus, a given crystal structure has a uniquely-defined lattice. In practice, experimental uncertainties could make it difficult to decide whether parts of a crystal structure are equivalent by translation or not. For the purposes of defining and understanding the crystalline state, however, we can say that the relationship between a crystal structure and its lattice is unique.

In Figure 2.2, two suitable lattice vectors, **a** and **b**, are indicated. Several alternatives are shown in Figure 2.3. Each of these pairs of vectors is equally suitable to define *exactly the same* lattice. What does this mean for the description of a crystal structure? It means that the relationship between a crystal structure and its lattice is unique, but the lattice could be described using many different sets of lattice vectors.

The three vectors used to describe a 3-dimensional lattice are often referred to as the *basis vectors*, or collectively just *the basis*. Although there may be numerous reasonable choices for the basis vectors of a given lattice, there is one absolute rule: the axis system must be *right-handed*. One way to visualise this is to consider a right hand with the index finger pointing naturally forward, the thumb pointing upwards and the middle finger pointing perpendicular to the palm (Figure 2.4(a)). In this pose, the thumb, index finger and middle finger indicate the correct relative directions of the lattice vectors **a**, **b**

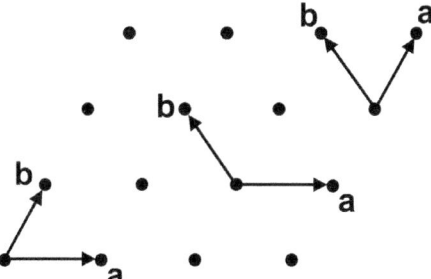

Figure 2.3 Three alternative choices of lattice vectors for the same lattice.

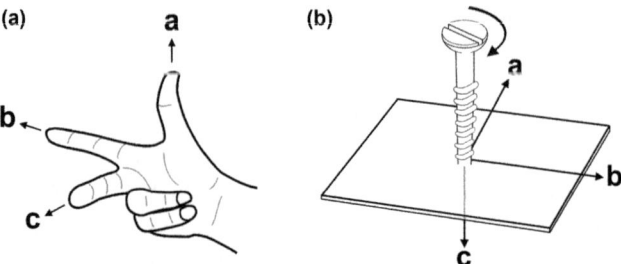

Figure 2.4   Illustration of right-handed axes: (a) using the thumb, index and middle fingers of the right hand; (b) when turning a conventional screw from **a** to **b**, the translation of the screw defines the positive direction for **c**.

and **c**. Alternatively, consider the action of a conventional screw, which moves into a plane if turned clockwise and out of a plane if turned anti-clockwise. Imagine a screw being screwed into the plane containing **a** and **b**: turning the screw from **a** towards **b** causes the screw to move in a *positive* direction along **c** (Figure 2.4(b)).

Additional rules have been devised in an effort to make a consistent choice of lattice vectors:

(1) The vectors should be the three shortest non-parallel, non-coplanar vectors.
(2) The vectors should be labelled so that their lengths are in the order: $a \leq b \leq c$.
(3) The angles between the vectors should be either all $<90°$ or all $\geq 90°$.

The result is called the *reduced basis.* The reduced basis defines the *reduced unit cell*, which must be primitive (see Section 2.3). Although the reduced unit cell should be uniquely defined, it may not always be the best choice to describe a crystal structure. Choices based on the orientations of symmetry elements usually take precedence (see Chapter 4). Nonetheless, the reduced unit cell remains useful to compare lattices, as illustrated in the Case Study at the end of this chapter.

2.3   Unit Cells

If we look again at the periodic paracetamol structure in Figure 2.1 and consider how much of the pattern must be described to be able to reconstruct the whole structure, it is sufficient to define a box that

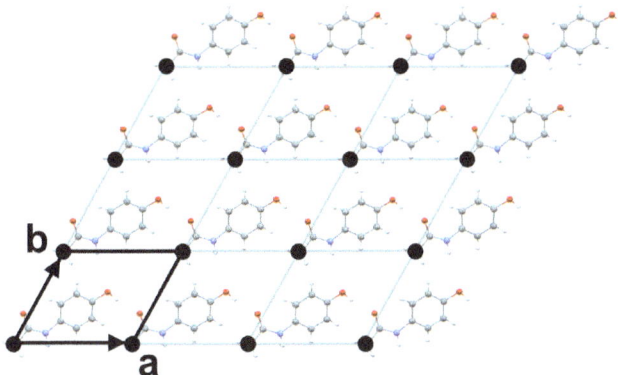

Figure 2.5 A primitive unit cell for the 2-D paracetamol structure in Figure 2.1. Each of the indicated boxes has identical contents. This is one of numerous possible choices for the lattice vectors and unit cell.

joins up adjacent lattice points (Figure 2.5). This building block does not need to contain more than one lattice point because we would needlessly describe points in the pattern that are known to be equivalent by translation. For crystal structures, the smallest possible building block is called the *primitive unit cell*. The word "primitive" means specifically that the unit cell contains only one lattice point. In Chapter 4, we will describe circumstances where it might be beneficial to define unit cells containing more than one lattice point, but for now we will use the intuitive concept that the unit cell should describe the smallest possible region between lattice points.

Commonly, unit cells are drawn so that lattice points lie at the corners. This can make it difficult to see how many lattice points the unit cell contains. One way to think about it is that the lattice points at the corners are "shared" between neighbouring unit cells, then to sum the appropriate fractions. For example, a lattice point at a corner of a three-dimensional unit cell is shared between 8 unit cells (Figure 2.6), so it contributes $\frac{1}{8}$ to any given unit cell. Since there are 8 corners, there is a total of one lattice point $(8 \times \frac{1}{8})$ per unit cell. Since the unit cell joins up adjacent lattice points, its metric dimensions are defined by the lengths of the lattice vectors and the angles between them. The unit-cell edges defined by the lattice vectors **a**, **b** and **c** have lengths written a, b and c, respectively. The angle between **a** and **b** is γ (gamma), the angle between **a** and **c** is β (beta) and the angle between **b** and **c** is α (alpha) (Figure 2.7). The angles can be remembered by noting the correspondence between the sets of symbols **abc** and $\alpha\beta\gamma$, as shown in Figure 2.7.

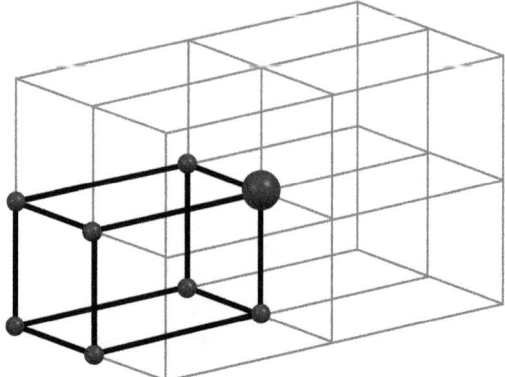

Figure 2.6 Counting the number of lattice points in a 3-D primitive unit cell. The central lattice point in the figure (enlarged) lies at the corner of the emphasised unit cell, and is seen to be shared between 8 unit cells (four in the lower tier and four in the upper tier). Thus, it contributes $\frac{1}{8}$ to the emphasised unit cell. Since there are eight corners, the unit cell contains $8 \times \frac{1}{8} = 1$ lattice point.

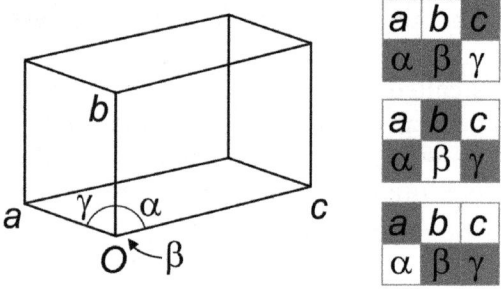

Figure 2.7 Conventional notation for the unit-cell lengths and angles. O represents the origin. The relationship between *abc* and $\alpha\beta\gamma$ is illustrated by the patterns on the right (*e.g.* the top pattern indicates that the angle between *a* and *b* is γ).

Since different lattice vectors could be chosen for the same lattice, the same crystal structure could be described using many different unit cells. Different unit-cell dimensions do not, therefore, immediately indicate different crystal structures. Similarly, there is no unique choice for the *position* of the unit cell with respect to the crystal structure. One corner of the unit cell is defined as the "origin", where the lattice vectors **a**, **b** and **c** are considered to originate, but deciding where to put the origin in a structure is arbitrary. We will see in Chapter 4 that some choices are more helpful than others, but it is nonetheless common for the same crystal structure to be described in the literature using different origins.

2.4 Fractional Coordinates

So far, we have seen how translational symmetry can be exploited to simplify the description of a crystal structure to just a description of the unit cell. It is implicit that the entire structure is built up by translating the unit cell along the three vectors that define its edges. Now we must describe the positions of atoms within the unit cell. An atom is considered to be located at a point defined by three co-ordinates x,y,z. Any definition of coordinates must refer to some set of axes and we can use the edges of the unit cell for this purpose. Frequently, we will refer to the a, b and c axes. Since the edges of the unit cell correspond to the lattice vectors, we could also say that the coordinates x,y,z are defined with respect to the lattice vectors $\mathbf{a}$, $\mathbf{b}$ and $\mathbf{c}$ (Figure 2.8). The coordinates are called *fractional coordinates* because they refer to fractions of the lattice vectors. Within a given unit cell, each coordinate spans the range 0–1 (with 0 being equivalent to 1). Since a crystal structure exhibits translational symmetry, it is always possible to add an integer to any of the fractional co-ordinates to describe an equivalent position in another unit cell. It is useful in several contexts to define the *position vector*, $\mathbf{r}$, which refers to the vector from the origin of the unit cell to the position with co-ordinates x,y,z. The formal definition is $\mathbf{r} = x\mathbf{a} + y\mathbf{b} + z\mathbf{c}$.

Since the fractional coordinates x,y,z are defined relative to the unit-cell edges, the coordinates assigned to a given point in the structure must depend on the unit cell that is chosen. If we choose a different unit cell, the same point in the same structure must be given different coordinates. A simple example of this could be that we choose to exchange the labels of the axes. In the 2-D example of Figure 2.2, we might have chosen to call the horizontal axis $\mathbf{b}$ and the

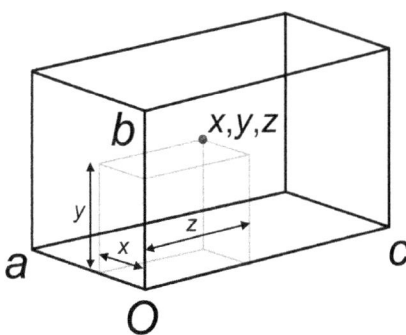

Figure 2.8 Fractional coordinates x,y,z defined relative to the unit-cell edges.

other axis **a**. The x and y coordinates of each atom would then be interchanged. We might also choose to define some other position for the origin of the unit cell, in which case *all* of the coordinates would be shifted by some specific values. This type of transformation does not change the crystal structure. The result describes the *same* crystal structure using a different unit cell.

2.5 Metric Properties

It has been deliberate so far to avoid any discussion of metric properties (*i.e.* the *values* of any distances or angles) because they are not directly important for the concepts of the lattice, unit cell and fractional coordinates. The concepts described in this chapter are derived solely from translational symmetry, which is the fundamental feature of a crystal and is independent of size and shape. Of course, the lattice of any real crystal structure must have a specific size and shape, and the unit cell used to describe the crystal structure will have some specified dimensions. Usually, the lengths of the unit-cell edges are given in angstrom units ($1 \text{ Å} = 10^{-10}$ m) because these provide a convenient range for typical molecular crystal structures and chemical bond lengths. For example, a C–C single bond length is typically *ca.* 1.54 Å and a C–H bond length is *ca.* 1 Å. A typical edge of a primitive unit cell for the types of APIs discussed in this book has length in the range 5–20 Å. The metric properties of a crystal relate the fractional coordinates of the atoms to the actual interatomic distances and angles. For example, the distance between two atoms with fractional coordinates 0,0,0 and $\frac{1}{2},\frac{1}{2},\frac{1}{2}$ must obviously depend on the dimensions of the unit cell.

2.6 Summary of Key Points

- The effective defining feature of the crystalline state is translational symmetry in three dimensions over a length scale much greater than the basic repeating unit of the structure.
- Equivalent points in a crystal structure, related by translational symmetry, define the lattice. The relationship between a crystal structure and its lattice is unique.
- A lattice is described by three lattice vectors (also known as basis vectors), denoted **a**, **b** and **c**. There are numerous choices for the lattice vectors, so the same lattice might be described by different vector sets. The lattice vectors must define a right-handed system.

- The unit cell is the fundamental building block of a crystal structure. The choice of unit cell is not unique, which means that the same crystal structure might be described using different unit cells. In general, it is most efficient to define a primitive unit cell, which contains only one lattice point, but there may be reasons to choose a larger unit cell.
- The lengths of the unit-cell edges (which correspond to the magnitudes of the lattice vectors) are denoted a, b and c. The angles between them are denoted α, β and γ.
- The positions of atoms within the unit cell are given as fractional coordinates, x,y,z, referring to the unit-cell edges, a, b and c. The coordinates have range 0–1 (with 0 being equivalent to 1). Any integer can be added to a fractional coordinate to give an equivalent position in another unit cell.
- If a different unit cell is chosen, the same structure will be represented by an entirely different set of atomic fractional coordinates. Different origins (*i.e.* the point in the structure considered to have fractional coordinates 0,0,0) might also be chosen.
- The definitions of the lattice, unit cell and fractional coordinates are independent of the metric properties of the crystal (*i.e.* the size and shape of the unit cell). However, the metric properties provide the link between the fractional coordinates and the actual distances and angles between atoms.

2.7 Case Study: Mefenamic Acid

Mefenamic acid (Figure 2.9) is a generic non-steroidal anti-inflammatory drug (NSAID). To date, crystal structures have been published for three polymorphs, labelled forms I, II and III. The structures of forms I and III are discussed here.

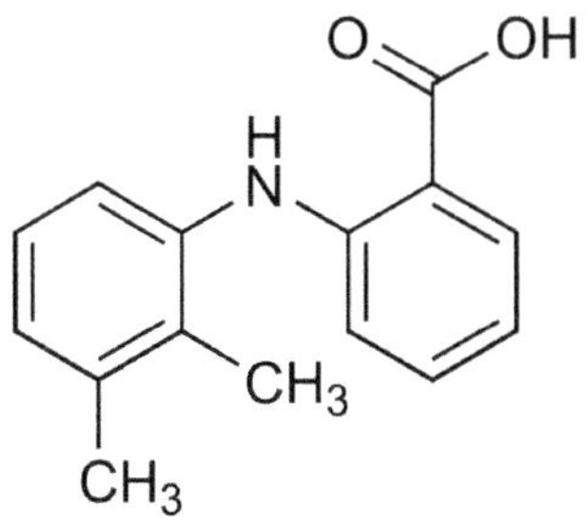

Figure 2.9 Mefenamic acid, $C_{15}H_{15}NO_2$.

The crystal structure of form I was reported in 1976,[1] and again in 2004.[2] The structure of form III was reported in 2012.[3] Both polymorphs adopt a primitive unit cell, containing two API molecules, which means that two molecules make up the basic structural unit repeated at each lattice point. The point of interest is the unit-cell parameters:

Reference	Poly-morph	a (Å)	b (Å)	c (Å)	α (°)	β (°)	γ (°)	Vol. (Å³)
1	I	14.556	6.811	7.657	119.57	103.93	91.30	631.8
2	I	7.337	14.306	6.790	101.01	114.64	76.05	625.4
3	III	7.723	7.934	11.232	83.59	80.94	67.51	627.0

The unit-cell volume is comparable in each case, consistent with the fact that each unit cell contains two mefenamic acid molecules, but all of the lengths and angles are different. We might reasonably conclude that these are three different crystal structures. However, we have said that the structures in ref. 1 and 2 are both form I. How can we see that? One way is to visualise the structure and its lattice points (Figure 2.10), and it can be seen that the alternative unit cells describe the same lattice. A more direct analytical method is to convert each unit cell to its reduced form, as discussed in Section 2.2. Since the reduced basis is unique for a given lattice, it can be used to compare

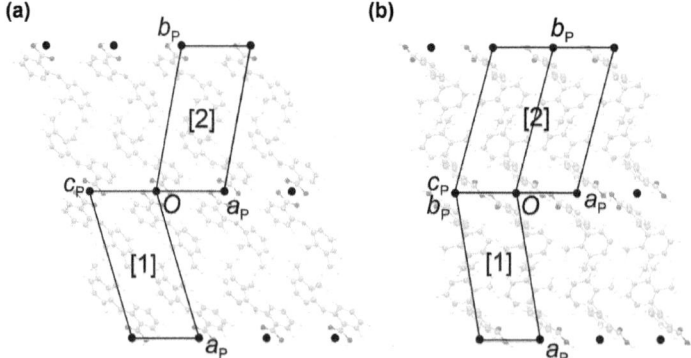

Figure 2.10 Two projections of the crystal structure of mefenamic acid form I, showing the lattice points as black dots and the two published choices for the unit cell. The subscript "P" indicates that the axes are shown in projection (*i.e.* they are not in the plane of the paper). In (b), the top unit cell is shown in projection along the *ac* body diagonal (it does not indicate two unit cells). The lattices are the same in (a) and (b).

lattice dimensions. The reduced unit cells, obtained using a standard computer algorithm, are as follows:

Reference	a (Å)	b (Å)	c (Å)	α (°)	β (°)	γ (°)	Vol. (Å³)
1	6.811	7.318	14.397	76.57	79.18	65.52	631.8
2	6.790	7.337	14.306	76.05	78.99	65.36	625.4

Now it is clear that the two lattices are the same (within experimental error). For form III in ref. 3, the published unit cell is already in its reduced form and it is clearly different.

References 1 and 2 report the same crystal structure with different unit-cell parameters (and therefore different sets of atomic coordinates), which could potentially be misinterpreted as different polymorphs. It is curious that neither reported structure uses the reduced unit cell because there is no apparent reason to make any other choice. Modern crystallographic software is programmed to use the reduced unit cell where possible, so consistency is more likely to be seen in more recent literature. However, this may not be the same unit cell that was used in earlier publications, so care must always be taken when comparing structures by their unit-cell parameters.

Note

Most of the examples in this book are retrieved from the Cambridge Structural Database (CSD; discussed in Chapter 14). The code following each reference below is the identifier for the CSD entry.

References

1. J. F. McConnell and F. Z. Company, *Cryst. Struct. Commun.*, 1976, **5**, 861, [CSD: XYANAC].
2. L. Fang, S. Numajiri, D. Kobayashi, H. Ueda, K. Nakayama, H. Miyamae and Y. Morimoto, *J. Pharm. Sci.*, 2004, **93**, 144, [CSD: XYANAC01 (no coordinates)].
3. S. SeethaLekshmi and T. N. Guru Row, *Cryst. Growth Des.*, 2012, **12**, 4283, [CSD: XYANAC03].

3 Symmetry in Crystals

Summary

In addition to their inherent translational symmetry, most crystals display other kinds of symmetry. Since a crystal is periodic by translation, a symmetry element such as a rotation axis, mirror plane or inversion centre within a crystal must be repeated at all positions that are equivalent by translation. Hence, these symmetry elements are present as sets distributed regularly through the crystal. Describing the positions of the symmetry elements within the unit cell is an efficient description of the symmetry elements through the entire crystal. It is shown in this chapter how the symmetry operations produce relationships between atomic coordinates, and how this leads to further efficiency in the description of a crystal structure. In addition to point symmetry operations (rotations, mirrors and inversions), space symmetry operations (screws and glides) are introduced.

3.1 Introduction

Chapter 2 showed how translational symmetry simplifies the description of a crystal structure to just a description of the unit cell. Translational symmetry is always present in a crystalline solid, so this simplification can always be made. Often, crystal structures display other types of symmetry, such as rotations, mirrors or inversions, which can be exploited to simplify structure descriptions even further. Crystals do not have to display non-translational symmetry, but most of them do. This chapter describes non-translational symmetry in crystals.

Pharmaceutical Crystallography: A Guide to Structure and Analysis
By Andrew Bond
© Andrew Bond 2019
Published by the Royal Society of Chemistry, www.rsc.org

3.2   Point Symmetry Operations

In general, a symmetry operation acts on an object to generate an identical copy of that object. A *point symmetry operation* generates the copy at the same location, which means that the geometrical centre of the object (its *centroid*) remains stationary (Figure 3.1). A collection of point symmetry operations is called a *point group*. All of the symmetry elements in a point group intersect at the centroid of the object. Here, the term *symmetry element* has been chosen instead of *symmetry operation*. There is a difference: a symmetry operation is an action, while a symmetry element is an object associated with that action. Specifically, a symmetry element is defined by the point, line or plane around which the symmetry operation acts. For example, the symmetry element for a mirror operation is a mirror plane and the symmetry element for a rotation operation is the rotation axis.

If we say that a crystal structure exhibits a certain type of symmetry, it means that the *whole structure* conforms to that symmetry. Since a crystal is periodic and contains points that are equivalent by translation, a symmetry element at some position within the crystal must be repeated at all positions related by translation. Hence, there must be a set of symmetry elements distributed through the crystal. In the case of mirror symmetry, parallel mirror planes must be spaced regularly through the crystal. As we have seen, describing such a situation is simplified by defining the unit cell. If we describe the positions of the symmetry elements within the unit cell, we efficiently describe the symmetry elements through the entire crystal. In the following sections, we will look in detail at the various point symmetry

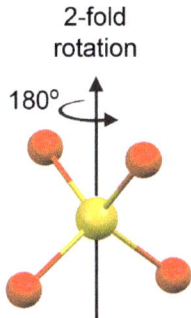

2-fold
rotation

180°

Figure 3.1   A point symmetry operation generates a copy of a molecule at the same location. A 2-fold rotation is shown for the sulfate anion, SO_4^{2-}. The central S atom remains stationary.

operations that are observed in crystals and consider how they can be described. In particular, it is vital to appreciate how symmetry operations produce relationships between atomic coordinates.

3.2.1   Mirrors

A mirror operation can be considered to move each point in the crystal onto the associated mirror plane, then out by the same distance on the other side. Points in the crystal related by mirror symmetry can be imagined to be joined by a line, with the mirror plane situated half-way along the line and perpendicular to it (Figure 3.2). For several pairs of points related by the same mirror plane, the imaginary lines joining each pair of points must all be parallel. For a chiral molecule, a mirror operation generates a copy of the molecule with the opposite handedness.

In order to exploit the existence of non-translational symmetry to describe crystal structures, we must consider the way that symmetry operations relate fractional coordinates. Figure 3.3 shows a primitive unit cell with a mirror plane perpendicular to the b axis, passing through the origin. The mirror operation acts on a point with fractional coordinates x,y,z to generate a point with coordinates $x, -y, z$. The positions x,y,z and $x, -y, z$ are said to be *symmetry equivalent*. When it comes to describing the structure, the simplifying feature is that it is necessary to specify only one coordinate out of each pair. If we know that a mirror operation relates x,y,z to $x, -y, z$, it is only

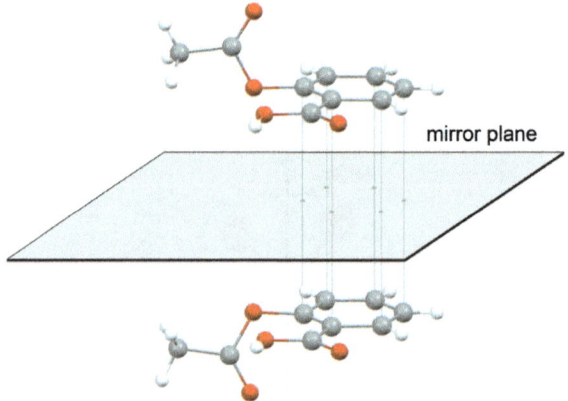

Figure 3.2   Mirror operation applied to a molecule of aspirin. For illustration, equivalent atoms in the benzene rings of the two molecules are joined by grey lines, which are all parallel and perpendicular to the mirror plane.

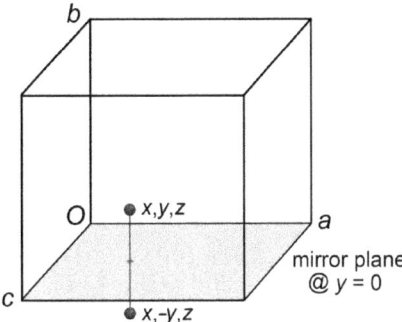

Figure 3.3 Primitive unit cell with a mirror plane perpendicular to the *b* axis, passing through the origin. The mirror operation relates a point with coordinates *x,y,z* to an equivalent point at *x, −y,z*.

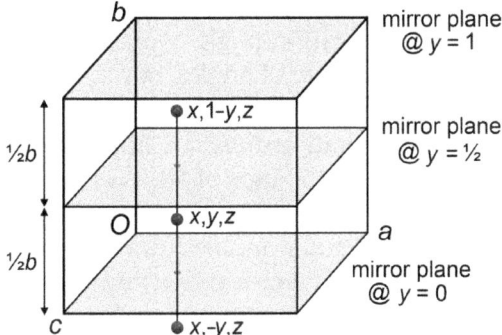

Figure 3.4 A mirror plane perpendicular to a unit-cell axis and situated at the origin must be accompanied by a parallel mirror plane half-way along the unit-cell edge. The translational symmetry of the crystal results in a set of parallel mirror planes with spacing equal to half of the unit-cell edge. A primitive unit cell is illustrated with mirror planes perpendicular to the *b* axis.

necessary to specify that there is an atom at x,y,z, and we know immediately that there is an equivalent atom at $x, -y,z$.

Now we can illustrate something vitally important about symmetry in crystalline solids. We know that fractional coordinates repeat themselves in the range 0–1 (Section 2.4). So, a position in a crystal with coordinates $x, -y,z$ lies outside of the unit cell. Since all unit cells are identical, there must be an equivalent position within the unit cell which is obtained by translating the point $x, -y,z$ by one unit along the *b* axis, *i.e.* there must be an equivalent position with coordinates $x,1 - y,z$. How is that position related to x,y,z? It is shown in Figure 3.4 that the point $x,1 - y,z$ is related to x,y,z by a mirror plane perpendicular to the b axis situated at $y = \frac{1}{2}$. Thus, if there is a mirror plane perpendicular to the b axis at $y = 0$, there must also be a parallel mirror plane half-way along the b axis of the unit cell. Since the unit cell is representative of the entire crystal, this means that there must be parallel mirror planes distributed through the entire crystal, separated by $\frac{1}{2}b$. This is an inevitable consequence of the way that point symmetry operations in crystals combine with the inherent translational symmetry. It is the first example of *the* key concept required to understand symmetry within crystal structures.

3.2.2   Rotations

For a rotation operation, the associated symmetry element is a *rotation axis*. The rotation operation can be considered to move each point in the crystal around the rotation axis, maintaining a fixed distance from that axis (Figure 3.5). The rotation angle could be specified in degrees (or radians, or any other measure of angle), but for crystals, the rotation type is most commonly specified by its *order*, which is the number of successive times that the rotation must be applied to complete a full rotation. For example, a rotation angle of 180° is usually referred to as a *2-fold rotation*, because two successive 180° rotations bring a point back to where it started. A 3-fold rotation is equivalent to 120°, a 4-fold rotation is equivalent to 90°, *etc.* For a chiral molecule, the action of a rotation operation does not change its handedness.

Figure 3.6 shows a primitive unit cell with a 2-fold rotation axis parallel to the b axis, passing through the origin. The rotation operation acts on a position with fractional coordinates x, y, z to generate an equivalent position with coordinates $-x, y, -z$. Just as we saw for mirror planes, the presence of translational symmetry must generate other rotation axes in the crystal. The positions of these can be

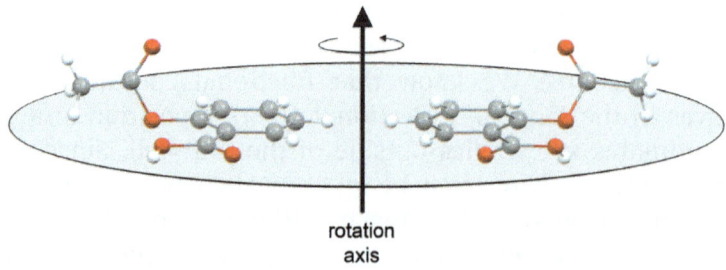

rotation
axis

Figure 3.5   2-Fold rotation operation applied to a molecule of aspirin. The grey disk is included solely for illustrative purposes.

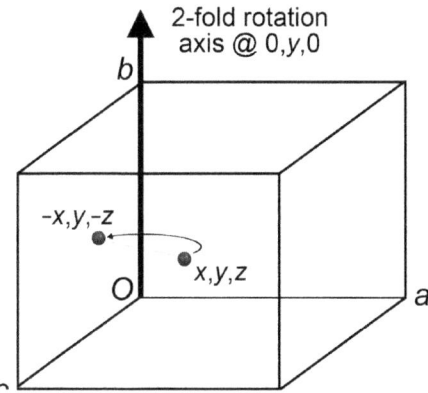

Figure 3.6 Primitive unit cell with a 2-fold rotation axis parallel to the *b* axis, passing through the origin. The rotation relates a point with coordinates *x,y,z* to an equivalent point at −*x,y,* − *z*.

deduced by adding 1 in sequence to both of the negative coordinates in −*x,y,* − *z*:

$$-x,y,-z \quad 1-x,y,-z \quad -x,y,1-z \quad 1-x,y,1-z$$

How are these positions related to *x,y,z*? For each, the symmetry operation corresponds to a 2-fold rotation parallel to the *b* axis. We can see this because the signs of *x* and *z* change, but the sign of *y* does not. However, the different translation parts $(1 - x, etc.)$ indicate that the rotation axes are not located at the unit-cell origin. The positions of the corresponding 2-fold rotation axes are:

$$0,y,0 \quad \tfrac{1}{2},y,0 \quad 0,y,\tfrac{1}{2} \quad \tfrac{1}{2},y,\tfrac{1}{2}$$

Here, *y* does not have any specific value because the rotation axes run parallel to the *b* axis. These coordinates correspond respectively to rotation axes passing through the origin, the middle of the unit-cell edge *a*, the middle of the unit-cell edge *c*, and the centre of the unit-cell face defined by *a* and *c* (Figure 3.7). Thus, 2-fold rotation axes are distributed through the entire crystal, all running parallel to the *b* axis, passing through and half-way between the lattice points in the plane perpendicular to the rotation axis.

By convention, a rotation operation has a defined direction, sometimes referred to as the *sense* of the rotation. For a 2-fold rotation axis, this is irrelevant because a rotation of 180° gives an identical result regardless of its direction. For other rotation orders, however, there is a difference between rotations in a positive or negative sense. By convention, an anticlockwise turn corresponds to a positive sense of rotation, while a clockwise turn corresponds to a negative sense of

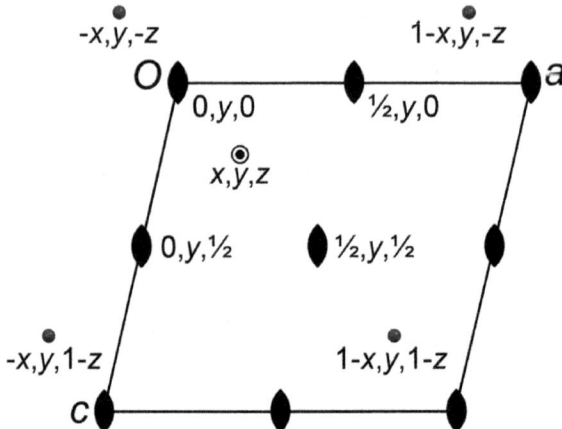

Figure 3.7 Projection of the primitive unit cell in Figure 3.6 along the *b* axis. If a 2-fold rotation axis passes through the origin, other 2-fold rotation axes pass through all lattice points and half-way between the lattice points in the *ac* plane. The diagram shows the standard symbol for a 2-fold rotation axis in projection and highlights equivalent positions as described in the main text.

rotation. For example, a positive 4-fold rotation means an anticlockwise rotation of 90°. A negative 4-fold rotation means a clockwise rotation of 90°, which is equivalent to an anticlockwise rotation of 270°. This is straightforward, but there is one further thing to consider: how do we define what is clockwise and what is anticlockwise? If we look along an axis in one direction and define a clockwise rotation, the same rotation will appear to be anticlockwise if we look from the opposite direction. The convention is that we look from the end of the specified axis *towards the origin* when considering whether a rotation takes place in a clockwise or anticlockwise direction.

3.2.3 Inversions

For an inversion operation, the associated symmetry element is an *inversion point*, or *inversion centre*. Sometimes this is referred to (rather imprecisely) as a *symmetry centre*. The inversion operation can be considered to move each point in a crystal to the inversion centre, then out by the same distance on the other side (Figure 3.8). Thus, two points in the crystal related by inversion symmetry can be imagined to be joined by a line with the inversion centre exactly half-way along its length. For several pairs of points related by the same inversion operation, the lines joining each pair of equivalent points intersect at the inversion centre. For a chiral molecule, an inversion operation generates a molecule with the opposite handedness.

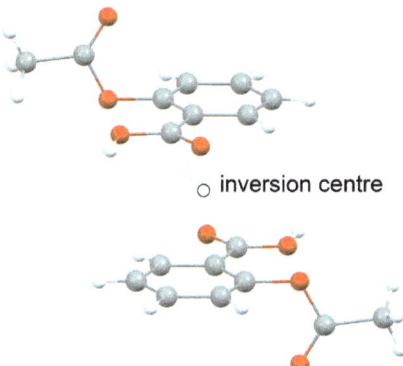

Figure 3.8   Inversion operation applied to a molecule of aspirin. For illustration, some equivalent atoms in the two molecules are joined by grey lines, which pass through the inversion centre.

An inversion operation at the origin acts on a position with fractional coordinates x,y,z to generate a position $-x, -y, -z$. In this case, the combination of the inversion centre with the crystal's translational symmetry generates a set of *eight* equivalent positions, seen by adding 1 in sequence to each of the negative coordinates:

$$-x,-y,-z \qquad 1-x,-y,-z \qquad -x,1-y,-z \qquad -x,-y,1-z$$
$$-x,1-y,1-z \quad 1-x,-y,1-z \quad 1-x,1-y,-z \quad 1-x,1-y,1-z$$

These coordinates are related to x,y,z by inversion centres located at:

$$0,0,0 \quad \tfrac{1}{2},0,0 \quad 0,\tfrac{1}{2},0 \quad 0,0,\tfrac{1}{2}$$
$$0,\tfrac{1}{2},\tfrac{1}{2} \quad \tfrac{1}{2},0,\tfrac{1}{2} \quad \tfrac{1}{2},\tfrac{1}{2},0 \quad \tfrac{1}{2},\tfrac{1}{2},\tfrac{1}{2}$$

These points correspond to the middle of each unit-cell edge, the middle of each unit-cell face, and the centre of the unit cell (Figure 3.9). Thus, a crystal with an inversion centre at the origin of the unit cell also has inversion centres situated on and half-way between all lattice points.

3.2.4   Improper Rotations

An improper rotation can be viewed as a combination of rotation and inversion (sometimes called a *rotoinversion*). This is not quite the definition usually used to describe the point symmetry of individual molecules, where an improper rotation is considered to be a

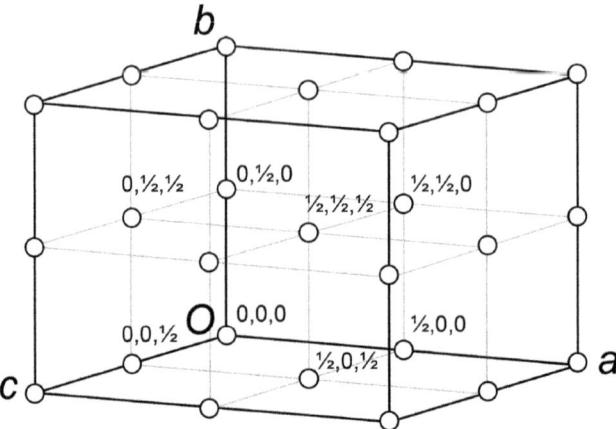

Figure 3.9 Inversion centres in a primitive unit cell. If there is an inversion centre at the origin, inversion centres must also exist on and half-way between all lattice points.

combination of rotation and reflection in the perpendicular plane. The difference comes from the existence of two systems commonly used to describe point symmetry. Chemists generally use the *Schoenflies* system, while crystallographers use the *Hermann–Mauguin* (H–M) system. These are just two different ways to describe the same symmetry operations. We will use the H–M system because it is the one that is invariably used for crystal structures. An improper rotation is indicated by the rotation order with a "bar" above the number. For example, $\bar{4}$ indicates a combination of a 4-fold rotation and inversion. The symmetry element for an improper rotation is an *improper rotation axis*, which comprises an axis for the rotation and a specified point on the axis through which inversion occurs.

Figure 3.10(a) shows a primitive unit cell with a $\bar{4}$ axis parallel to the c axis, with its inversion point at the origin. The $\bar{4}$ operation acts on a position with fractional coordinates x,y,z to generate a set of four equivalent positions:

$$x,y,z \quad y,-x,-z \quad -x,-y,z \quad -y,x,-z$$

Although the improper rotation can be visualised in terms of a combination of rotation and inversion, this does not necessarily mean that the constituent rotation and inversion operations are present individually. For example, the set of four coordinates listed above does not contain $y,-x,z$ (which corresponds to a 4-fold rotation around the c axis) or $-x,-y,-z$ (which corresponds to inversion through the origin). Thus, a crystal containing a $\bar{4}$ axis does not

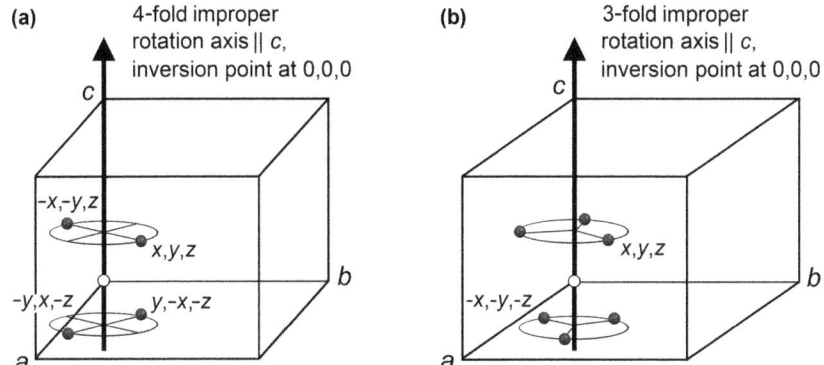

Figure 3.10 (a) Primitive unit cell with a $\bar{4}$ axis parallel to c, with its inversion point at the origin. The four equivalent positions are indicated. These do not exhibit either a 4-fold rotation around the c axis or inversion through the origin. (b) Primitive unit cell with a $\bar{3}$ axis parallel to c, with its inversion point at the origin. The six equivalent points exhibit both a 3-fold rotation parallel to the c axis and inversion through the origin (only two inversion-related points are labelled).

display either 4-fold rotation symmetry around that axis or inversion symmetry through its defined inversion point. Actually, this depends on whether the improper rotation has even or odd order. For example, Figure 3.10(b) shows the equivalent positions generated by a $\bar{3}$ axis parallel to the c axis, with its inversion point at the origin. In this case, the set of positions *does* exhibit both 3-fold rotation symmetry and inversion symmetry. This is true in general: for even-order improper rotations, the inherent rotation and inversion operations are not individually present, while for odd-order improper rotations, both the inherent rotation and inversion operations are also present.

As we saw for mirrors, rotations and inversions, the translational symmetry within crystals must reproduce improper rotation axes through the entire crystal. In this case, three further $\bar{4}$ axes lie within the unit cell, running parallel to the c axis, half-way between the lattice points in the ab plane of the unit cell. For each axis, the inversion point lies at $z = 0$. The sense of an improper rotation (positive or negative) is defined by the sense of the constituent rotation operation, as discussed in Section 3.2.2.

3.2.5 The Identity Operation

One further symmetry operation to be defined is the *identity operation*, which acts to generate a copy of the object at exactly the same position—essentially, it does nothing. In terms of fractional co-ordinates, it relates point x,y,z to point x,y,z. The identity operation is defined for logical and mathematical completeness. One important feature of symmetry in a crystal is that any symmetry operation followed by another symmetry operation must always generate a third symmetry operation that is present in the crystal. For example, a positive 4-fold rotation $(+90°)$ followed by a 2-fold rotation $(180°)$ around the same axis is equivalent to a negative 4-fold rotation $(-90°)$. Some combinations of operations will return an object back to where it started: for example, a positive 4-fold rotation followed by a negative 4-fold rotation, or two consecutive 2-fold rotations. The identity operation is required to describe this possibility. We return to this in more detail in Chapter 4.

3.2.6   A Brief Recap

While discussing the various symmetry operations in crystals, it is possible to lose focus amongst the jumble of operation types, fractional coordinates and symbols. It is therefore worth reminding ourselves of what we are trying to do: we are trying to describe a crystal structure, which means that we must describe the positions of a huge number of atoms. Within a crystal, the atomic positions are periodic by translation, which enables us to define the unit cell and fractional coordinates x,y,z that are periodic in the range 0–1. This dramatic simplification is always possible for a crystal. More often than not, crystals also display non-translational symmetry, which establishes relationships between the fractional coordinates of atoms *within the unit cell*. Since the unit cell is representative of the entire crystal structure, we only need to describe symmetry within the unit cell to account for the symmetry of the entire crystal. A crucial point is that the arrangement of the atoms is not a *result* of the symmetry. Rather, the atoms are arranged how they are and we are developing tools to describe that. We are not discussing *why* atoms commonly arrange themselves in a symmetrical manner. We are just observing that they do, and exploiting it to develop an efficient framework to describe crystal structures.

3.3   Point Symmetry in Crystals

In principle, there is no limit to the variety of point symmetry operations. However, there is a significant simplification to be made for crystals because any point symmetry operation in a crystal must be consistent with its inherent translational symmetry. This means that the rotation orders that can be seen in crystals are limited to just a few possible values. In fact, the *only* point symmetry operations that can be applied to a crystal structure are:

$$\text{Proper rotations}: \quad 1, 2, 3, 4, 6$$

$$\text{Improper rotations}: \quad \bar{1}, \bar{2}(=m), \bar{3}, \bar{4}, \bar{6}$$

This is probably best illustrated by considering an example that is *not* allowed. Imagine a situation with a 5-fold rotation axis perpendicular to a plane of lattice points, passing through one of the lattice points (Figure 3.11). If this 5-fold axis were to apply to the whole crystal structure, it must be surrounded by a regular pentagon of lattice points. Each of these lattice points must also be in a 5-fold symmetric environment, which means that we should be able to lay a new pentagonal motif onto each of them. If we try that, however, we see that the corners of the various motifs do not overlay. Indeed, there is no way that pentagonal motifs can be arranged consistently to cover all of the space, so the 5-fold rotation axis cannot apply to the whole crystal structure. Basically, we are demonstrating that pentagons do not tessellate. This does not mean that 5-fold symmetric objects cannot arrange themselves into crystals, or even that there cannot be some *local* 5-fold symmetric arrangement within a crystal. But 5-fold symmetry cannot be a feature of the entire crystal structure. The same can be shown for any rotation order higher than 6-fold.

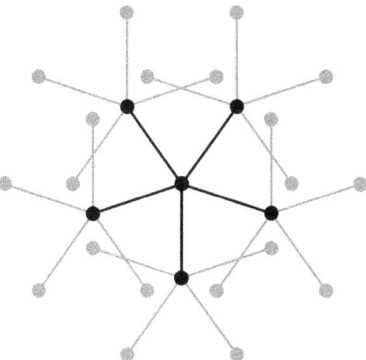

Figure 3.11 5-fold rotational symmetry is not consistent with the translational symmetry of a crystal. Repeating the central (bold) pentagonal motif on each of the lattice points cannot generate a genuine periodic lattice. We are not dealing with "quasicrystals" in this book.

3.4 Space Symmetry Operations

The existence of translational symmetry within crystals gives rise to additional types of symmetry operations, where a copy of the object is generated at a location different from the original object. Because these symmetry operations act to move the object through space, they are called *space symmetry* operations. Translation is a space symmetry operation. The following sections describe two other types of space symmetry operations that are seen in crystals.

3.4.1 Screws

A screw operation is a combination of rotation and translation. The corresponding symmetry element is a *screw axis*, where the translation occurs parallel to the axis. The inherent translational symmetry within the crystal means that the translation part of a screw operation can only take certain values. For example, consider a 2-fold screw parallel to the b axis passing through the unit-cell origin (Figure 3.12). The screw operation comprises a 2-fold rotation around the screw axis, plus a translation parallel to that axis. The rotation part relates fractional coordinates x,y,z to $-x,y,-z$, then the translation generates a position $-x,\delta+y,-z$, where δ is the fractional translation along the b axis. Applying the screw operation for a second time moves the position $-x,\delta+y,-z$ to $x,2\delta+y,z$. If this is to be consistent

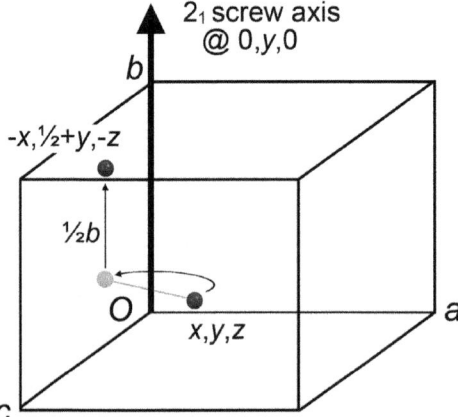

Figure 3.12 A primitive unit cell with a 2_1 screw axis parallel to the b axis, passing through the origin. If the 2_1 operation were applied again to the point $-x,\frac{1}{2}+y,-z$, it would generate a point $x,1+y,z$ in the next unit cell.

with the translational symmetry along y, 2δ must be 1, so the value of δ must be $\frac{1}{2}$. Thus, the 2-fold screw operation generates the following sequence of equivalent positions:

$$x,y,z \quad \rightarrow \quad -x,y+\tfrac{1}{2},-z \quad \rightarrow \quad x,y+1,z \quad \rightarrow \quad \ldots$$

In the H–M system, this is labelled a 2_1 screw. The 2 refers to the rotation order and the subscript indicates the translation by defining a fraction with the subscript in the numerator and the rotation order in the denominator, *i.e.* $\frac{1}{2}$.

Each of the other possible rotation orders in crystals can be combined with translation to generate screw operations. There are two possible types of 3-fold screw, namely 3_1 and 3_2. A 3_1 screw combines a 3-fold rotation with a translation of $\frac{1}{3}$ parallel to the screw axis. For example, a 3_1 screw parallel to the c axis of a primitive unit cell, passing through the origin (Figure 3.13(a)), generates the positions:

$$x,y,z \quad \rightarrow \quad -y,x-y,\tfrac{1}{3}+z \quad \rightarrow \quad -x+y,-x,\tfrac{2}{3}+z \quad \rightarrow \quad x,y,1+z \quad \rightarrow \quad \ldots$$

In this sequence, the translation occurs in steps of $\frac{1}{3}$ in a positive direction along the c axis. A 3_2 screw combines a 3-fold rotation with a translation of $\frac{2}{3}$ parallel to the screw axis (Figure 3.13(b)). This looks unusual at first sight because it does not seem to be consistent with the translational symmetry within the unit cell. A 3_2 screw axis parallel to the c axis generates the positions:

$$x,y,z \quad \rightarrow \quad -y,x-y,\tfrac{2}{3}+z \quad \rightarrow \quad -x+y,-x,\tfrac{4}{3}+z \quad \rightarrow \quad x,y,2+z \quad \rightarrow \quad \ldots$$

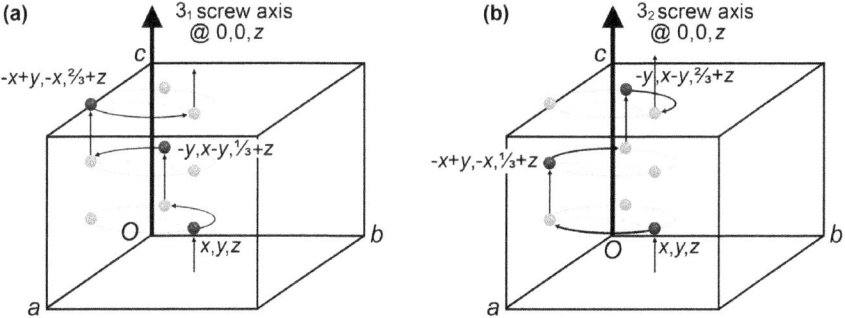

Figure 3.13 Primitive unit cells with 3-fold screw axes running parallel to the c axis, passing through the origin. (a) A 3_1 screw combines an anti-clockwise rotation of 120° with a translation of $+\frac{1}{3}\mathbf{c}$; (b) A 3_2 screw combines an anti-clockwise rotation with a translation of $-\frac{1}{3}\mathbf{c}$. In the diagram, this can be viewed as a clockwise rotation with a translation of $+\frac{1}{3}\mathbf{c}$. The 3_1 and 3_2 screw axes are enantiomers.

which spans two unit cells. However, all of these points can be moved into the same unit cell by subtracting integers from the z coordinates. Then, the points generated by a 3_2 screw can be written:

$$x,y,1+z \quad \rightarrow \quad -y,x-y,\tfrac{2}{3}+z \quad \rightarrow \quad -x+y,-x,\tfrac{1}{3}+z \quad \rightarrow \quad x,y,z \quad \rightarrow \quad \cdots$$

Now it can be seen that the z coordinate *decreases* by $\tfrac{1}{3}$ in sequence, which means that the translation part of the 3_2 screw acts in the opposite direction compared to the 3_1 screw. Thus, if the 3-fold rotation is considered to act in the same sense, 3_1 and 3_2 screw axes are associated with opposite translation directions. A 3_1 screw along a given direction represents a right-handed helix, while a 3_2 screw along the same direction represents a left-handed helix. The screw is a chiral object, and the $3_1/3_2$ pair are enantiomers. Similar considerations apply to $4_1/4_3$ and $6_1/6_5$ screws. A 2_1 screw axis is not chiral because a rotation of $+180°$ is identical to a rotation of $-180°$.

For 4-fold and 6-fold rotations, there are other possibilities, each of which can be viewed as a 2-fold rotation axis coincident with the screw axes that have already been described. For example, a 4_2 screw axis can be viewed as a 2-fold rotation axis coincident with a 4_1 or 4_3 screw axis (it doesn't matter which). Similarly, a 6_3 screw axis is a 2-fold rotation axis coincident with either a 6_1 or 6_5 screw axis. Finally, 6_2 and 6_4 screw axes arise when a 2-fold rotation axis is coincident with a 3_1 or 3_2 screw axis, respectively. The details of these relationships are not important for this discussion, and we are again in danger of losing focus. We just need to be aware of the symbols that describe the different types of screw operations and to understand that screws combine rotation with translation parallel to the rotation axis.

3.4.2 Glides

A glide operation combines a mirror and translation. The corresponding symmetry element is a *glide plane* and the translation occurs parallel to that plane. As for screws, the inherent translational symmetry in the crystal means that the direction and magnitude of the translation part can only take certain values. For example, Figure 3.14 shows a glide plane perpendicular to the b axis of a primitive unit cell, passing through the origin. Since two successive applications of the glide operation must be consistent with the translational symmetry in the ac plane of the unit cell, the

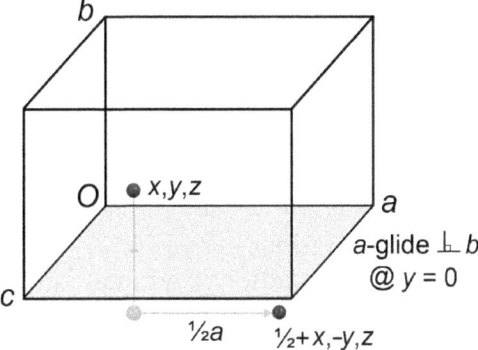

Figure 3.14   Example of the action of a glide plane. The glide operation combines reflection in the glide plane with translation parallel to the glide plane. An a-glide is shown, which has a translation component of $\frac{1}{2}\mathbf{a}$. Translation of $\frac{1}{2}\mathbf{c}$ or $\frac{1}{2}(\mathbf{a}+\mathbf{c})$ would also be possible for a glide plane perpendicular to the *b* axis.

translational part of the glide can be either $\frac{1}{2}\mathbf{a}$ or $\frac{1}{2}\mathbf{c}$. These two possibilities are referred to respectively as an *a*-glide perpendicular to the *b* axis or a *c*-glide perpendicular to the *b* axis. There is one further possibility, which is to combine both translations, $\frac{1}{2}\mathbf{a}$ and $\frac{1}{2}\mathbf{c}$, thereby moving across the diagonal of the *ac* unit-cell face. This is referred to as an *n*-glide perpendicular to the *b* axis. For a glide plane perpendicular to the *a* axis of a primitive unit cell, the possible associated translations are $\frac{1}{2}\mathbf{b}$, $\frac{1}{2}\mathbf{c}$ or $\frac{1}{2}(\mathbf{b}+\mathbf{c})$. For a glide plane perpendicular to the *c* axis, the possible associated translations are $\frac{1}{2}\mathbf{a}$, $\frac{1}{2}\mathbf{b}$ or $\frac{1}{2}(\mathbf{a}+\mathbf{b})$. There are some additional possibilities that arise in centred unit cells, which we will meet in Section 4.4.3.

3.5   Summary of Key Points

- In addition to the translational symmetry that defines the crystalline state, crystal structures frequently display non-translational symmetry.
- Non-translational symmetry defines relationships between fractional coordinates within the unit cell, which can be exploited to simplify descriptions of crystal structures.
- Point symmetry operations act to generate a copy of an object at the same location. These include mirrors, rotations, inversions and improper rotations. In crystals, improper rotations are defined as a combination of rotation and inversion through a defined point on the rotation axis.

- The types of point symmetry operations that can exist in crystals are limited by the requirement for them to be consistent with the translational symmetry. Allowed rotation orders are restricted to 1, 2, 3, 4 and 6.
- Space symmetry operations act to create a copy of an object at a different location from the original object. Space symmetry operations in crystals include screws and glides. Translation is also a space symmetry operation.
- Due to the crystal's translational symmetry, all symmetry elements appear in sets distributed through the crystal. An efficient representation of the symmetry through an entire crystal is given by a description of the symmetry elements within the unit cell.

3.6   Case Study: Formoterol Fumarate Dihydrate

Formoterol is a long-acting β2 agonist (LABA) used to control asthma. The API is delivered by inhalation of a powdered fumarate salt, which is universally referred to as formoterol fumarate dihydrate (Figure 3.15). The contents of the unit cell, including the symmetry elements of the crystal, are shown in Figure 3.16. There are two protonated formoterol molecules in the unit cell, related by an inversion centre at $\frac{1}{2}, \frac{1}{2}, \frac{1}{2}$. There are also two water molecules, related to each other by the same inversion centre. Fumarate molecules (present as dianions) occupy sites halfway along the unit-cell edges, which must also correspond to inversion centres, according to the discussion in Section 3.2.3.

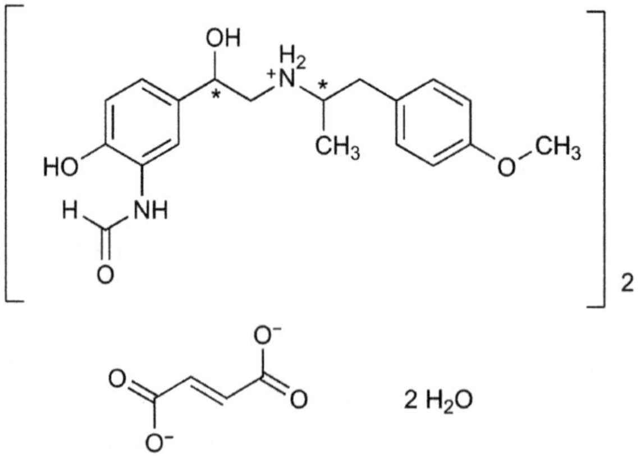

Figure 3.15   Formoterol fumarate dihydrate, $2(C_{19}H_{25}N_2O_4{}^+) \cdot (C_4H_2O_4{}^{2-}) \cdot 2(H_2O)$. The two chiral centres are indicated by asterisks.

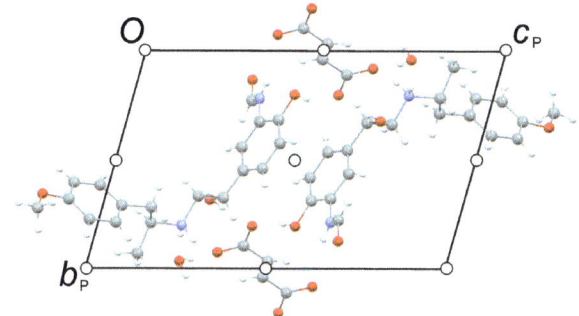

Figure 3.16 Unit-cell contents in formoterol fumarate dihydrate, projected along the *a* axis. Inversion centres are shown as circles. There are no other symmetry elements in the unit cell.

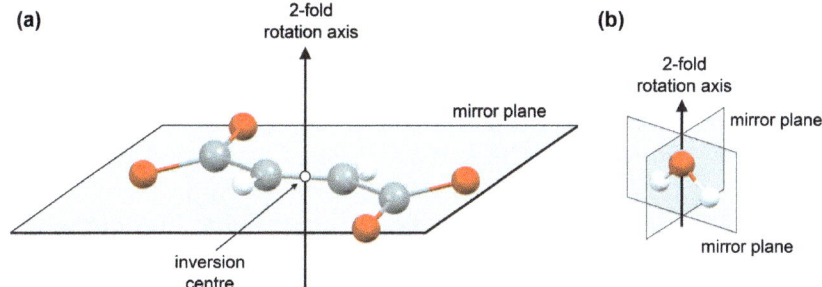

Figure 3.17 Symmetry elements in (a) fumarate molecules; (b) water molecules.

Thus, the fumarate molecules themselves display inversion symmetry. Actually, they display more symmetry than that: a 2-fold rotation axis lies perpendicular to the plane of each molecule and the molecular plane constitutes a mirror plane (Figure 3.17(a)). The point group of each individual molecule is therefore 2/*m* in Hermann–Mauguin notation (C_{2h} in Schoenflies notation). However, it is only the molecule's inversion centre that is retained as a symmetry operator that applies to the whole crystal, so the molecules are said to have *crystallographic inversion symmetry*. The other symmetry elements of the molecule apply only in a local sense. They do not apply to the entire crystal structure. Incidentally, the water molecules have point group *mm*2 (C_{2v}) (Figure 3.16(b)), but none of their symmetry elements apply to the whole crystal structure.

There is an important point here: it is tempting to think that the fumarate molecules display inversion symmetry *because* they lie on inversion centres within the unit cell. This is certainly the wrong way to think about it. The molecules have the point symmetry that they have because of the various physical/chemical factors that control molecular structure. The symmetry elements of a molecule may become symmetry elements of the entire crystal if the molecules are suitably arranged. For example, the fumarate molecules would have to be arranged so that their 2-fold rotation axes are parallel if the 2-fold rotation is to be consistent with the translational symmetry of the crystal. Similarly, the molecular planes must all be parallel and equally spaced if the mirror symmetry is to apply to the whole crystal structure. If they are not, then these symmetry operations apply only locally to the molecules. With this view, it is easy to see why centrosymmetric molecules quite frequently retain their inversion symmetry as part of the symmetry of the crystal, because inversion symmetry does not place any constraints on the relative orientations of molecules. Water molecules rarely retain their 2-fold or mirror symmetry as part of the crystal's symmetry because it is unlikely that all molecules in the crystal will be suitably aligned.

The formoterol molecule contains two chiral centres, indicated by asterisks in Figure 3.15. The fact that the crystal structure contains inversion centres means that the crystal must comprise a racemic mixture of two enantiomers, *i.e.* it contains an equal number of (R,R) and (S,S) molecules. Again, it is not the inversion symmetry in the crystal that forces the molecules to be opposite enantiomers. Rather, the (R,R) and (S,S) enantiomers were present in the crystallisation solution and they have come together to make a crystal where (R,R) and (S,S) molecules are present in equal quantities and are related by inversion symmetry. This is driven by the kinetic and thermodynamic aspects of the crystallisation process; we are just describing the result.

The stoichiometry of the crystal is $2(\text{formoterol}^+) \cdot (\text{fumarate}^{2-}) \cdot 2(\text{H}_2\text{O})$, which can be seen by looking at the contents of the unit cell. To count the number of fumarate molecules in the unit cell, consider that there is a molecule at the middle of four of the unit-cell edges, and that each edge is shared between four unit cells. Thus, there are $4 \times \frac{1}{4} = 1$ fumarate molecule per unit cell. Another way is to imagine shifting the unit cell half a unit along the a and c axes (remembering we are free to define the origin where we like). Then there would clearly be one fumarate molecule in the middle of each unit cell. One final question: is this a dihydrate? That depends on how we choose to designate the formoterol fumarate part. In the literature paper reporting this crystal structure,[1] the authors state explicitly that they consider "formoterol fumarate" to mean $2(\text{formoterol}^+) \cdot (\text{fumarate}^{2-})$, which therefore makes the structure a dihydrate. Alternatively, we might consider the number of water molecules per formoterol molecule: then the compound would be formoterol hemifumarate monohydrate. All of which changes nothing about the crystal structure but reminds us to be very careful with names!

Reference

1. K. Jarring, T. Larsson, B. Stensland and I. Ymén, *J. Pharm. Sci.*, 2006, **95**, 1144, [CSD: WAXZIG].

4 Space Groups

Summary

This chapter describes how the symmetry operations in a crystal combine to form a space group and how the space group provides a general framework to describe a crystal structure. The simplest space group comprises just the three translational symmetry operations of the crystal. Other space groups also include non-translational symmetry operations such as rotations, mirrors, inversions, screws and glides. The requirement for all of the symmetry operations to be consistent with each other, and in particular with the translational periodicity of the crystal, means that the number of possible space group types is limited to exactly 230. The relationships between the symmetry operations and the atomic coordinates are expressed by the general equivalent positions, which provide the information to generate the complete contents of the unit cell from a minimal set of coordinates called the asymmetric unit. Various ways are discussed in which crystals are classified by their symmetry, with an especially important distinction made between centrosymmetric, non-centrosymmetric and chiral space groups.

4.1 Introduction

Just as a collection of point symmetry operations is called a point group, a collection of space symmetry operations is called a *space group*. The simplest space group comprises just three translation operations (plus the identity operation), so the description of the unit cell in Chapter 2 is in fact a description of the simplest crystallographic space group, $P1$. This chapter will describe how all of the

Pharmaceutical Crystallography: A Guide to Structure and Analysis
By Andrew Bond
© Andrew Bond 2019
Published by the Royal Society of Chemistry, www.rsc.org

symmetry operations in a crystal combine to form a space group, and how the space group provides a general framework to describe a crystal structure. Perhaps one of the most surprising aspects of symmetry in crystals is that the number of space group types is limited to exactly 230. This is a consequence of the fact that all of the symmetry operations must be consistent with each other, and in particular with the translational periodicity of the crystal. Each of the 230 space group types has a common symbol, such as $P2_1/c$ or $P2_12_12_1$. They are catalogued and illustrated in the *International Tables for Crystallography*, published and maintained by the International Union of Crystallography (IUCr).[1]

4.2 General Concepts of Space Groups

All crystallographic space groups contain the three translation operations that define the crystal's periodicity. This fact is so fundamental that it is generally taken for granted. For example, space group $P1$ is often stated (erroneously) to have "no symmetry" or "only the identity operation". The translational symmetry operations are just as much part of the space group as any other symmetry operation, even if we do not continue to stress that explicitly. We have also discussed how a description of the unit cell represents a description of the entire crystal structure, so a space group can be described fully by describing only the symmetry elements within the unit cell. We have seen that a given crystal lattice can be described in principle by many different unit cells and that the choice of unit cell has an influence on how the fractional coordinates are represented. Thus, the symmetry operations of a given space group type, which express relationships between coordinates, can also appear to change if the unit cell should be redefined. We will see that the same set of symmetry operations can be described in numerous ways, which are called different *settings* of the space group. An example is given in the Case Study at the end of the chapter.

4.2.1 Combining Symmetry Operations

We have seen how symmetry operations within the unit cell combine with translational symmetry to produce other symmetry operations. In Section 3.2.2, for example, we saw how a 2-fold rotation axis running parallel to the b axis passing through the origin requires that there must also be 2-fold rotation axes at positions $0,y,\frac{1}{2}$; $\frac{1}{2},y,0$ and $\frac{1}{2},y,\frac{1}{2}$. This is an illustration of a crucial general concept: *any symmetry operation in a space group combines with any other to produce another operation of the space group.* As an example, consider a 2_1 screw axis running parallel to the b axis, passing through the origin (Figure 4.1). Applying this operation to a point with coordinates x,y,z generates another point with coordinates $-x,\frac{1}{2}+y,-z$. Now consider that there is also a mirror plane perpendicular to the b axis at $y=0$. If the mirror operation is applied to the previously generated point, the result is a new point at $-x,-\frac{1}{2}-y,-z$. How is this related to the point x,y,z? Since the signs of all three coordinates have changed compared to x,y,z, the resultant operation must be an inversion centre. The addition of $-\frac{1}{2}$ on the y coordinate places the inversion centre at $0,-\frac{1}{4},0$ (Figure 4.1). Thus, we see that the combination of the specified 2_1 screw axis and mirror plane generates an inversion centre. Put another way: if a space group contains 2_1 screw axes and perpendicular mirror planes, it must also contain inversion centres. If we add all of the symmetry elements within the unit cell (in the manner of Section 3.2.3), we find that inversion centres lie on the 2_1 axes, halfway between the mirror planes (Figure 4.2). We can apply this approach to visualise any combination of symmetry operations in any space group and thereby understand how the symmetry operations of a group fit together. The example should not give the impression that there are some "primary" operations that generate "secondary" operations in any particular sequence. Rather, the space group comprises a self-consistent set of symmetry operations, all of which can be applied in any sequence to generate other operations of the group.

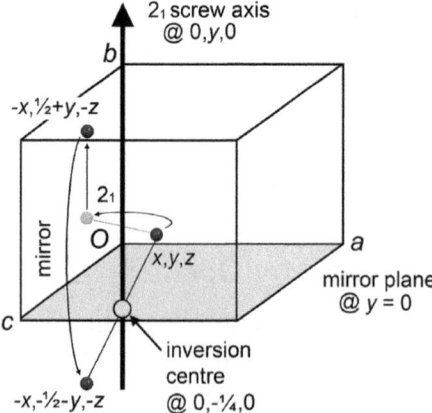

Figure 4.1   The combination of a 2_1 screw axis parallel to the b axis passing through the origin and a mirror plane perpendicular to the b axis, also passing through the origin, generates an inversion centre at $0,-\frac{1}{4},0$.

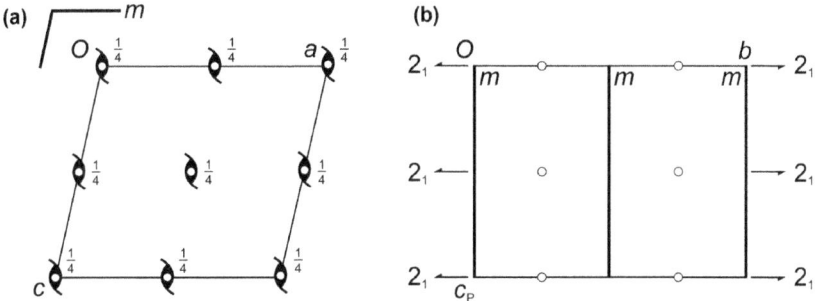

Figure 4.2 Distribution of symmetry elements in the unit cell for the case developed in Figure 4.1: (a) projection along the *b* axis; (b) projection along the *a* axis. In (a), the symbol *m* denotes mirror planes parallel to the plane of the page at $y = 0$ and $\frac{1}{2}$. The other symbol denotes a 2_1 screw axis running perpendicular to the plane of the page, plus inversion centres at $y = \frac{1}{4}$ and $\frac{3}{4}$. In (b), the 2_1 screw axes and inversion centres are at $x = 0$ and $\frac{1}{2}$. The label c_P refers to the *c* axis of the unit cell drawn in projection (*i.e.* not in the plane of the page).

The situation we have just described corresponds to space group type $P2_1/m$. The diagrams in Figure 4.2 are the key illustrations of the space group, which show how the symmetry elements are distributed within the unit cell. These types of diagrams are given for all 230 crystallographic space group types in *International Tables*. Electronic representations also exist, which can demonstrate the distribution of the symmetry elements in three dimensions. The space group types in *International Tables* are assigned a number (1–230) which is often quoted in the literature, *e.g.* space group $P2_1/c$ (no. 14). There is little value to this because the numbering scheme does not add anything to the information already given by the space group symbol, especially for the more common and easily recognisable space groups.

In fact, Figure 4.2 is not quite the diagram that appears in *International Tables* for space group type $P2_1/m$. The difference is that *International Tables* shows an inversion centre at the origin of the unit cell, with the mirror planes positioned at $y = \frac{1}{4}$ and $\frac{3}{4}$ (Figure 4.3). This corresponds to "pushing" all of the symmetry elements by $\frac{1}{4}$ along the *b* axis, which amounts to changing the origin of the unit cell. We have already seen that the choice of origin in a crystal structure is arbitrary. It is the *relative* positions of the symmetry elements that characterize the space group type. The choice made in *International Tables* is convenient because it gives a tidy relationship between the coordinates related by the inversion centre, namely x,y,z and $-x,-y,-z$. For this reason, space groups containing inversion centres are preferably described with an inversion centre at the origin.

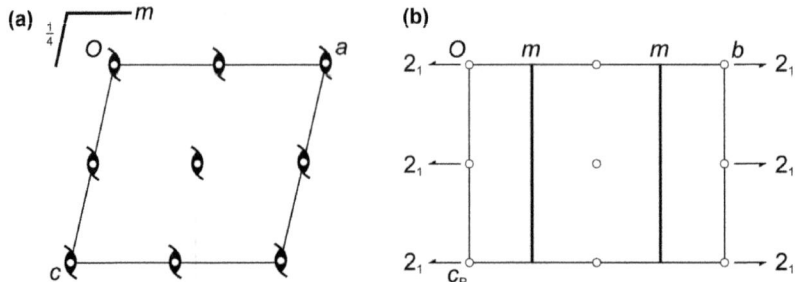

Figure 4.3   Two projections of space group $P2_1/m$ as they appear in the *International Tables for Crystallography*. The origin is defined at an inversion centre (denoted by open circles). The label c_P refers to the c axis of the unit cell drawn in projection (*i.e.* not in the plane of the page). The mirror plane perpendicular to b is indicated at $y = \frac{1}{4}$; the additional mirror plane at $y = \frac{3}{4}$ is implied.

4.2.2   Equivalent Positions

Applying all of the symmetry operations of a space group to a position x,y,z generates a set of coordinates known as the *general equivalent positions* (GEPs). Formally, the equivalence refers to points that are *equivalent under the action of the symmetry operations of the space group*. For space group $P2_1/m$, represented as in Figure 4.3, the GEPs are:

$$x,y,z \quad -x,\tfrac{1}{2}+y,-z \quad -x,-y,-z \quad x,\tfrac{1}{2}-y,z$$

The list shows that atoms appear in groups of four, so we only need to specify coordinates for one atom in the group to know the coordinates of all four. This is the key simplifying feature of the space group for describing a crystal structure because it reduces the number of parameters that must be specified. For example, if we specify the GEPs above and we know that there is a C atom at fractional coordinates 0.1,0.1,0.1, we know that there must also be C atoms at fractional coordinates:

$$0.1,0.1,0.1 \quad -0.1,0.6,-0.1 \quad -0.1,-0.1,-0.1 \quad 0.1,0.4,0.1$$

Moving all of these into the unit cell (by adding 1 to the negative coordinates) gives:

$$0.1,0.1,0.1 \quad 0.9,0.6,0.9 \quad 0.9,0.9,0.9 \quad 0.1,0.4,0.1$$

Every space group has a constituent set of GEPs (which will depend on the unit-cell setting and origin choice), which are derived from the set of symmetry operations. The simplest space group $P1$ has only one GEP, x,y,z, while the most complicated space groups have 192 GEPs. All of these lists are tabulated in *International Tables*.

There is one more thing to note on this topic: there can be some specific values of x,y,z that produce a smaller set of positions when they are substituted into the list of GEPs. For example, consider what happens to the list of coordinates in space group $P2_1/m$ when $x,y,z = 0,0,0$:

$$x,y,z \quad -x,\tfrac{1}{2}+y,-z \quad -x,-y,-z \quad x,\tfrac{1}{2}-y,z$$
$$0,0,0 \quad\quad 0,\tfrac{1}{2},0 \quad\quad\quad 0,0,0 \quad\quad\quad 0,\tfrac{1}{2},0$$

Some of the coordinates are duplicated, so an atom at 0,0,0 would actually have only one other equivalent at $0,\tfrac{1}{2},0$. The physical reason for this is obvious: an atom at 0,0,0 lies on the inversion centre, so application of the inversion operation does not generate another atom. Any values of x,y,z that produce duplicates when inserted into the list of GEPs are referred to as *special equivalent positions* (SEPs), or commonly just *special positions*. SEPs arise on point symmetry elements, which could be inversion centres, mirror planes, rotation axes or the inversion point of an improper rotation axis. For mirror planes, any point on the plane is an SEP, while for rotation axes, any point on the axis is an SEP. Symmetry operations with an inherent translation component, *i.e.* screws or glides, cannot have associated SEPs because there is no way that these operations can map a point onto itself.

4.2.3 The Asymmetric Unit

Once the symmetry operations of the space group are specified, we only need to specify the coordinates of one atom in each group of GEPs. These are sometimes referred to as the "unique" atoms, and collectively as the *asymmetric unit*. Formally, the asymmetric unit is the minimal fraction of the unit-cell contents that must be described to generate the entire unit-cell contents by applying the symmetry operations of the space group (Figure 4.4). For single-component molecular crystals, the asymmetric unit frequently comprises one molecule, but it does not have to. It could comprise only a fraction of a molecule if that molecule exhibits point symmetry that is retained within the crystal's space group. For example, the fumarate dianions in the Case Study of Chapter 3 are situated on inversion centres within the space group so the asymmetric unit contains only half of each fumarate molecule. Alternatively, there could be more than one molecule in the asymmetric unit. Obviously, this must be the case for a multi-component crystal, but it could also be the case for a single-component crystal if there are molecules that are not related by any

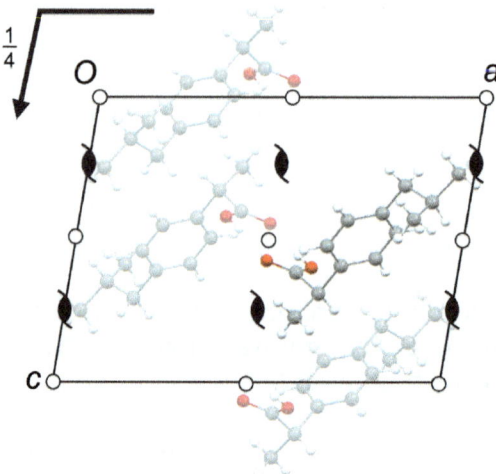

Figure 4.4 Unit-cell contents for the crystal structure of ibuprofen ($C_{13}H_{18}O_2$) in space group $P2_1/c$. The symmetry elements are shown and the asymmetric unit is highlighted. The other three molecules within the unit cell are generated by application of the symmetry operations. The symbol in the top-left denotes the c-glide parallel to the plane of the page at $y=\frac{1}{4}$ and $\frac{3}{4}$.

symmetry operation of the crystal's space group. In that case, the molecules are said to be *crystallographically distinct* or *independent*. For single-component crystals, the number of molecules in the unit cell is usually given the symbol Z and the number of molecules in the asymmetric unit is given the symbol Z'. An example of a structure with $Z'=2$ is form 2 of ropivacaine hydrochloride, described in the Case Studies of Chapters 6 and 7. Practically, there is no unique representation of the asymmetric unit for any structure. For example, if there are four symmetry-related molecules in the unit cell, we could pick any of the four molecules to be the asymmetric unit. In principle, we could even pick an appropriate set of atoms from different molecules, but it is sensible for the asymmetric unit to comprise a complete molecule where possible.

4.3 Crystal Systems

Space groups are commonly classified into different categories according to some feature of their symmetry. Most prevalent is the division into *crystal systems*, based on the types of symmetry operations that each group contains. There are seven crystal systems, described in Table 4.1. As indicated in the table, the crystal system

Table 4.1 Description of the seven crystal systems.[a,b]

Crystal system	Essential symmetry	Unique axis	Unit-cell dimensions
Triclinic	—	—	$a \neq b \neq c,\ \alpha \neq \beta \neq \gamma$
Monoclinic	2-fold along one axis	b	$a \neq b \neq c,\ \alpha = \gamma = 90°,\ \beta \neq 90°$
Orthorhombic	2-folds along three perpendicular axes	—	$a \neq b \neq c,\ \alpha = \beta = \gamma = 90°$
Trigonal	3-fold along one axis	[111]	$a = b = c,\ \alpha = \beta = \gamma$
		c	$a = b \neq c,\ \alpha = \beta = 90°,\ \gamma = 120°$
Tetragonal	4-fold along one axis	c	$a = b \neq c,\ \alpha = \beta = \gamma = 90°$
Hexagonal	6-fold along one axis	c	$a = b \neq c,\ \alpha = \beta = 90°,\ \gamma = 120°$
Cubic	3-folds along four axes mutually disposed in a tetrahedral arrangement	—	$a = b = c,\ \alpha = \beta = \gamma = 90°$

[a]2-fold, 3-fold, *etc.*, means a proper or improper rotation or screw with order 2, 3, *etc.* The 2-fold case includes the possibility of a mirror or glide plane perpendicular to the axis.
[b]The trigonal system includes two different lattice types. [111] denotes the body-diagonal of the unit cell.

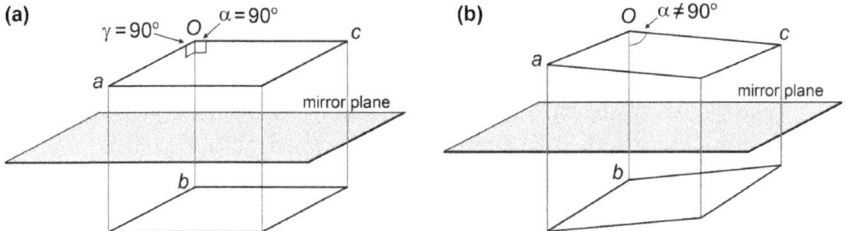

Figure 4.5 Mirror symmetry perpendicular to *b* imposes $\alpha = \gamma = 90°$, as shown in (a). Diagram (b) shows the upper and lower faces not parallel to the mirror plane, thereby producing an unacceptable unit cell.

provides a link between the space group type and the metric properties of the crystal because the shape of the unit cell must conform to the symmetry of the space group. As an example, consider again space group $P2_1/m$ as defined in Figure 4.3. If the structure has mirror symmetry perpendicular to the *b* axis of the unit cell, the *b* axis must be perpendicular to the *ac* face of the unit cell (Figure 4.5). If it was not, the *ac* face at one end of the unit cell would not be properly reflected onto the *ac* face at the other end of the unit cell. Hence, the unit-cell angles α and γ *must* be 90°. Crystal systems with higher symmetry can impose further constraints on the unit-cell angles, and also on the lengths, as summarised in Table 4.1. In the table, " $=$ " means "is required to be equal to" and " $\neq$ " means "is not required to be equal to", rather than "does not equal". For example, a triclinic crystal could have a lattice that can be described with a unit cell having $a = b = c$ and $\alpha = \beta = \gamma = 90°$, but this is not a *requirement*. By contrast, a cubic crystal *must* have $a = b = c$ and $\alpha = \beta = \gamma = 90°$. If we know the crystal system, we know any constraints on the metric properties; but if we know the metric properties, we do not necessarily know the crystal system. The metric constraints are therefore a consequence of the crystal system, not *vice versa*.

Another thing to note in Table 4.1 is the conventional choice of the "unique axis" for the monoclinic, trigonal, tetragonal and hexagonal crystal systems. This is the direction along which the principal symmetry element of the crystal system is aligned. For the monoclinic crystal system, the characteristic 2-fold is generally chosen to be aligned with the b axis. This could mean a 2-fold rotation or screw axis parallel to the b axis, and/or a mirror or glide plane perpendicular to the b axis. The consequence is that the angles α and γ must be 90°, while β is not required to be 90°. For the trigonal, tetragonal and hexagonal systems, the unique axis is conventionally chosen to be the c axis, which gives $a = b$ in each case, and $\gamma = 120°$ for the trigonal and hexagonal systems. This glosses over a few details for the trigonal crystal system, which is actually compatible with two possible lattice types with associated constraints as indicated in Table 4.1. Trigonal space groups are quite rare for typical pharmaceutical compounds, so they are not discussed here in any further depth. There is no obvious logic in choosing a different unique axis for the monoclinic crystal system compared to the trigonal, tetragonal and hexagonal systems, but this convention is applied more-or-less universally in the contemporary literature and it will not change.

4.4   Centred Unit Cells

We saw in Chapter 2 that a given lattice can be described by any number of unit cells. It seems reasonable to choose a unit cell that is as small as possible, *i.e.* a primitive unit cell, because we would otherwise include redundant translational symmetry within the unit cell. If we only wish to define the positions of the *lattice points*, this is the most sensible approach. When we consider the full symmetry of a crystal structure, however, there are situations where it is better to choose a unit cell encompassing more than one lattice point. For example, consider the circumstance shown in Figure 4.6. The structure belongs to the monoclinic crystal system with its 2-fold rotation axes running conventionally parallel to the b axis. Perpendicular to the b axis, the lattice points lie in planes within which

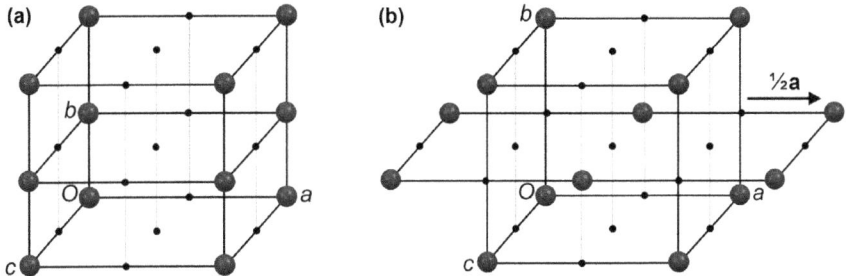

Figure 4.6 Alternative arrangements of layers of lattice points with perpendicular 2-fold axes. The lattice points are indicated by the larger spheres. The small dots show the locations of the 2-fold axes between lattice points. The 2-fold axes run parallel to the *b* axis. Arrangement (a) produces a monoclinic *P* unit cell, with the *b* axis spanning between adjacent layers (two unit cells are shown in the vertical direction). Arrangement (b) shows an offset of $\frac{1}{2}\mathbf{a}$, producing a monoclinic *C* unit cell.

the *a* and *c* axes are defined. All of these lattice planes must be parallel, but it is possible for them to be *offset* with respect to each other while still maintaining the 2-fold rotation symmetry of the structure. We have seen on a few occasions that symmetry elements positioned on lattice points generate additional symmetry elements between lattice points due to combination with the crystal's translational symmetry. So, each *ac* plane of the lattice could accommodate the structure's 2-fold rotation axes passing either through its lattice points or between its lattice points. If all *ac* planes are aligned so that a given 2-fold axis passes through lattice points, we can define a primitive unit cell with the *b* axis spanning the distance between adjacent *ac* planes (Figure 4.6(a)). Alternatively, a given 2-fold axis might pass alternately through and between lattice points in neighbouring *ac* planes, so that the *b* axis defined parallel to the 2-fold axes spans *two ac* planes (Figure 4.6(b)). The resulting unit cell has lattice points at its corners (as always), but it will have an additional lattice point in the central *ac* plane.

For the 2-fold axes to extend properly through the crystal structure, the offset of the adjacent *ac* planes must be specifically defined. However, there are several options: in this case, the middle *ac* plane could be offset by $\frac{1}{2}\mathbf{a}$, $\frac{1}{2}\mathbf{c}$, or $\frac{1}{2}(\mathbf{a}+\mathbf{c})$. The lattice point in the central plane then lies at either $\frac{1}{2},\frac{1}{2},0$; $0,\frac{1}{2},\frac{1}{2}$ or $\frac{1}{2},\frac{1}{2},\frac{1}{2}$, which correspond respectively to the centre of the *ab* face of the unit cell, the centre of the *bc* face, or the centre of the entire unit cell. In each case, the unit cell is said to be *centred*, and the three cases are labelled *C*-centred, *A*-centred and *I*-centred, respectively. In general, centred unit cells arise when symmetry elements of the space group do not align with any primitive lattice vector, so it is not possible to define a primitive unit cell in which the symmetry elements are aligned with the unit-cell edges. There is nothing unusual about this. It is just another way in which the symmetry operations in a crystal structure can combine with the crystal's inherent translational periodicity.

4.4.1 Bravais Lattices

For the 230 possible space group types, it turns out that there are only 14 lattice types, which are called the *Bravais lattices*. Bravais lattices tend to be quite heavily emphasised in crystallographic texts, but really the concept is secondary. We already understand that the symmetry operations of a space group must combine with the crystal's inherent translational periodicity. The Bravais lattices merely catalogue the various lattice types that permit that. A Bravais lattice is labelled according to the crystal system to which it belongs and the centring type of the conventional unit cell, *e.g.* monoclinic P or orthorhombic C. A summary is given in Table 4.2. The labelling scheme relies as ever on certain conventions. For the monoclinic crystal system, Table 4.2 indicates specifically that the centred Bravais lattice is monoclinic C. We saw previously that there are a few potential ways in which a centred monoclinic lattice might be defined. With the b axis unique, we could define an A-centred, C-centred or I-centred unit cell. These possibilities can be transformed into one another by re-defining the unit-cell edges within the ac plane.

Table 4.2 The 14 conventional Bravais lattice types.

	Centring points	Lattice points per unit cell
Triclinic P	—	1
Monoclinic P	—	1
Monoclinic C	$\frac{1}{2},\frac{1}{2},0$	2
Orthorhombic P	—	1
Orthorhombic C	$\frac{1}{2},\frac{1}{2},0$	2
Orthorhombic F	$0,\frac{1}{2},\frac{1}{2}$; $\frac{1}{2},0,\frac{1}{2}$; $\frac{1}{2},\frac{1}{2},0$	4
Tetragonal P	—	1
Tetragonal I	$\frac{1}{2},\frac{1}{2},\frac{1}{2}$	2
Trigonal P	—	1
Trigonal R	Rhombohedral axes: —	1
	Hexagonal axes: $\frac{2}{3},\frac{1}{3},\frac{1}{3}$; $\frac{1}{3},\frac{2}{3},\frac{2}{3}$	3
Hexagonal P	—	1
Cubic P	—	1
Cubic I	$\frac{1}{2},\frac{1}{2},\frac{1}{2}$	2
Cubic F	$0,\frac{1}{2},\frac{1}{2}$; $\frac{1}{2},0,\frac{1}{2}$; $\frac{1}{2},\frac{1}{2},0$	4

By convention, monoclinic C is the standard choice. Similar conventions are chosen to label the unique Bravais lattice types in the other crystal systems. It is seen in Table 4.2 that the conventional unit cells for the various Bravais lattice types contain one, two, three or four lattice points. Thus, centred unit cells have a volume up to four times that of a primitive unit cell for the same lattice. This should be remembered in circumstances where it might be necessary to compare unit-cell volumes for different crystal structures.

4.4.2 Centred Unit Cells and Equivalent Positions

For a space group described with a centred unit cell, the translation operation within the unit cell must appear in the list of equivalent positions. A centring operation is easily recognizable because it does not change the sign or otherwise rearrange any of the coordinates x,y,z. For example, consider a C-centred monoclinic unit cell with the 2-fold rotation axis parallel to b and passing through the origin (space group type $C2$). The GEPs and corresponding symmetry elements are:

$$x,y,z$$
$$-x,y,-z \qquad \text{2-fold rotation axis at } (0,y,0)$$
$$\tfrac{1}{2}+x,\tfrac{1}{2}+y,z \qquad C\text{-centring operation}$$
$$\tfrac{1}{2}-x,\tfrac{1}{2}+y,-z \qquad 2_1 \text{ screw axis at } (\tfrac{1}{4},y,0)$$

The fourth coordinate in the list is obtained by applying the 2-fold rotation to the point generated by the C-centring operation, *i.e.* $\tfrac{1}{2}+x,\tfrac{1}{2}+y,z \rightarrow -(\tfrac{1}{2}+x),\tfrac{1}{2}+y,-z$, then adding 1 to the resulting x coordinate to bring it into the unit cell. Compared to the general point x,y,z, it corresponds to the action of a 2_1 screw axis parallel to the b axis at $(\tfrac{1}{4},y,0)$. Thus, if there are 2-fold rotation axes parallel to the b axis with a monoclinic C lattice, there must also be 2_1 screw axes parallel to the b axis. This concept is no different from the earlier discussion on how to combine symmetry operations, and centred space groups should just be viewed in the same way.

In the above list of GEPs for space group $C2$, notice that the coordinates can be split into two sets of two, related to each other by addition of $\tfrac{1}{2},\tfrac{1}{2},0$. An efficient description of these GEPs needs only to list the first two coordinates, with a note that the centring operation $\tfrac{1}{2}+x,\tfrac{1}{2}+y,z$ should also be applied. In the example given, this does not

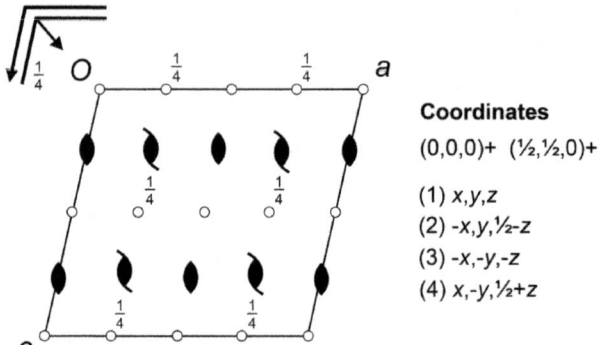

Coordinates

$(0,0,0)+$ $(\frac{1}{2},\frac{1}{2},0)+$

(1) x,y,z
(2) $-x,y,\frac{1}{2}-z$
(3) $-x,-y,-z$
(4) $x,-y,\frac{1}{2}+z$

Figure 4.7 Symmetry elements (projected down the *b* axis) and coordinates for space group *C2/c*. The symbols at the top left represent a *c*-glide parallel to the plane of the page at $y=0$ and $\frac{1}{2}$ and an *n*-glide parallel to the plane of the page at $y=\frac{1}{4}$ and $\frac{3}{4}$. The complete list of GEPs comprises 0,0,0 plus each of (1)–(4) and $\frac{1}{2},\frac{1}{2},0$ plus each of (1)–(4). Hence, there are eight GEPs in total.

achieve much in the way of abbreviation but it can simplify matters considerably for higher-symmetry groups with many more GEPs. An example for the common monoclinic space group *C2/c* is illustrated in Figure 4.7.

4.4.3 Glides in Centred Unit Cells

We saw in Section 3.4.2 that the inherent translation part of a glide is limited in a primitive unit cell to half a unit-cell edge in order to be compatible with the inherent crystal translations. For example, a glide perpendicular to the *b* axis can only have translation $\frac{1}{2}\mathbf{a}$, $\frac{1}{2}\mathbf{c}$ or $\frac{1}{2}(\mathbf{a}+\mathbf{c})$. In a centred unit cell, there are other possibilities because the translational component of the glide could be half of a *centring* translation. Again, there is no new concept: it is simply that the relationship between the unit cell and the inherent crystal translations has changed, so there are some new possibilities to define the glide translation relative to the unit cell. We will not attempt to list every possibility. A representative example is the orthorhombic space group *Fdd2*, which has an *F*-centred unit cell that includes "*d*-glides" perpendicular to both the *a* and *b* axes (Figure 4.8). The *d*-glides perpendicular to the *a* axis have translational component $\frac{1}{4}(\mathbf{b}+\mathbf{c})$, while the *d*-glides perpendicular to the *b* axis have translational component $\frac{1}{4}(\mathbf{a}+\mathbf{c})$. Since the *F*-centred unit cell has centring points at $0,\frac{1}{2},\frac{1}{2}$ and $\frac{1}{2},\frac{1}{2},0$ (and also $\frac{1}{2},0,\frac{1}{2}$), the translational components of the *d*-glides clearly correspond to half of the centring translations.

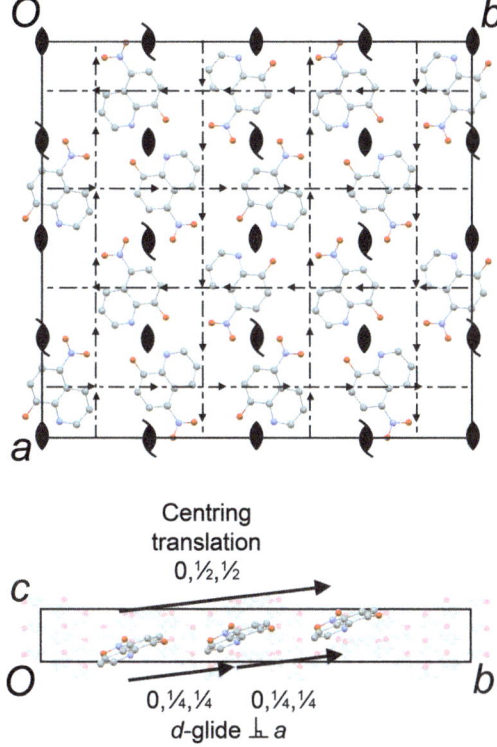

Figure 4.8 Illustration of *d*-glides in space group *Fdd2*. In the upper diagram, the *d*-glides are indicated by the dot-dash lines, with the arrows showing the positive direction of the glide translation. The lower diagram highlights three molecules related by a *d*-glide. The outer molecules of the three are related by the *F*-centring translation $0, \frac{1}{2}, \frac{1}{2}$ and the central molecule is reflected in the plane of the page. The structure shows the antibiotic nitroxoline (CSD: YOKQAQ).

4.5 Space Group Symbols

For ease of communication, each of the 230 space group types is given a symbol. Most common are the Hermann–Mauguin (H–M) symbols, which comprise a letter to describe the centring type and an indication of the key symmetry elements lying parallel to a specified set of directions. We have already used the "short" H–M symbols $P1$, $P\bar{1}$, $P2_1/m$, $C2$, *etc.* An example of a "long" H–M symbol, corresponding to the situation discussed in Section 4.2.1, is $P\,1\,2_1/m\,1$. For the monoclinic crystal system, the order of the operations listed in the full symbol corresponds to the symmetry elements aligned with the unit-cell axes. So, $P\,1\,2_1/m\,1$ indicates the identity operation "1" parallel to the *a* and *c* axes (*i.e.* no specific symmetry elements other than translation), and "$2_1/m$" parallel to the *b* axis. The "$/m$" part of "$2_1/m$" indicates a mirror plane perpendicular to the *b* axis. In making the contraction to the short symbol $P2_1/m$, we clearly must know that the symmetry elements are conventionally aligned along the *b* axis. Short symbols are generally used for conventional cases (which are overwhelmingly most common), while long symbols are used if there is a need to demonstrate that the description deviates from convention.

4.6 Other Classifications of Space Groups

In addition to the seven crystal systems, there are numerous other ways in which the 230 space group types are classified according to some aspect of their symmetry. For pharmaceutical crystals, two of the most relevant refer to centrosymmetry and chirality. An overview is given in Figure 4.9.

4.6.1 Centrosymmetric and Non-centrosymmetric Space Groups

A centrosymmetric space group contains inversion centres, while a non-centrosymmetric space group does not. Of the 230 space group types, 91 are centrosymmetric and 139 are non-centrosymmetric. The primary practical importance (as far as we are concerned) is that, under the right conditions, X-ray diffraction can distinguish between structures that are related to each other by an inversion operation. For a centrosymmetric crystal, this has no relevance because the

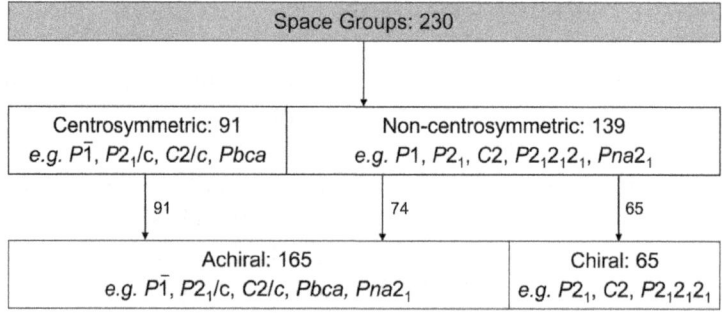

Figure 4.9 Overview of relevant classifications for space groups, considering centrosymmetry (containing inversion centres) and chirality (containing only proper symmetry operations).

structure is identical to its inversion-related counterpart. For a non-centrosymmetric crystal, however, there is a difference between inversion-related structures. Specifically, the two cases have different *absolute structure*, which refers to their "handedness". If we can determine absolute structure using X-ray diffraction, it is possible to distinguish between different enantiomers of chiral molecules in a non-centrosymmetric crystal. This will be discussed further in Chapter 7.

4.6.2 Chiral Space Groups

A chiral space group is one that does not contain any symmetry operation that acts to change the handedness of an object. This means that it does not contain any inversion centres, mirrors/glides, or improper rotations. Thus, chiral space groups contain only proper operations. It is vital to distinguish chiral space groups from centrosymmetric or non-centrosymmetric space groups. A centrosymmetric space group cannot be chiral, but a non-centrosymmetric space group may or may not be chiral (Figure 4.9). For example, space group *Pm* (GEPs: x,y,z and $-x,y,-z$) is non-centrosymmetric, but it is not chiral because it contains mirror planes. Of the 230 space group types, 65 are chiral. These groups are sometimes referred to (rather obtusely) as the *Sohncke groups*. An enantiomerically pure sample of chiral molecules must crystallise in a chiral space group because there is no way that molecules with the same handedness could be related by any symmetry operation that acts to change the hand of the molecule—*i.e.* the space group used to describe the crystal cannot possibly contain any improper operation. Since chiral space groups must be non-centrosymmetric, their handedness can be determined using single-crystal X-ray diffraction. Thus, X-ray diffraction provides a means to determine the absolute configuration of chiral API molecules.

4.7 Summary of Key Points

- All symmetry operations within a crystal, including its inherent translational symmetry, combine to make the space group. The space group is a *closed* set of symmetry operations, which means that any symmetry operation in the group combines with any other to produce another operation of the group. The number of space group types is exactly 230.

- Coordinates related by symmetry operations of the space group are called general equivalent positions (GEPs). Coordinates located on point symmetry elements are called special equivalent positions (SEPs). The minimal set of coordinates that must be described to generate the entire unit-cell contents, by applying the symmetry operations of the space group, is called the asymmetric unit.
- Space groups are classified into seven crystal systems, based on the symmetry operations that they contain. The symmetry operations that characterise each crystal system can impose constraints on the metric features of the unit cell.
- In circumstances where symmetry elements do not align with the shortest (primitive) lattice vectors, a centred unit cell might be defined, containing additional lattice points.
- The combination of the crystal system and the unit-cell centring type is called the Bravais lattice. There are only 14 possible Bravais lattice types. For circumstances where a given Bravais lattice might be described by several alternative centred unit cells, conventions are defined to establish a standard setting.
- Structures in non-centrosymmetric space groups have an absolute structure, which refers to their "handedness". Under certain circumstances, absolute structure can be determined using single-crystal X-ray diffraction.

4.8 Case Study: The Monoclinic Form of Paracetamol (Acetaminophen)

Paracetamol (shown in Figure 1.1) is polymorphic and has been the subject of numerous crystallographic studies under a variety of conditions. In this case study, we are interested in the monoclinic form, which was first reported in 1974 in space group $P2_1/a$.[2] The same structure was reported in 1998 in space group $P2_1/n$.[3] The unit-cell parameters are as follows:

CSD identifier	Temp (K)	Space group	a (Å)	b (Å)	c (Å)	α (°)	β (°)	γ (°)
HXACAN01	298	$P2_1/a$	12.93(4)	9.40(1)	7.10(2)	90	115.9(2)	90
HXACAN07	123	$P2_1/n$	7.094(1)	9.232(1)	11.620(1)	90	97.82(1)	90

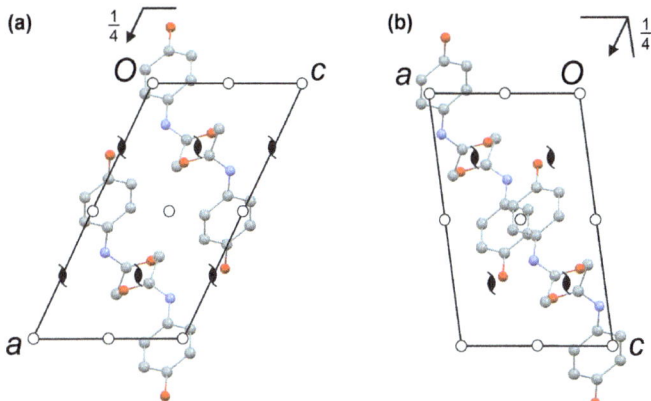

Figure 4.10 Monoclinic form of paracetamol projected onto the *ac* plane, showing the unit cell and symmetry elements: (a) HXACAN01 described in space group $P2_1/a$; (b) HXACAN07 described in space group $P2_1/n$. The similarity between the orientations of the molecules and their positions relative to the symmetry elements should be apparent.

These are two descriptions of the same crystal structure, so what must be done to the structure described in $P2_1/a$ to produce the description in $P2_1/n$? Firstly, note that both space groups are monoclinic with 2_1 screw axes running parallel to the *b* axis and that the length of the *b* axis is comparable in the two cases (within experimental error and changes due to temperature). This is a vital indicator when comparing monoclinic crystal structures: if the structures are the same, the unique axes must be comparable. Both space groups have a glide plane perpendicular to *b*. In $P2_1/a$, the translation component of the glide is $\frac{1}{2}\mathbf{a}$, while in $P2_1/n$ it is $\frac{1}{2}(\mathbf{a}+\mathbf{c})$. Since we know that these crystal structures are the same, these glides must be equivalent, which means that **a** in the first structure must correspond to $\mathbf{a}+\mathbf{c}$ in the second structure. The situation in projection onto the *ac* plane is shown in Figure 4.10. To transform the $P2_1/a$ description into the $P2_1/n$ description, it is apparent that the new **a** vector should be obtained from the old $-\mathbf{c}$, and the new **c** vector should be obtained from the old $\mathbf{a}+\mathbf{c}$. After re-defining the HXACAN01 unit cell in this way, the result is:

CSD identifier	Temp (K)	Space group	*a* (Å)	*b* (Å)	*c* (Å)	α (°)	β (°)	γ (°)
HXACAN01	298	$P2_1/n$	7.10	9.40	11.72	90	97.1	90
HXACAN07	123	$P2_1/n$	7.094(1)	9.232(1)	11.620(1)	90	97.82(1)	90

The space group is now $P2_1/n$ because the translation component of the glide runs along the *ac* diagonal of the new unit cell. Note that the structure and its constituent symmetry elements have not changed; all that has changed is the description of the lattice vectors within the *ac* plane. The lattice vectors now match those of HXACAN07.

The transformation must also be applied to the reported atomic coordinates of HXACAN01 to give a new set of coordinates for the structure referring to the new unit cell. Practically, this is best achieved within a suitable crystallographic software package. After making the transformation, the resulting *x* and *z* coordinates are identical to those reported for HXACAN07 but the *y* coordinates are still different. Turning the two structures to look along the *a* axis (Figure 4.11) shows that the molecules appear to be mirrored perpendicular to the *b* axis. Should we have inverted the *b* axis when making our previous transformation of the axes? Actually, we could not have done that (and we can't do it now), because the result would be a left-handed set of axes. So perhaps we could just "reflect" our transformed HXACAN01 structure in a plane perpendicular to the *b* axis? This could be achieved simply by multiplying all of the *y* coordinates by -1. But this also is not allowed because it would be *changing the structure*. We are not trying to change the structure of HXACAN01 to make it match HXACAN07. We know that the structures are the same, and we are trying to find corresponding *descriptions*.

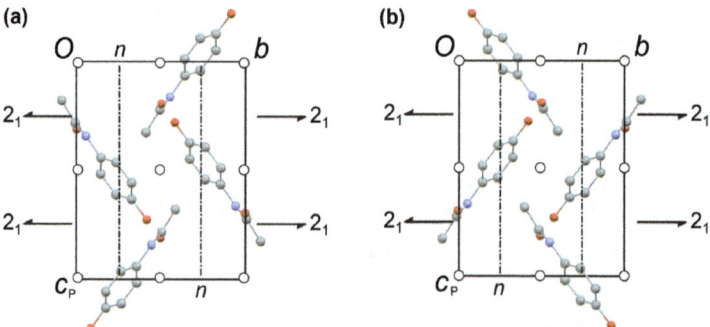

Figure 4.11   View of (a) HXACAN01 and (b) HXACAN07 along the *a* axis after transforming HXACAN01 as described in the text. Both structures are now described in space group $P2_1/n$. The dot-dash lines indicate the *n*-glide viewed parallel to its plane. It appears as if the two structures are mirrored perpendicular to the *b* axis, but they can be brought into coincidence by choosing a different origin for HXACAN01, as described in the text.

One thing that we are allowed to do is change the origin of the unit cell. This might seem unnecessary, because the origin of the unit cell is already on an inversion centre in both HXACAN01 and HXACAN07. However, we saw in Section 3.2.3 that an inversion centre at the origin of a primitive unit cell must be accompanied by inversion centres half-way between all of the lattice points. These inversion centres are not all related by the translational symmetry of the crystal structure, so they are not all equivalent origin choices. The final step to match the descriptions of HXACAN01 and HXACAN07 is to apply an origin shift of $\frac{1}{2},\frac{1}{2},0$. This brings the unit cells and fractional coordinates of the atoms into complete coincidence.

References

1. International Union of Crystallography, *International Tables for Crystallography, Volume A*, ed. T. Hahn, 2005.
2. M. Haisa, S. Kashino, R. Kawai and H. Maeda, *Acta Crystallogr., Sect. B: Struct. Crystallogr. Cryst. Chem.*, 1974, **B32**, 1283, [CSD: HXACAN01].
3. G. Nichols and C. S. Frampton, *J. Pharm. Sci.*, 1998, **87**, 684, [CSD: HXACAN07].

5 Planes and Crystal Morphology

Summary

Defining planes in a crystal is necessary to describe morphology and for visualisation and description of a diffraction pattern. Planes in crystals are defined relative to the unit cell using a set of three integers called Miller indices. Since a crystal is periodic by translation, Miller indices actually describe sets of planes through the crystal with a perpendicular separation known as the d-spacing. Each set of planes can be efficiently represented by its normal vector, which has its direction perpendicular to the planes. If the length of the normal vector is defined as the reciprocal of the d-spacing, the ends of the vectors representing many sets of planes define a lattice, called the reciprocal lattice. The crystal habit can be constructed by placing each external crystal face along the line of its associated normal vector at some defined distance from the crystal's centre. The resulting habit is dominated by the faces that lie closest to the origin.

5.1 Introduction

There are a few reasons why we need to describe planes in crystals. Firstly, it will become useful when we discuss X-ray diffraction. Secondly, it is necessary to describe the *morphology* of crystals, which refers to their external shape. Morphology has practical importance for pharmaceutical crystals because it can influence, for example, whether a crystalline powder flows easily (perhaps from a hopper feeding a tablet press), or whether it might have a tendency to stick. Crystals usually have flat faces which can be described as planes with a defined relationship to the unit cell. We begin this chapter by

Pharmaceutical Crystallography: A Guide to Structure and Analysis
By Andrew Bond
© Andrew Bond 2019
Published by the Royal Society of Chemistry, www.rsc.org

showing how to label planes with respect to the unit cell then move on to consider crystal faces. In the process, the important concept of the *reciprocal lattice* will be introduced, which will be useful later to make an efficient description of a diffraction pattern.

5.2 Defining Planes in Crystals

In geometrical terms, there are several ways to define planes. When dealing with crystals, we are most interested in two of them (Figure 5.1):

(a) A plane can be defined by specifying three (non-collinear) points on the plane.
(b) A plane can be defined by specifying a line perpendicular to the plane (called the *normal*) and one point on the plane.

We will use method (a) to develop an intuitive picture of planes defined with reference to the unit cell, then apply method (b) to introduce the reciprocal lattice.

5.2.1 Miller Indices

Although we could use any three points to define a plane in a crystal structure, a sensible choice is to define the points where the plane cuts the three axes of the unit cell. This is simplified by the fact that a point on one of the unit-cell axes has two fractional coordinates that are zero. So we need to define the three points $x,0,0$; $0,y,0$ and $0,0,z$, which amounts to specifying only three values, xyz. For reasons we will see in a moment (Section 5.2.2), the xyz values are not used directly to label the plane. Instead, we specify the *reciprocals* of xyz, which are denoted hkl and called *Miller indices*. If a plane does not cross a unit-cell edge, it is given a Miller index of 0 with reference to that edge. When describing planes, Miller indices are enclosed in parentheses, (hkl). It is possible to specify a negative intercept on an axis. In that case, the Miller index is usually written with the negative sign above (an

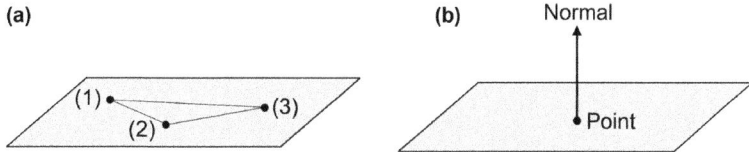

(a) **(b)** Normal

(1) (3)
(2)

Point

Figure 5.1 Two ways to define planes: (a) specifying three (non-collinear) points; (b) specifying the normal (perpendicular to the plane) and one point on the plane.

"overbar"). For example, Figure 5.2 shows the difference between (111) (spoken "one, one, one") and (11$\bar{1}$) (spoken "one, one, bar one").

Since the unit cell is the fundamental building block of the crystal structure, defining a plane with respect to the unit cell actually defines a set of parallel planes through the entire structure. If we consider the (001) plane, for example, it crosses the c axis of the unit cell at $z = 1$ and lies parallel to the a and b axes (Figure 5.3(a)). Thus, it corresponds to the upper ab face of the unit cell. There must be an equivalent plane at the lower ab face of the unit cell, and in every other unit cell, so the (001) planes actually comprise a parallel set with separation defined by the c axis. Note that this is not necessarily the *perpendicular* separation between the planes because the c axis may not lie perpendicular to the ab face. We return to this subject in a moment.

Now consider the (002) plane. According to our definition, this also lies parallel to the ab face of the unit cell, passing through $z = \frac{1}{2}$ (Figure 5.3(b)). How is this different from (001)? As we have described it so far, it isn't. The set of (002) planes in all unit cells would have the same separation as the (001) planes, just intersecting the c axis at

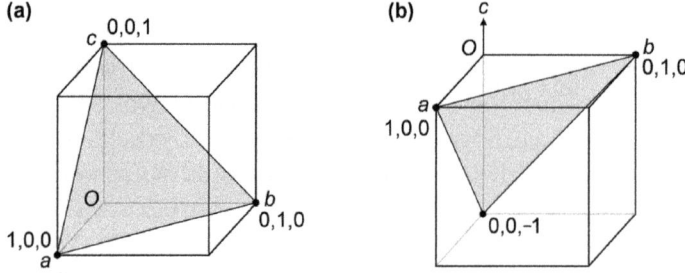

Figure 5.2 Defining Miller indices: (a) (111) planes; (b) (11$\bar{1}$) planes. The angle formed between the two planes can be visualised by imagining diagram (a) stacked on top of diagram (b).

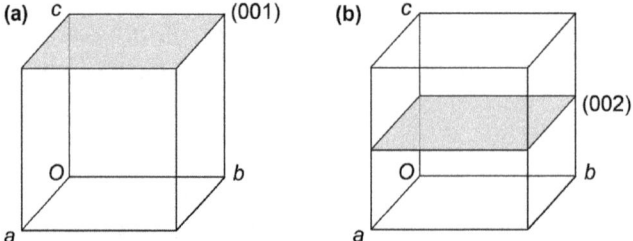

Figure 5.3 Planes defined relative to the unit cell: (a) (001); (b) (002). Individual planes are shown for illustration only (as described in the text). In crystals, sets of parallel planes run through the entire structure, as shown in Figure 5.4.

$z = \frac{1}{2}$ instead of $z = 1$. This corresponds only to choosing a different origin for the unit cell. So the definition must be updated: *Miller indices define a set of parallel and equally-spaced planes in a crystal, with the indices referring to the first plane out from a plane passing through the origin.* On this basis, the (002) planes comprise a set of planes parallel to the *ab* face of the unit cell, passing through $z = 0, \frac{1}{2}, 1$, *etc.* They are parallel to the (001) planes, but have a separation of only $\frac{1}{2}c$. The (003) planes would also be parallel, with a separation of $\frac{1}{3}c$, and so on. This picture leads to a common alternative description of Miller indices, by which they specify the number of planes in a parallel set that intersect each unit-cell edge in the range 0–1 (counting *either* the plane at 0 *or* the plane at 1). Thus, the (001) planes cut the *c* axis once, the (002) planes cut the *c* axis twice, *etc.*

5.2.2 *d*-Spacing

Geometrically, the distance between two parallel planes is defined by their perpendicular separation, which is the shortest distance between any two points on neighbouring planes. For a set of planes in a crystal, this distance is called the *d-spacing*. We cannot immediately know the *d*-spacing from the Miller indices because the calculation depends on the metric properties of the unit cell. In some crystal systems, the relationship can be defined quite easily in terms of some or all of the unit-cell parameters, but it becomes cumbersome for the triclinic case and we will not attempt to list any explicit mathematical expressions. Instead, we will consider another way to think about *d*-spacing, based on method (b) for defining planes.

5.2.3 Plane Normals and the Reciprocal Lattice

To show how method (b) can be used to define planes within a crystal structure, consider again the situation shown in Figure 5.3. Since the (001) and (002) planes are parallel, they have the same normal line. We can therefore describe the *orientation* of both the (001) and (002) planes by specifying the same normal line, but we need something more to distinguish the two cases. Following the same argument developed for Miller indices, we can complete the description if we also define a point on the first plane out from the origin. There is a neat way to achieve this: if we consider the normal line to extend from the origin and specify the required point where the normal line intersects the first plane out from the origin, we define the *normal vector*, which has length equal to the *d*-spacing. In this way, the normal vector uniquely

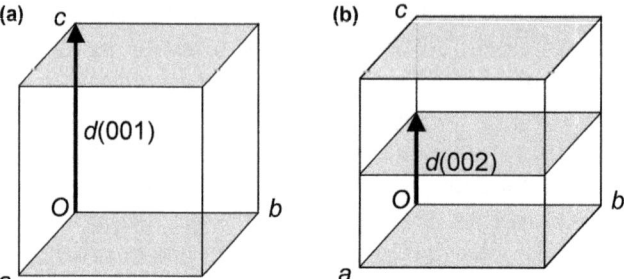

Figure 5.4   Representing the (001) and (002) sets of planes by their normal vectors. The vectors are parallel but are distinguished by their lengths, equal to the *d*-spacing of the planes they represent.

identifies each set of planes (Figure 5.4). In any circumstance where we need to represent several sets of planes, it is far easier to draw a collection of lines representing the normal vectors than it is to draw the planes themselves.

If we consider representing the planes (001), (002), (003), *etc.* in this way, we would define parallel normal vectors with relative length 1, $\frac{1}{2}$, $\frac{1}{3}$, *etc.*, as the planes become progressively more closely spaced. This would rapidly become inconvenient as the vectors become shorter and shorter. A more effective way is to define the normal vector length as the *reciprocal* of the *d*-spacing, as we did when we defined Miller indices. Then, the planes (001), (002), (003), *etc.*, are represented by parallel normal vectors with evenly-spaced relative lengths 1, 2, 3, *etc.* This is why the Miller indices are defined as they are: the *h* indices in this set of parallel planes become integral multipliers of the *reciprocal d*-spacing. If we draw the reciprocal vectors for planes (001), (002), (003), *etc.* at the same origin, their ends define a regular row of points (Figure 5.5). We can extend this picture by considering a few sets of planes with other orientations. For example, the normal vectors representing the planes (010), (020), (030), *etc.* define a second row of points with spacing equal to the reciprocal of the *d*-spacing for the (010) planes and with the position of each point along the row defined by the Miller index *k* (Figure 5.5). The vectors representing the planes (100), (200), (300), *etc.* define a third row, with the Miller index *l* specifying the multiple of the reciprocal of the *d*-spacing for the (100) planes. The result is that we build a lattice, exactly like the lattice described for a crystal structure in Chapter 2. Because the lattice is built up from normal vectors with lengths proportional to the reciprocals of the *d*-spacings of the sets of planes they represent, it is called the *reciprocal lattice*. The reciprocal vector for any set of planes can be defined by the indices *hkl*, which

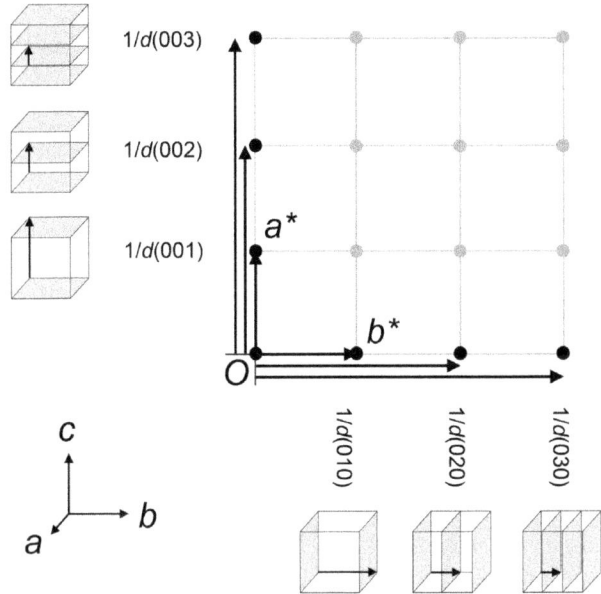

Figure 5.5 Building the reciprocal lattice from the (0*k*0) and (00*l*) planes. The Miller index *k* gives the position along the horizontal axis and the Miller index *l* gives the position along the vertical axis. The grey points correspond to the sets of planes with Miller indices (0*kl*).

serve as coordinates referred to three *reciprocal lattice axes*. The reciprocal vector immediately identifies the orientation of the plane normal and provides the *d*-spacing for the set of planes by taking the reciprocal of its length. Hence, the reciprocal lattice is a convenient and efficient way to represent many sets of planes.

5.2.4 Formalising the Reciprocal Lattice

The axes of the reciprocal lattice are called $\mathbf{a}^*$, $\mathbf{b}^*$ and $\mathbf{c}^*$, and are normal to the (100), (010) and (001) planes, respectively (Figure 5.6). The asterisk is used to denote that the vectors refer to the reciprocal lattice rather than the real lattice. Since the (100), (010) and (001) planes have a definite relationship to the unit cell, the axes $\mathbf{a}^*$, $\mathbf{b}^*$ and $\mathbf{c}^*$ must have a definite relationship to the unit cell. Specifically, $\mathbf{a}^*$ is perpendicular to both $\mathbf{b}$ and $\mathbf{c}$ in the real lattice, $\mathbf{b}^*$ is perpendicular to both $\mathbf{a}$ and $\mathbf{c}$, and $\mathbf{c}^*$ is perpendicular to both $\mathbf{a}$ and $\mathbf{b}$. The reciprocal lattice is a geometrical construction that exists for any real lattice. Whenever we think of a lattice in a crystal structure, we can consider that there is a reciprocal lattice tied to it. Let us remind ourselves why we want to do this: for any crystal lattice, a vector from the origin

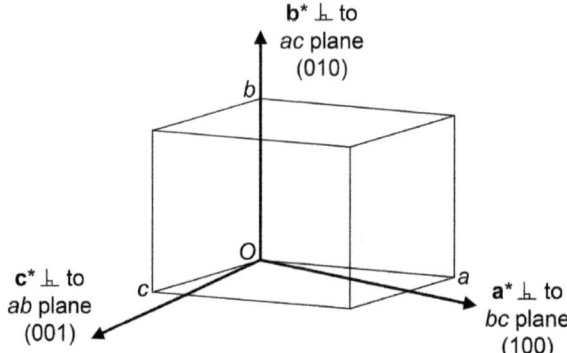

Figure 5.6   Directions of the reciprocal lattice vectors **a***, **b*** and **c*** illustrated for a monoclinic unit cell with **b** unique. For each label, the relevant plane is denoted both by a descriptive phrase and the equivalent Miller indices.

to any reciprocal lattice point neatly defines the orientation and *d*-spacing of a set of planes within the crystal structure. Formally, any vector in the reciprocal lattice can be defined as $\mathbf{r}^* = h\mathbf{a}^* + k\mathbf{b}^* + l\mathbf{c}^*$, where *hkl* are the Miller indices of the corresponding set of planes. This will become extremely useful when we think about the geometry of diffraction patterns in Chapter 6, and we will use it immediately to describe crystal morphology.

5.3   Crystal Planes and Crystal Faces

The orientation of the external faces of a crystal can be described by Miller indices in exactly the same way as we described for sets of crystal planes. Since we are dealing with a single plane rather than a set of planes, only the orientation is relevant (not the *d*-spacing), so we usually choose the smallest possible Miller indices to define a given face. For example, we would choose (100) rather than (200) or (300). If we consider the origin of the lattice to lie at the centre of a crystal, we need to describe crystal faces on either side of the origin, so (100) and ($\bar{1}$00) would describe opposite faces. Using the idea that the orientation of a plane can be neatly described by the direction of its normal, we can define the centroid of a crystal and draw normal lines from it through the centre of each crystal face (Figure 5.7). This simplifies the representation of several crystal faces to a set of lines intersecting at the origin. We have seen that these lines correspond to vectors in the reciprocal lattice, and we have seen that the reciprocal lattice is tied to the real lattice. Thus, we have an effective way to describe the external faces of a crystal *with reference to the crystal's*

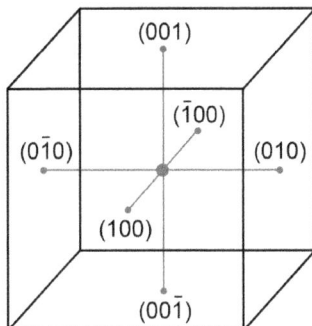

Figure 5.7 Representing crystal faces by their normal vectors. The lines represent the directions of the normals to each crystal face and the dots show the intersection of the faces with their normals. The labels give the Miller indices of each face, which are the same as the *hkl* values of the corresponding reciprocal lattice vector.

unit cell. Conceptually, we could also just imagine the crystal planes directly within the unit cell and deduce *hkl* in the manner of Section 5.2.1. But the reciprocal lattice vectors make it far easier to quantify the situation, particularly to define angles between crystal faces. As ever, we note that the labelling of crystal faces will change if we choose a different unit cell.

5.3.1 Crystal Morphology and Crystal Habit

Describing the morphology of a crystal requires that Miller indices are assigned to each crystal face. To do this, we need to know the details of the crystal lattice and we must have chosen a unit cell. We will describe in Chapter 8 how this can be achieved practically using a single-crystal X-ray diffractometer. For the moment, we will assume that we know the lattice of the crystal, and therefore that we can deduce the reciprocal lattice, which allows us to assign Miller indices to any crystal face. In geometry, the angle between two planes is defined as the (acute) angle between the plane normals. Thus, the angle between two crystal faces is simply the angle between the associated reciprocal lattice vectors. Since the reciprocal lattice is tied to the real lattice, and there is a unique relationship between a crystal structure and its lattice (Chapter 2), the angles between crystal faces are always the same for a given crystal structure. This is a profound link between a microscopic crystal structure and the macroscopic external appearance of a crystal.

There is one more thing to consider: we understand the relative orientations of the crystal faces, but we don't immediately know where each face should be placed relative to the crystal's origin.

For example, Figure 5.7 showed an orthogonal system ($\alpha = \beta = \gamma = 90°$) with all crystal faces at the same distance from the origin. This defines a cube. Assuming that opposing faces (*e.g.* (100) and ($\bar{1}$00)) remain equidistant from the crystal's origin but that the three pairs of faces could adopt different distances from the origin, the crystal could look quite different. If one of the distances is much longer than the other two, the crystal will appear to be a needle (Figure 5.8(a)). If one of the distances is shorter than the other two, the crystal will appear to be a plate (Figure 5.8(b)). The angles between the faces remains the same, but the crystals have different shapes, referred to as the crystal *habit*. The terms *morphology* and *habit* are often used interchangeably. We will use *morphology* to describe the underlying crystal faces and their Miller indices (which are a consistent property of the crystal structure as defined by the specified unit cell), and *habit* when we mean crystal shape (which can be variable as described). The three cases shown in Figure 5.7 and 5.8 are thus different crystal habits based on the same crystal morphology.

To describe a crystal's habit, we must describe the orientation of the crystal faces with respect to the unit cell and the distance of each face from the origin. Imagine constructing a framework of normal vectors then placing each face at some position along its normal vector. The crystal habit is defined by the intersection of all of the faces. As seen in Figure 5.8, it is the faces that are *closest* to the origin that have the largest relative area on the surface of the crystal. The faces furthest from the origin are either very small or not observed at all. As an example, consider the cubic habit in Figure 5.7, and imagine the (111) face (Figure 5.9). If it is placed at a similar distance from the origin as the (100), (010) and (001) faces, it serves to truncate the edges where these faces intersect (Figure 5.9(a)). If the (111) face moves further out

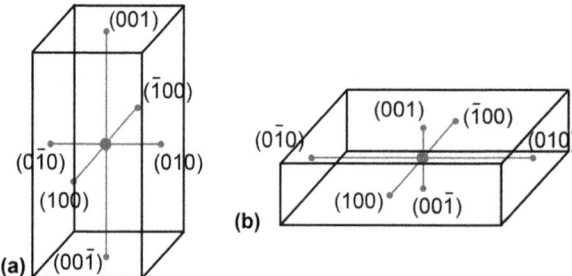

Figure 5.8 Different crystal habits based on the same crystal morphology: (a) needle; (b) plate. If all faces are equidistant from the origin, a cube is defined as in Figure 5.7. The angles between the faces are identical in all three diagrams.

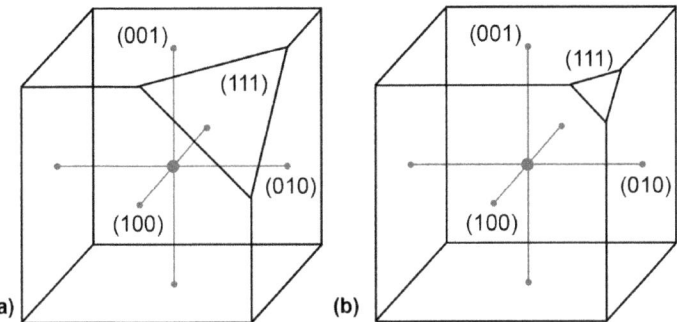

Figure 5.9   Truncation of a cube by the (111) face: (a) with (111) placed close to the origin; (b) with (111) placed further from the origin. If (111) were to be placed further out than the intersection of the (100), (010) and (001) faces, it would not be seen at all in the final habit.

compared to the (100), (010) and (001) faces, the extent of the truncation becomes less (Figure 5.9(b)). At a certain point, where the (111) face moves further out than the intersection of the (100), (010) and (001) faces, the (111) face is not seen at all in the final crystal habit. Hence it is the faces closest to the origin that dominate the crystal habit.

5.3.2   Modelling Crystal Habit

To model crystal habit, we need to know how far to place each face from the origin. Is there any information from the structure that enables us to do that? One piece of information that is always available is the d-spacing for the associated set of crystal planes. We might make the simple assumption that the position of each crystal face along its normal line is proportional to the reciprocal d-spacing. There is actually some physical justification to this, based on the idea that a crystal grown under equilibrium conditions should have minimum (*i.e.* most stable) surface energy. It means that the crystal faces with the *largest* associated d-spacings dominate the observed crystal habit, while faces with the smallest associated d-spacings have only a small relative area or are not seen at all. This approach to predict crystal habit is called the BFDH (Bravais–Friedel–Donnay–Harker) method. It provides a rough and rapid way to relate crystal habit to the underlying crystal *lattice*, under the assumption that the crystal grows under thermodynamic (equilibrium) conditions. An example is given in the Case Study at the end of this chapter. There are many more sophisticated methods to model crystal habit, for example based on the kinetics of crystal growth. Since each crystal face is chemically different, the faces grow at different rates, depending on the rate at which molecules join the crystal surfaces. By the arguments developed earlier, the observed habit under kinetic growth conditions will be dominated by the faces that grow most slowly outwards from the crystal's centre.

5.4 Summary of Key Points

- Planes in crystal structures can be defined using Miller indices, *hkl*, which are reciprocals of the coordinates where the planes cut the unit-cell axes. If a plane does not cut a unit-cell axis, it has a corresponding Miller index of 0.
- Due to the periodicity of the crystal structure, Miller indices describe sets of parallel equally-spaced planes, with the *hkl* indices referring to the first plane out from the origin. The perpendicular separation between neighbouring planes is called the *d*-spacing.
- The orientation of a set of planes is concisely described by the direction of the line perpendicular to the planes. Drawing this line from the origin to the point where it intersects the first plane out from the origin defines the normal vector, with length equal to the *d*-spacing.
- Plotting the normal vectors for many sets of planes, and representing them by the reciprocals of their lengths, defines the reciprocal lattice. Vectors in the reciprocal lattice are defined by $\mathbf{r}^* = h\mathbf{a}^* + k\mathbf{b}^* + l\mathbf{c}^*$, where $\mathbf{a}^*$, $\mathbf{b}^*$ and $\mathbf{c}^*$ are the reciprocal lattice vectors and *hkl* are the Miller indices. The reciprocal lattice is a geometrical construction that has a specific relationship to the real lattice.
- The reciprocal lattice provides a convenient framework to describe crystal morphology: the orientation of each crystal face is defined by its reciprocal lattice vector and the habit of the crystal is constructed by placing each face along this vector at some defined distance from the origin. The crystal habit is defined by the intersecting faces. The faces closest to the origin are most prominent in the defined habit.

5.5 Case Study: BFDH Morphology of Famotidine Form A

Famotidine (Figure 5.10) is used to relieve symptoms of acid reflux and heartburn. Crystal structures are known to date for two polymorphs, both of which adopt space group $P2_1/c$. This case study considers form A, which has unit-cell parameters $a = 11.986$ Å,

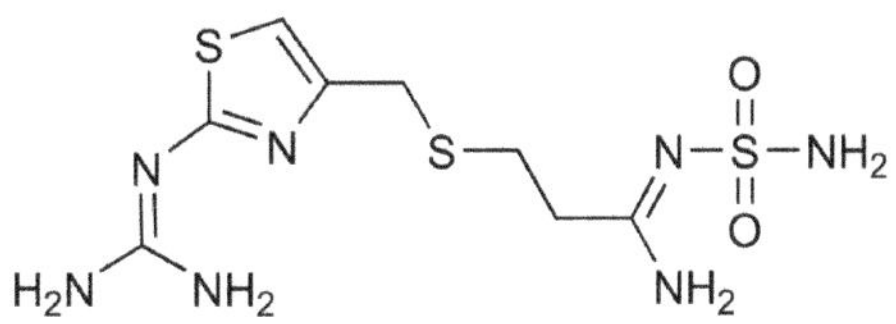

Figure 5.10 Famotidine, $C_8H_{15}N_7O_2S_3$.

$b = 7.200$ Å, $c = 16.818$ Å, $\alpha = 90°$, $\beta = 99.82°$, $\gamma = 90°$.[1] To visualise the crystal morphology, the first task is to construct the reciprocal lattice. Axis $\mathbf{a^*}$ is defined perpendicular to the bc plane of the unit cell, $\mathbf{b^*}$ is defined perpendicular to the ac plane, and $\mathbf{c^*}$ is defined perpendicular to the ab plane. For a monoclinic unit cell in the conventional setting, the b axis of the unit cell must be perpendicular to the ac plane to conform to the 2-fold symmetry (Section 4.3), which means that $\mathbf{b^*}$ must be parallel to the b axis (Figure 5.11). The metric parameters of the reciprocal lattice are $a^* = 0.0847$ Å, $b^* = 0.1389$ Å, $c^* = 0.0603$ Å, $\alpha^* = 90°$, $\beta^* = 80.18°$, $\gamma^* = 90°$.

According to the BFDH model, the faces with the largest d-spacing dominate the crystal habit:

Label	Faces equivalent by symmetry	d (Å)	d^* (Å^{-1})
{100}	(100), ($\bar{1}$00)	11.810	0.0847
{002}	(002), (00$\bar{2}$)	8.286	0.1207
{10$\bar{2}$}	($\bar{1}$02), (10$\bar{2}$)	7.402	0.1351
{011}	(011), (0$\bar{1}$1), (0$\bar{1}\bar{1}$), (01$\bar{1}$)	6.604	0.1514
{110}	(110), (1$\bar{1}$0), ($\bar{1}\bar{1}$0), ($\bar{1}$10)	6.148	0.1627
{11$\bar{1}$}	(11$\bar{1}$), (1$\bar{1}\bar{1}$), ($\bar{1}$11), ($\bar{1}\bar{1}$1)	5.938	0.1684

Here, the "curly bracket" notation in the left-hand column indicates sets of crystal faces that are equivalent by symmetry, according to point group $2/m$. The general relationship is $\{hkl\} = (hkl)$, $(h\bar{k}l)$, $(\bar{h}k\bar{l})$, $(\bar{h}kl)$. Three of the sets listed in the table have only two equivalent faces because the four combinations produce only two unique results. The physical interpretation is that the faces are perpendicular to the symmetry elements of the point group (Figure 5.12), so application of those symmetry elements does not generate any new face. {002} appears in the list rather than {001} because the c-glide in the space group produces equivalent planes of molecules interleaved along the c axis, which effectively halves the d-spacing (Figure 5.12).

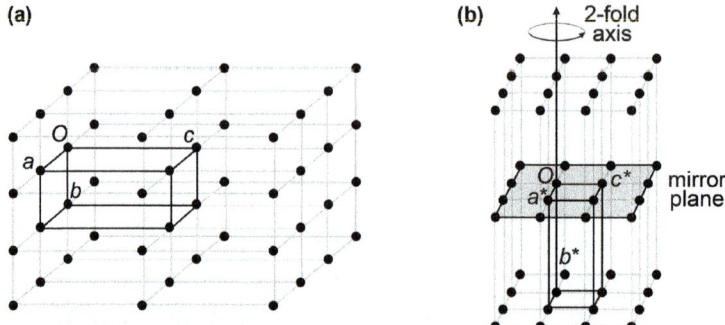

Figure 5.11 Real and reciprocal lattices for famotidine form A. The two diagrams are drawn in comparable orientations. The symmetry elements of point group 2/*m* are indicated for the reciprocal lattice in (b).

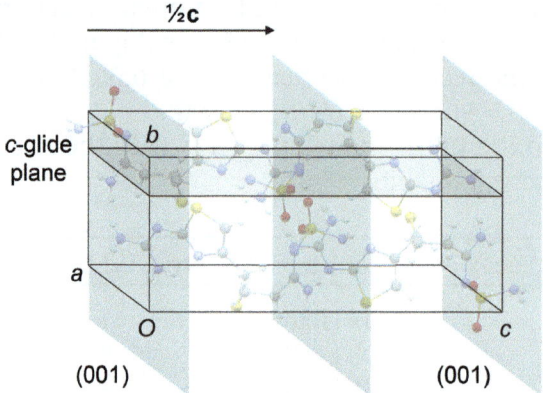

Figure 5.12 The presence of the *c*-glide in space group $P2_1/c$ effectively halves the *d*-spacing of equivalent planes perpendicular to the *c* axis. Each (001) plane can be imagined to be reflected by the glide plane (which has no consequence for the plane) then translated by $\frac{1}{2}$**c**. The resulting set of planes is spaced equivalently to (002).

Looking along the *b* axis (Figure 5.13(a)), the crystal habit appears almost rectangular, bounded principally by the {100} and {002} faces. The angle between these faces is 80.18° ($= \beta^*$). The {100} faces are bigger than the {002} faces because they lie closer to the origin (they have a smaller *d**). The small truncations visible in the bottom-left and top-right corners are the {10$\bar{2}$} faces, which are at a distance from the origin just less than the intersection of the {100} and {002} faces. Looking in the perpendicular direction onto (100) (Figure 5.13(b)), the bounding faces are {011} and {002}. {011} comprises *four* faces, as listed in the table. Since *d** is similar for {011} and {002}, all faces are similar distances from the origin and the projection appears roughly

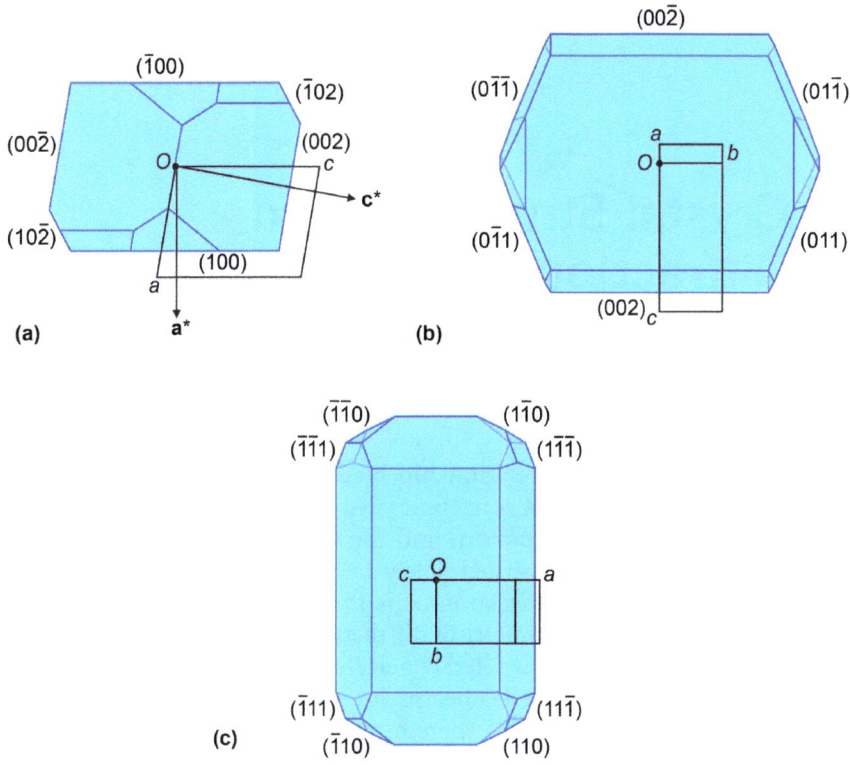

Figure 5.13 BFDH morphology of famotidine form A: (a) projection onto (010); (b) projection onto (100); (c) projection onto (001). The directions of the reciprocal lattice vectors are indicated in (a), and it can be seen that they are normal to the {100} and {002} faces.

hexagonal. The {002} faces are actually slightly larger than {011}, on account of their smaller d^*. In the final projection onto (001) (Figure 5.13(c)), the smaller {110} and {11$\bar{1}$} faces are clearly visible.

Reference

1. L. Golic, K. Djinovic and M. Florjanic, *Acta Crystallogr., Sect. C: Struct. Chem.*, 1989, **C45**, 1381, [CSD: FOGVIG01].

6 Crystal Structures and Diffraction Patterns

Summary

This chapter describes the relationship between a crystal structure and its X-ray diffraction pattern. A conceptual distinction is made between the geometry of the diffraction pattern and the intensities of the diffracted beams. The geometry is governed solely by the translational symmetry of the crystal. Geometrical measurements therefore provide information on the crystal lattice. The geometry of diffraction can be understood using Bragg's description and represented concisely using the reciprocal lattice. The intensities of the diffracted beams are determined by the electron density. Intensity measurements therefore reveal the positions and types of atoms within the crystal. The quantitative link is the structure factor equation, which enables the amplitudes and relative phases of the diffracted beams to be calculated from a crystal structure.

6.1 Introduction

Chapters 2–5 have provided the concepts required to describe crystals and crystal structures. There are a few details still to be mentioned, but we are ready to discuss the relationship between a single crystal and its X-ray diffraction pattern. The aim is to develop a physical understanding of the diffraction process together with a mathematical framework to describe it. We are working towards the practical description of single-crystal and powder X-ray diffraction in Chapters 8–10.

Pharmaceutical Crystallography: A Guide to Structure and Analysis
By Andrew Bond
© Andrew Bond 2019
Published by the Royal Society of Chemistry, www.rsc.org

6.2   X-ray Diffraction

X-ray diffraction encompasses two processes: (1) scattering of incident X-rays by electrons in the sample and (2) interference of the X-rays scattered from each point in the sample to produce a diffraction pattern. X-rays are scattered by any type of sample (solid, liquid or gas), and therefore any type of sample will produce a diffraction pattern. The importance of crystals is that their periodic structure leads to a well-ordered diffraction pattern that can be interpreted in a straightforward and meaningful way. An amorphous solid also displays a diffraction pattern, but its lack of translational symmetry makes the pattern much harder to analyse and it is far less clear what actually constitutes "the structure" (Section 1.3).

Put simply, an X-ray diffraction measurement comprises an X-ray beam entering the sample (the *incident beam*) and X-ray beams emerging from the sample (the *diffracted beams*). We can measure the angles between the incident beam and diffracted beams and also the relative intensities of the diffracted beams. It is useful to keep these two aspects separate: the geometrical measurements contain information about the crystal lattice while the intensity measurements contain information about the atom types and positions. A crucial point is that X-rays are scattered by electrons. When we analyse a crystal using X-rays we therefore get information on the *electron density*. In principle, the electron density is a continuous function with a value at every point in the crystal. However, we usually assume that the electron density is gathered into spherical atoms and is zero everywhere else. This is generally a fair approximation for small-molecule pharmaceutical crystals and it is universally applied for standard crystal-structure determination.

6.3   The Geometry of Diffraction from a Single Crystal

The geometry of a diffraction pattern reflects the translational symmetry of the crystal. To appreciate this, we can visualise how a diffraction pattern is generated from a regular array of unit cells. Each unit cell is identical and therefore scatters X-rays in exactly the same way. We must consider the interference of X-rays scattered from each unit cell in a given direction. We will take a simplistic view of an X-ray as a sinusoidal wave with its *amplitude* (A) equal to the maximum absolute displacement of the wave (Figure 6.1). The *intensity* of the wave (I) is equal to the

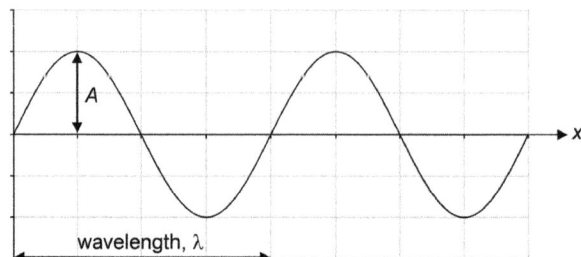

Figure 6.1   Schematic representation of an X-ray as a sinusoidal wave. The amplitude (A) and wavelength (λ) are indicated. The axis "x" represents the direction of travel. The intensity of the wave (I) is equal to the square of its amplitude.

square of its amplitude. Waves passing simultaneously through the same space sum together so that the total wave displacement at a given point is equal to the sum of the displacements of the individual waves. Fully constructive interference occurs when waves travel exactly in phase (Figure 6.2(a)), while fully destructive interference occurs when waves travel exactly out of phase (Figure 6.2(b)). Here, *phase* refers to the position of the wave through its wave cycle as it travels along its propagation path. This is usually expressed as an angle, where a complete wavelength (λ) corresponds to 2π radians (360°) and the phase angle at a position x along the propagation path is $2\pi(x/\lambda)$ radians. Two waves travelling exactly in phase have zero phase difference, while two waves travelling exactly out of phase have a phase difference of π radians (180°). Intermediate cases produce a total wave amplitude somewhere between zero and the maximum observed for fully constructive interference (Figure 6.2(c)). It can also be seen for the intermediate case in Figure 6.2(c) that the phase of the resultant wave is different from the phases of either of the initial waves.

6.3.1   A Trigonometric Construction

For simplicity, start by considering a one-dimensional row of unit cells, with an incident beam arriving from an arbitrary direction relative to the row (Figure 6.3). The phase difference between X-rays scattered from equivalent points in each unit cell in a given direction depends on the different path lengths travelled by the X-rays. In Figure 6.3, there are two things to consider:

(1) Path differences for the incident X-rays before they are scattered by each unit cell.
(2) Path differences for the X-rays scattered by each unit cell in the specified direction.

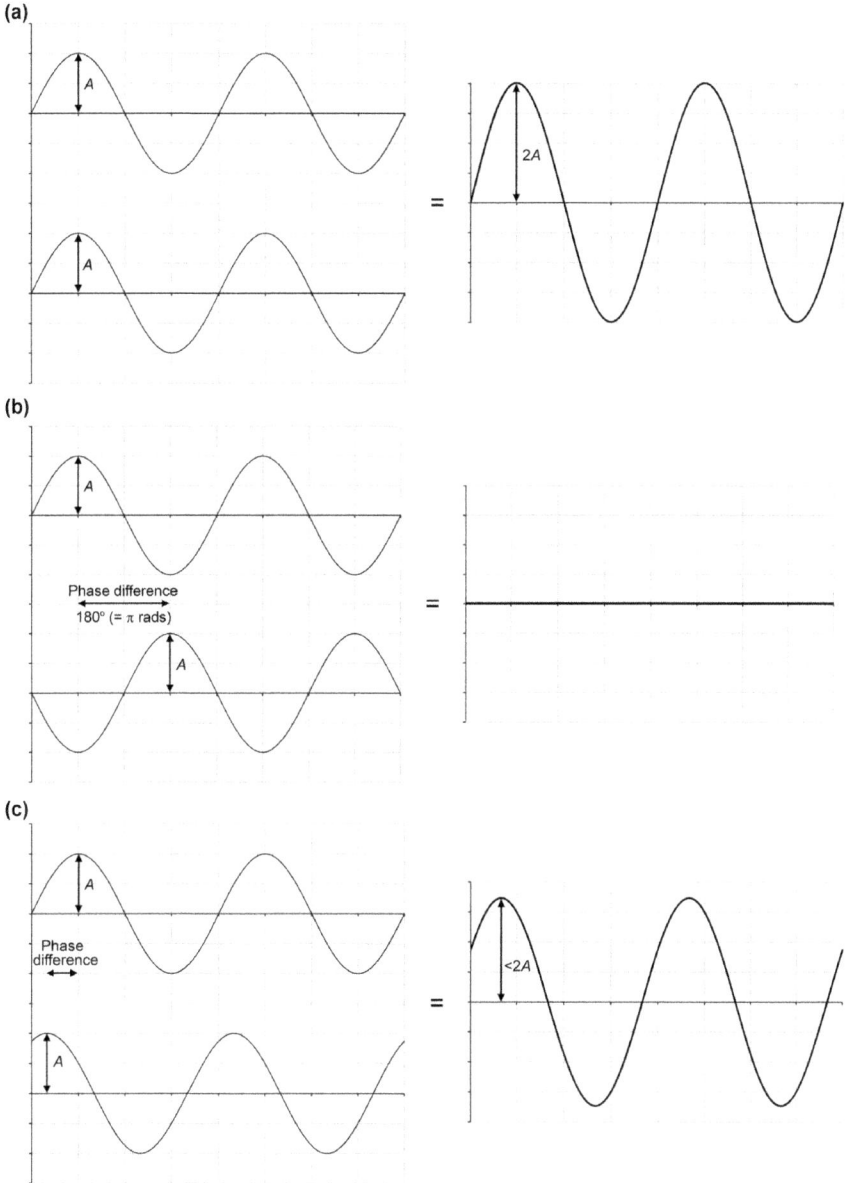

Figure 6.2 Interference of waves: (a) fully in phase; (b) fully out of phase; (c) intermediate case. In (a), the resultant wave has amplitude 2*A* and is fully in phase with the two contributing waves. In (b), there is no resultant wave. In (c), the resultant wave has amplitude less than 2*A* and phase different from either of the contributing waves.

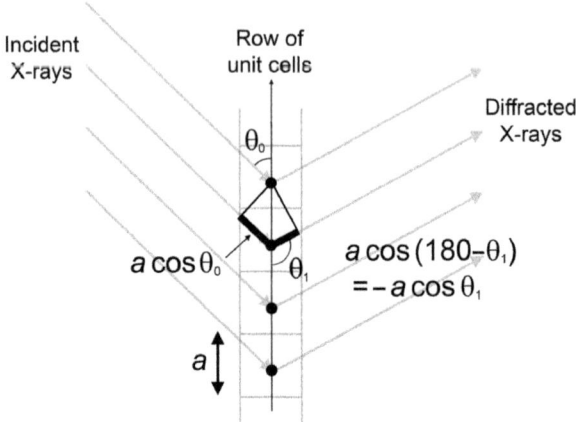

Figure 6.3 X-ray scattering from a 1-D row of unit cells. The black dots indicate equivalent points within each unit cell. The thick black line shows the path difference between X-rays scattered from equivalent points in neighbouring unit cells. Defining the angles of the incident and diffracted X-ray beams as shown, the path difference is $a \cdot (\cos \theta_0 - \cos \theta_1)$.

The path differences can be calculated using the geometrical construction shown in Figure 6.3. For incident X-rays forming an angle of θ_0 to the row of unit cells (with cell dimension a), the path difference between beams hitting adjacent unit cells is $a \cdot \cos \theta_0$. For diffracted X-rays forming an angle θ_1 to the row of unit cells, the path difference between beams emerging from adjacent unit cells is $-a \cdot \cos \theta_1$. Hence, the total path difference between the X-rays emerging from adjacent unit cells is $a \cdot (\cos \theta_0 - \cos \theta_1)$. Maximum diffracted intensity will be seen along directions where this path difference corresponds to an integral multiple of the X-ray wavelength because this ensures that the X-rays are travelling exactly in phase. Referring to Figure 6.3, we arrive at the following description for the directions of these maxima:

$$n\lambda = a \cdot (\cos \theta_0 - \cos \theta_1)$$

Note that there are numerous directions that satisfy this equation, corresponding to different integral values of n. If we consider a similar argument for complete destructive interference, we can expect to see zero diffracted intensity along directions where the total path difference corresponds to half-integral multiples of the X-ray wavelength.

These conditions for fully constructive and fully destructive interference are clear. But what about everything between? The exact details actually depend on the number of unit cells (which can become crucial for diffraction from nanoparticles or thin films, for example). For a single crystal, however, we can assume an "infinite" number of unit cells in perfect alignment. If we move slightly away from the direction required for complete constructive interference, the path difference between X-rays scattered from neighbouring unit cells deviates slightly from an integral multiple of the wavelength. For next nearest neighbours, the path difference is slightly different again, and for next, next neighbours, slightly different again, and so on. When this situation is summed over an infinite number of unit cells, the X-rays travelling in a given direction exhibit all possible phase differences and the result is complete destructive interference. The only directions where this does not occur are those corresponding to complete constructive interference. So, for an infinite number of unit cells in a single crystal, the situation becomes "all or nothing". Discrete diffracted beams are seen along directions corresponding to complete constructive interference, with no intensity between.

6.3.2 Bragg's Description

The trigonometric description above is physically realistic (because it stresses the importance of equivalent scattering from points related by translation), but it is cumbersome to apply, especially when extended to three dimensions. A more intuitive link between the crystal lattice and the geometry of diffraction is provided by Bragg's description, which is based on the simple fact that the incident X-ray beam enters the crystal and a diffracted X-ray beam emerges from the crystal with some angular relationship between them. This behaviour is familiar from day-to-day experience of visible light reflected in a mirror. We can easily imagine holding a mirror and aligning it to reflect a beam of visible light in any direction. Bragg's description allows X-ray diffraction to be visualised in the same way. It leads to the common description of diffracted X-ray beams as "reflections". This is solely a geometrical construction. Bragg's description does not reveal *why* the diffracted beams emerge in the directions that they do (this is shown in Section 6.3.1), but it does give a powerful method to determine those directions.

For the situation shown in Figure 6.3, we can instinctively draw a mirror line that would act to reflect the incident X-ray beam in the direction of the diffracted beam (Figure 6.4(a)). We can draw a set of these mirror lines, one through each unit cell, and note that they must

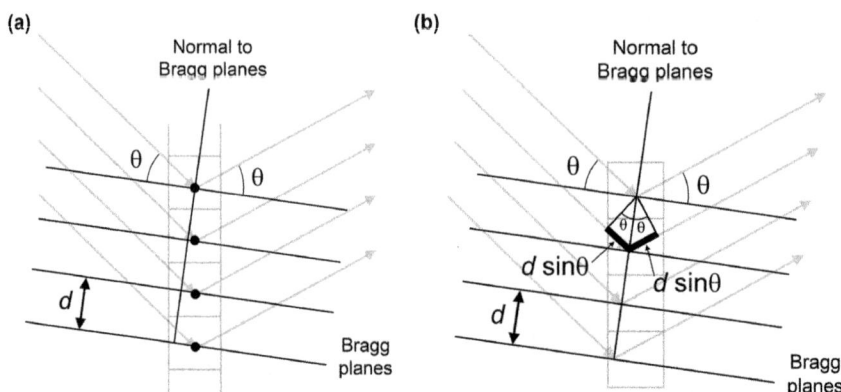

Figure 6.4   Bragg's description of diffraction geometry. The directions of the incident and diffracted beams are identical to Figure 6.3. Bragg planes are
drawn through the equivalent points in each unit cell. In (a), the
equivalent points do not align with the plane normal, which makes it
quite difficult to calculate the path differences between X-rays reflected
from neighbouring planes. In (b), the incident and diffracted beams are
shown reflected from points aligned along the normal, which gives the
path difference clearly as $2d \sin \theta$. In general, the result does not depend
on where the reflecting points are considered to lie in each plane—it
only depends on d and θ, as shown.

all be parallel. In three dimensions we consider mirror planes, which
we call *Bragg planes* to remind us that we are using them in relation to
Bragg's description. We can draw the normal to this set of planes and
define the d-spacing. Now the path difference between X-rays scattered
from neighbouring unit cells can be calculated *with reference to the set
of Bragg planes*. Look carefully at Figure 6.4(a): the dots indicating
the equivalent points in each unit cell from which the X-rays are being
scattered are not aligned with the normal to the Bragg planes. This
makes it quite difficult to give a trigonometric description of the path
differences in terms of d and θ. It can be shown, however, that the path
difference does not depend on where each point is situated within each
Bragg plane—it only depends on d and θ. Qualitatively, we can imagine
that if the X-ray has to travel further before it hits the reflecting
point in the next Bragg plane (as it does in Figure 6.4(a)), it will travel
correspondingly less far after it is reflected. Given that this is the case,
the simplest illustration of the path difference is obtained if the X-rays
are considered to be reflected from points within each Bragg plane
aligned along the normal (Figure 6.4(b)). Expressing the path difference
in terms of the d-spacing now gives the famous Bragg equation for
directions of fully constructive interference:

$$n\lambda = 2d \sin \theta$$

where n is an integer, λ is the X-ray wavelength, d is the d-spacing of the Bragg planes, and θ is half of the angle formed between the incident and diffracted beams (equal to the angle of incidence and angle of reflection at the Bragg planes). We are not doing anything different than we did in Section 6.3.1; we are just quantifying the X-ray path differences using an equation that refers to a different geometrical construction. To extend the description to a 3-dimensional array of unit cells, we describe the Bragg planes by their Miller indices. When referring to a diffracted beam, there are no parentheses around hkl.

The n in Bragg's equation indicates that the path difference for constructive interference can be any integral multiple of the X-ray wavelength. When we come to describe the diffraction pattern from a crystal, n is usually eliminated by dividing it through on both sides:

$$\lambda = 2(d/n)\sin\theta$$

This means that any direction corresponding to $n > 1$ for a set of Bragg planes with d-spacing d will be identical to the direction corresponding to $n = 1$ for a set of parallel Bragg planes with d-spacing d/n. For example, we don't need to consider the direction with $n = 2$ for the (100) planes, because it will be the same as the direction with $n = 1$ for the (200) planes; this is discussed further in the Case Study at the end of the chapter. It can also be seen that a specific X-ray wavelength (λ) gives a minimum value of the d-spacing for which diffraction can occur. Since $\sin\theta$ has a maximum value of 1, d has a minimum value of $\frac{1}{2}\lambda$. This imposes a limit on the *resolution* of the structure that can be obtained using a given X-ray wavelength (see Section 11.2.3).

A final thing to note about Bragg's description is that the incident beam, diffracted beam and normal to the Bragg planes all lie in a common plane, but Bragg's equation does not specify what that plane is. Imagine taking hold of the normal vector to the Bragg planes and rotating the entire construction around the incident beam: the angular relationship described by Bragg's equation will still be satisfied, but the diffracted beam could emerge anywhere on a cone around the incident beam (Figure 6.5(a)). Bragg's equation describes the diffraction angle (2θ) for a specified set of Bragg planes, hkl, but the information on where the diffracted beam emerges in the 3-D diffraction pattern is tied up in the specification of hkl, which we discussed in Chapter 5. Also, Bragg's equation does not put any absolute restriction on the orientation of the crystal with respect to the incident beam because the crystal could be rotated around the

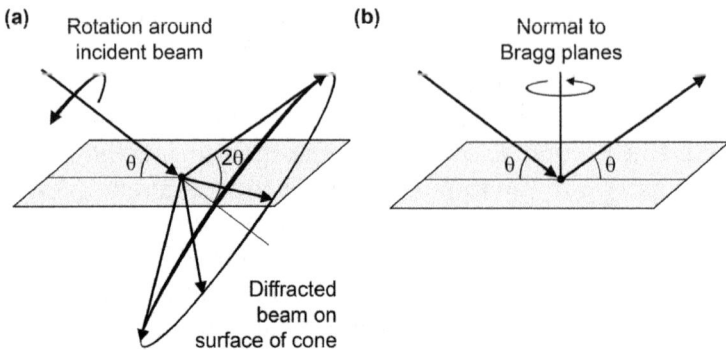

(a) Rotation around incident beam

(b) Normal to Bragg planes

Diffracted beam on surface of cone

Figure 6.5   Bragg's law does not put an absolute restriction on the orientation of the crystal relative to the incident beam: (a) rotation around the incident beam means that the diffracted beam could emerge anywhere on the surface of a cone with half-angle 2θ; (b) rotation around the normal to the Bragg planes does not change the construction at all.

normal to the Bragg planes without changing anything about the directions of the incident and diffracted beams (Figure 6.5(b)).

6.3.3   The Role of the Reciprocal Lattice

The reciprocal lattice was introduced in Chapter 5 as a concise way to represent crystal planes. It is exactly what we need now to describe the orientation and d-spacing of Bragg planes. Knowing that a set of Bragg planes in a crystal structure can be represented by a vector in the reciprocal lattice, each diffracted beam in a diffraction pattern can be associated with a specific reciprocal lattice point, which is labelled using the corresponding Miller indices. This is the most effective way to describe the geometry of a diffraction pattern but it can also be a source for confusion and apprehension. Diffraction patterns are often said to exist "in reciprocal space", which conjures up an image of some strange alternative dimension. This is not the case at all. The reciprocal lattice is just a geometrical construction related to the real crystal lattice in the way that we saw in Chapter 5. Physically, we measure diffraction angles, which can be visualised and understood using Bragg's description. The numerous sets of Bragg planes can then be represented efficiently using the reciprocal lattice.

When we come to measure a single-crystal X-ray diffraction pattern, we can see from Bragg's description that the reciprocal lattice vector (normal to the Bragg planes) must bisect the incident and diffracted beams (Figure 6.4). For each diffracted beam, the bisector of the incident and diffracted beams thereby gives the direction of the associated reciprocal lattice vector. The length (d^*) of that reciprocal

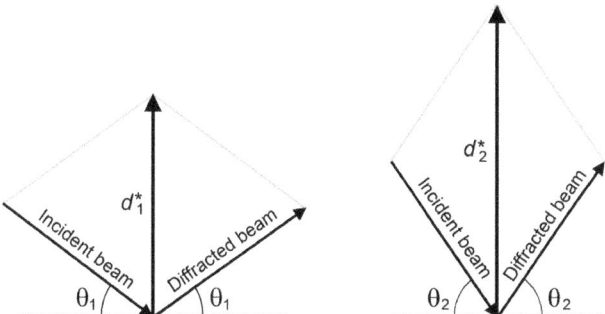

Figure 6.6 Relationship between the reciprocal lattice vector and the incident/ diffracted beams. If the vectors representing the incident/diffracted beams are considered to have unit length, the length of the reciprocal lattice vector (d^*) is given by $(2 \sin \theta)/\lambda$, according to Bragg's equation. Longer reciprocal lattice vectors represent larger diffraction angles.

lattice vector can be obtained from Bragg's equation by calculating the d-spacing, and taking its reciprocal:

$$d^* = 1/d = (2 \sin \theta)/\lambda$$

A useful way to visualise this is to consider the incident and diffracted beams as rigid rods with unit length, physically linked to both ends of the reciprocal lattice vector (Figure 6.6). Stretching the reciprocal lattice vector pulls the incident and diffracted beams upwards so that θ gets larger. Compressing the reciprocal lattice vector does the opposite. This reiterates that diffracted beams with larger diffraction angles are represented by longer reciprocal lattice vectors.

Now we have a route to construct the reciprocal lattice from a set of measured diffraction angles. And we know how the reciprocal lattice is linked to the real lattice (Section 5.2). So to determine the crystal lattice from a diffraction pattern, we measure the angles between the incident beam and an adequate number of diffracted beams, convert each one to a reciprocal lattice vector, then convert the resulting reciprocal lattice into the real lattice.

6.3.4 Bringing Each Beam into the "Diffracting Position"

For practical measurement of an X-ray diffraction pattern, each diffracted beam is observed only when the crystal adopts the correct orientation with respect to the incident beam. We cannot simply illuminate a crystal with a fixed incident beam and see all diffracted beams emerge at once. Considering Bragg's description, we know that the incident beam must make an appropriate angle θ to a set of Bragg

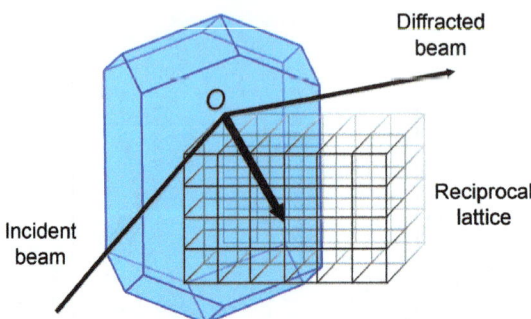

Figure 6.7 Schematic illustration of the link between a crystal, its associated reciprocal lattice and the direction of the incident and diffracted beams in a single-crystal measurement. We can imagine moving the reciprocal lattice vector to each reciprocal lattice point in turn, pulling the incident and diffracted beams with it. If the incident beam stays static, the crystal must be rotated.

planes and that θ is restricted to specific values expressed by Bragg's equation. Hence, the crystal must be aligned in a specific orientation to observe each diffracted beam. It is commonly said that the crystal must be in an appropriate *diffracting position*. If we want to observe each diffracted beam in turn, imagine a stationary crystal with its associated lattice and reciprocal lattice (Figure 6.7). We can select any reciprocal lattice vector and draw in the associated directions of the incident and diffracted beams using the construction in Figure 6.6. Now imagine dragging the reciprocal lattice vector sequentially from reciprocal lattice point to reciprocal lattice point, pulling the incident and diffracted beams along with it. This shows how the incident beam direction must change relative to the crystal in order to observe each diffracted beam. Practically, this is not quite how we measure a single-crystal diffraction pattern. Rather than changing the direction of the incident beam with respect to a static crystal, we change the orientation of the crystal with respect to a fixed incident beam. This is considered further in the Case Study at the end of the chapter. The practical procedure will be described in Chapter 8.

6.3.5 Angular Width of the Diffracted Beams

So far, we have considered that the incident and diffracted beams travel along perfectly sharp lines. In practice, diffracted X-ray beams have a certain width, which is governed partly by the width of the incident beam and partly by a physical feature of the crystal known as its *mosaicity*. Basically, crystals are never perfect but actually have more of a mosaic structure where there is some degree of misalignment

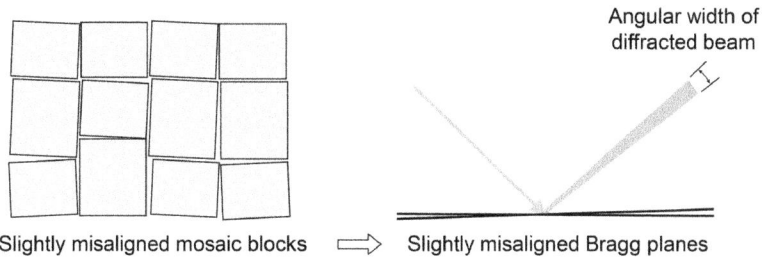

Figure 6.8 Schematic illustration of mosaicity, whereby a crystal comprises blocks with some degree of misalignment between them. This leads to slightly misaligned Bragg planes which result in some angular width for the diffracted beams.

between constituent *mosaic blocks* (Figure 6.8). The extent of this misalignment governs the crystal's contribution to the angular width of the diffracted beams. We can imagine that the Bragg planes within the mosaic blocks are slightly misaligned so that each diffracted beam is distributed over a small range of diffraction angles. A greater degree of misalignment gives a greater angular width. A typical angular width for a single crystal is in the region of 1°, but angular beam widths up to a few degrees can be managed in a practical single-crystal analysis.

6.3.6 Influence of the X-ray Wavelength

Bragg's equation also demonstrates how the diffraction geometry depends on the X-ray wavelength. For measurements on a given single crystal, the diffraction angle θ (or 2θ) increases for a particular diffracted beam as the wavelength increases. For the commonly used laboratory X-ray wavelengths, Mo ($\lambda \approx 0.71$ Å) and Cu ($\lambda \approx 1.54$ Å), diffraction patterns are spread over a wider angular range for Cu radiation than for Mo. This can have disadvantages for practical measurement because it generally takes longer to measure a diffraction pattern with Cu radiation. But it can also have advantages, particularly for powder X-ray diffraction, where it is desirable to spread the diffraction pattern to minimise peak overlap. Further details are described in Chapters 8–10.

6.4 The Intensities of Diffraction from a Single Crystal

We have considered so far that a crystal behaves as a collection of unit cells, each of which scatters X-rays identically. By considering interference between X-rays scattered from an effectively infinite number of unit cells related by the crystal's translational symmetry, we can describe the geometry of the diffraction pattern. Diffracted beams are produced in directions where the scattered X-rays from all unit cells are fully in phase, with no intensity seen in other directions. For X-rays that combine fully in phase, the total amplitude is simply the sum of the amplitudes of all individual X-rays (as in Figure 6.2(a)). Hence, the total amplitude of each diffracted beam in a single crystal must be the amplitude scattered from one unit cell multiplied by the number of unit cells. The total intensity is the square of this total amplitude. To determine crystal structures, we are interested in the *relative* intensities of the diffracted beams (not the absolute values, which will vary from crystal to crystal and instrument to instrument). Hence, we only need to quantify the diffracted intensities for X-rays scattered by one unit cell.

6.4.1 Scattering from an Individual Unit Cell

To determine the amplitudes of the X-rays scattered by one unit cell in a given direction, we must consider that the incident X-ray beam will be scattered from every point within the unit cell, then sum the scattered X-rays to produce the resultant X-ray beam. We adopt an atomic picture, which means that the electron density is gathered into spherical regions corresponding to the atoms and is zero everywhere else. Eventually, we will have to account for the fact that atoms have volume, but for now we can consider that all electron density in each atom is located at the specified atom position, *xyz*. How do X-rays scattered from these points interfere to produce diffracted beams from the entire unit cell? For this calculation, we cannot take the "all or nothing" approach that applies to an infinite array of unit cells. We must make a proper calculation of the amplitude and phase of all interfering X-rays along each direction.

6.4.2 Summing the Scattered X-rays

A general method to sum waves with the same wavelength is to represent them as vectors on a 2-D diagram (Figure 6.9). The length of each vector is equal to the amplitude of the wave and the vector is aligned so that the angle to the horizontal axis of the diagram is equal to the phase angle at the point of measurement. Waves can then be summed by adding the components of each vector along each axis of the diagram to produce a total. This corresponds to adding the vectors in a "head-to-tail" manner. The resultant vector reveals the amplitude and phase angle of the resultant wave. It is easy to confirm

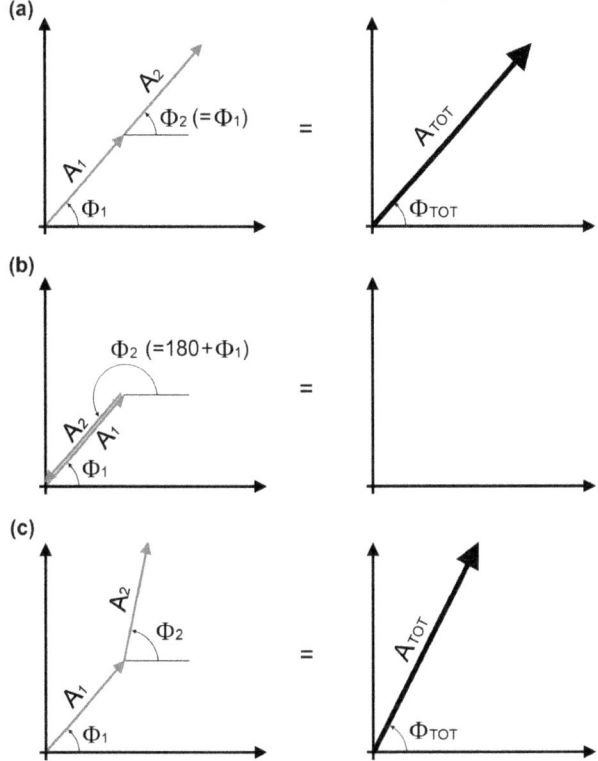

Figure 6.9 Summing X-rays by representing them as vectors on a 2-D diagram. Cases (a), (b) and (c) correspond to the situations shown in Figure 6.2. In (a), representing fully constructive interference, $A_{TOT} = A_1 + A_2$ and $\Phi_{TOT} = \Phi_1 = \Phi_2$. In (b), representing fully destructive interference, there is no resultant vector. In the intermediate case (c), $A_{TOT} < A_1 + A_2$ and $\Phi_{TOT} \neq \Phi_1 \neq \Phi_2$.

that this works for two waves exactly in phase (Figure 6.9(a)): the resultant vector points in the same direction (*i.e.* it has the same phase) and its length is equal to the sum of the lengths of the vectors representing the input waves, exactly as in Figure 6.2(a). It is also easy to see that it works for fully destructive interference (Figure 6.9(b)), since two vectors representing waves with the same amplitude but a phase difference of 180° point in opposite directions and cancel each other out. An intermediate scenario is shown in Figure 6.9(c), where the vectors have the same length but different phase angles. The resultant vector has length less than the sum of the two contributing waves and phase angle different from either wave. This represents the scenario shown in Figure 6.2(c).

A convenient mathematical way to describe this is to represent X-rays by complex numbers. A complex number has two parts, called the real and imaginary parts, which are kept separate. These correspond to the components of the vector relative to the two axes of the diagram described above. In the language of complex numbers, the diagram is called the *Argand diagram*, the horizontal axis is called the real axis and the vertical axis is called the imaginary axis. For an X-ray with amplitude A and phase Φ (phi), the required vector has component $A\cos\Phi$ along the real axis and $A\sin\Phi$ along the imaginary axis (Figure 6.10). This is written $A(\cos\Phi + i\sin\Phi)$, where the symbol i denotes that $\sin\Phi$ is the component along the imaginary axis. A convenient contraction is referred to as Euler's notation:

$$A(\cos\Phi + i\sin\Phi) = A\exp\{i\Phi\}$$

This enables the interference between X-rays to be expressed symbolically as a concise summation. For example, $A_1\exp\{i\Phi_1\} + A_2\exp\{i\Phi_2\}$ sums two waves with amplitudes A_1 and A_2 and phases Φ_1 and Φ_2. The notation should be interpreted simply as the head-to-tail summation of vectors on the 2-D (Argand) diagram, as we have shown.

When multiplying complex numbers, the amplitudes are multiplied and the phases are added:

$$A_1\exp\{i\Phi_1\} \times A_2\exp\{i\Phi_2\}$$
$$= A_1 \times A_2 \times \exp\{i\Phi_1\} \times \exp\{i\Phi_2\}$$
$$= (A_1 \times A_2)\exp\{i(\Phi_1 + \Phi_2)\}$$

An important application for X-ray diffraction is multiplication of a complex number by its *complex conjugate*, which has the same amplitude but negative phase:

$$A_1\exp\{i\Phi_1\} \times A_1\exp\{i(-\Phi_1)\} = (A_1 \times A_1)\exp\{i(\Phi_1 - \Phi_1)\} = (A_1)^2$$

The result is a positive real number (*i.e.* it has zero phase), equal to the square of the amplitude. The relevance is that the intensity of a

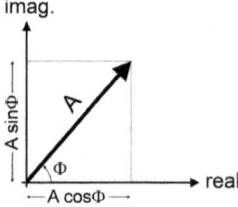

Figure 6.10 The Argand diagram formalises the link between summing X-rays and summing complex numbers. The horizontal and vertical axes are called the real and imaginary axes, respectively. The phase of the wave (Φ) is the angle formed to the real axis, defined as an anti-clockwise rotation. A negative value of Φ corresponds to a clockwise rotation.

diffracted beam, which corresponds physically to the square of its amplitude, can be obtained by multiplying a complex number by its complex conjugate.

6.4.3 Structure Factors

To sum the X-rays scattered from all atoms in the unit cell, we must consider both the amplitudes and phases of the scattered X-rays. We will assert for now (and clarify later) that the amplitude of the X-rays scattered from an atom is given by a quantity called the *atomic scattering factor*, which is related to the number of electrons in the atom. Phase differences between the scattered X-rays arise from path differences between X-rays scattered from atoms at different positions within the unit cell. These path differences must be measured relative to some zero point, and it is obviously most convenient to choose the origin of the unit cell. Thus, we describe the path differences of X-rays scattered in a particular direction from each atom in the unit cell *relative to an X-ray that would be scattered from the unit-cell origin.* There does not need to be an atom at the origin, or any X-ray scattered from the origin—we are just using it to define a reference point to express relative path differences.

The position of each atom within the unit cell is given by its fractional coordinates, *xyz*. Recall from Section 2.4 that this can be expressed as the position vector $\mathbf{r} = x\mathbf{a} + y\mathbf{b} + z\mathbf{c}$. The path differences travelled by the incident and diffracted X-ray beams are seen by drawing lines perpendicular to the direction of those beams, as in Figure 6.11. This defines two right-angled triangles, each having $\mathbf{r}$ as the hypotenuse and the required path difference as one of its sides. We could describe this with some trigonometric construction involving sines or cosines. However, a more general way is to view each path difference as the *projection* of the vector $\mathbf{r}$ onto the incident or

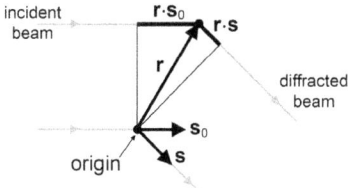

Figure 6.11   Expressing the path difference between X-rays scattered from a point specified by position vector $\mathbf{r}$ and X-rays scattered from the origin. The two parts of the path difference correspond to the projection of $\mathbf{r}$ onto the directions of the incident and diffracted beams, defined by unit vectors $\mathbf{s_0}$ and $\mathbf{s}$, respectively.

diffracted beam. Mathematically, the projection of one vector onto another is achieved by taking the *scalar product* (also called the *dot product*). If we define vectors with unit length along the directions of the incident and diffracted beams, labelled s_0 and s, respectively, the magnitudes of the path differences are given by the scalar product of **r** with the relevant unit vector. The total path difference is obviously the sum. We have to be careful with signs because a scalar product is positive if the angle between the vectors is acute (as for **r** and s_0 in Figure 6.11), but negative if the angle is obtuse (as for **r** and **s** in Figure 6.11). So, the total path difference shown in Figure 6.11 is given by $\mathbf{r} \cdot \mathbf{s}_0 - \mathbf{r} \cdot \mathbf{s}$. This result is true for any relative orientation of **s** and s_0, and would emerge however we draw the diagram. To convert the path difference to a phase difference in radians, we multiply by 2π. We also have to note (perhaps counterintuitively) that a wave travelling further has its phase *behind* that of a wave travelling a shorter distance. We are interested to sum the waves arriving at the same time at a particular measuring point along the diffracted beam. If a wave arriving at that point at a given time has travelled further, it must have started out earlier so its phase will correspond to that of the incident beam at an *earlier* point in time. The conclusion is that the phase difference is given by $-2\pi(\mathbf{r} \cdot \mathbf{s}_0 - \mathbf{r} \cdot \mathbf{s})$, which is the same as $2\pi(\mathbf{r} \cdot \mathbf{s} - \mathbf{r} \cdot \mathbf{s}_0)$, or $2\pi(\mathbf{r} \cdot (\mathbf{s} - \mathbf{s}_0))$.

This description is starting to look complicated, but it can be simplified by returning to the idea introduced in Section 6.3.3. The incident and diffracted beams can be considered to be tied to the reciprocal lattice vector, as in Figure 6.6. In the vector notation that has just been introduced, the reciprocal lattice vector is seen to equal $\mathbf{s} - \mathbf{s}_0$, so the phase difference is given by $2\pi(\mathbf{r} \cdot \mathbf{r}^*)$. In other words, the phase difference for an X-ray scattered from an atom with position vector **r** relative to the origin is given by the scalar product of **r** and the reciprocal lattice vector $\mathbf{r}^*$ associated with the specific diffracted beam. Again, this may sound complicated, but it can be simplified by the fact that the real and reciprocal lattices have a specific geometrical relationship. Since $\mathbf{r} = x\mathbf{a} + y\mathbf{b} + z\mathbf{c}$ and $\mathbf{r}^* = h\mathbf{a}^* + k\mathbf{b}^* + l\mathbf{c}^*$, the various relationships between **a**, **b**, **c** and $\mathbf{a}^*$, $\mathbf{b}^*$, $\mathbf{c}^*$ give the following result: the total path difference for an X-ray scattered from an atom at position *xyz* relative to the unit-cell origin in the direction of the diffracted beam labelled *hkl* is given by the simple relationship $hx + ky + lz$. Hence, the phase difference in radians is $2\pi(hx + ky + lz)$. To obtain the resultant X-rays scattered from the unit cell, we sum over all *N* atoms in the unit cell using the "head-to-tail" summation method. The result is called the structure factor, denoted by the symbol $F(hkl)$. Using Euler's notation, the summation is as follows:

$$F(hkl) = \sum_{n=1}^{N} f_n \exp\{i(2\pi(hx_n + ky_n + lz_n))\}$$

where f_n is the atomic scattering factor for the nth atom at fractional coordinates x_n, y_n, z_n. This summation is basically all that we need. It gives the amplitude and phase of the diffracted beam produced by the unit cell in the direction specified by the indices hkl. The units are electrons (because f is expressed in electrons, as we shall see below). The phase is relative to the phase of an X-ray beam that would be scattered from the origin.

6.4.4 Atomic Scattering Factors

To complete our understanding, we must look more closely at the atomic scattering factor. The idea of "point atoms" is obviously not realistic because the electron density in an atom must be distributed over some volume. Conceptually, we need to consider interference of X-rays scattered from *all* points within the unit cell, not just the points considered to be the centres of atoms. However, we would like to retain the atomic picture because it allows us to describe crystal structures in the straightforward way presented in Chapters 2–4. The atomic scattering factor reconciles these two views. It is a function that encapsulates the scattering behaviour of all electron density associated with an atom so that the atom can be represented at a single point. As well as dealing with atomic volume, the function accounts for the fundamental physical effects that actually reveal the value of the electron density at each position. The full details are usually presented in more physical/mathematical descriptions of X-ray diffraction. We will simply note that the atomic scattering factor f produces a scalar value that gives the length of the vector on the Argand diagram representing the contribution of each atom to the resultant diffracted beam (Figure 6.12).

At zero diffraction angle (*i.e.* considering a diffracted beam emerging in the same direction as the continuing incident beam) the value of f is equal to the number of electrons in the atom. This means that there is fully constructive interference between X-rays scattered from all f electrons in the atom so the total scattered amplitude from the atom is f times the scattered amplitude from one electron. As the scattering angle moves away from zero, path differences are

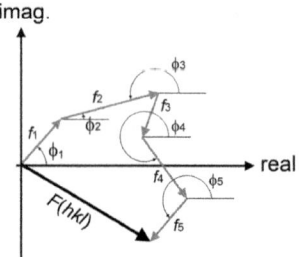

Figure 6.12   Representation of the structure factor equation on the Argand diagram.
Each atom in the summation contributes a vector with length equal to
its atomic scattering factor, f (at the given diffraction angle) with phase
$\phi = 2\pi(hx + ky + lz)$. The resultant vector gives the amplitude and phase
of the structure factor $F(hkl)$.

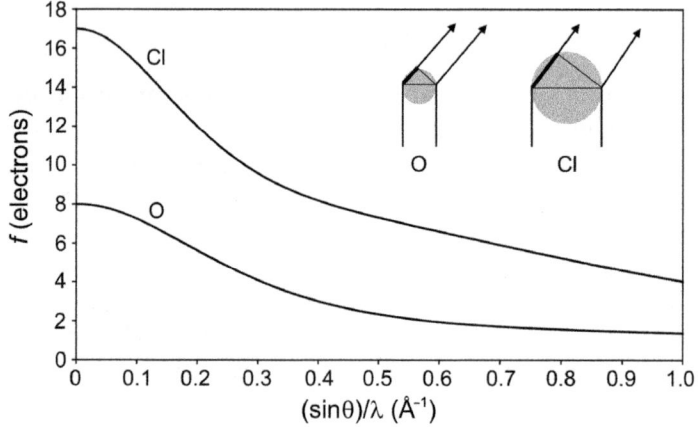

Figure 6.13   Atomic scattering factors for O and Cl. At zero scattering angle, the
values correspond to the total number of electrons in each atom. The
values decrease as scattering angle increases because path differences
are introduced between X-rays scattered from different positions
within each atom (shown in the inset). The function shows a more
rapid drop-off for Cl than for O because Cl occupies a greater volume
than O.

introduced between X-rays scattered from electron density at different
positions within the atom, which introduces some degree of destructive interference (Figure 6.13). The value of the atomic scattering
factor decreases as the scattering angle increases because these path
differences become progressively greater. The rate at which an atomic
scattering factor decreases with diffraction angle depends on the
volume occupied by the atom. For example, Figure 6.13 shows atomic
scattering factors for O and Cl. The drop-off in the value of f is more
rapid for Cl than for O because Cl occupies a larger volume. We will

return to this point in Chapter 12 when we discuss the influence of temperature. Because we assume that atoms are spherical, the atomic scattering factor is a spherical function. Its value (for a given atom and X-ray wavelength) depends only on the scattering angle θ. In the structure factor equation, each atom has a different value of f for each diffracted beam hkl, depending on the associated value of θ. In practice, the values are obtained from a parameterised equation, where the parameters are tabulated for each atom type. This is coded within crystallographic software for all commonly-used laboratory wavelengths. Methods exist to extrapolate to non-standard wavelengths, as might be used at a synchrotron source.

6.5 A Review

This chapter has presented a lot of information, which is worth reviewing concisely in preparation for the coming description of practical single-crystal X-ray analysis. When a crystal is illuminated with an incident X-ray beam, we can first visualise X-rays scattered from just one unit cell (Figure 6.14). The X-rays emerge from this single unit cell in *all* directions, and the resulting amplitudes and (relative) phases depend on the positions and types of atoms within the unit cell. For a single crystal comprising an effectively infinite number of unit cells, the emerging X-rays are eliminated in most directions because of destructive interference between X-rays scattered from different unit cells. Diffracted X-rays are observed only along specific

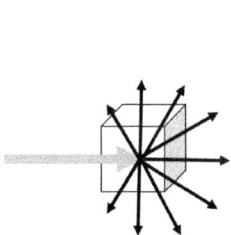

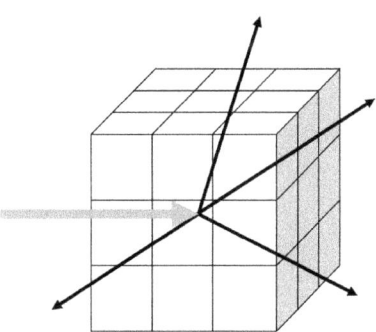

(1) Scattering from all atoms in one unit cell produces diffracted X-rays in all directions. The diffracted amplitude and phase along a given direction is quantified by the structure factor equation.

(2) Translational periodicity in the crystal means that diffracted X-rays emerge only along directions of complete constructive interference between unit cells. These directions can be visualised with Bragg's construction.

Figure 6.14   Schematic review of the information presented in this chapter.

directions where there is fully constructive interference between X-rays scattered from all unit cells. The diffraction angles that remain are determined by the geometry of the crystal lattice. In the directions where diffracted beams are seen, the X-ray amplitude from one unit cell is multiplied by the number of unit cells because all unit cells act in concert. Since the scaling is linear and the same for all diffracted beams, it is sufficient just to describe the X-rays scattered from one unit cell. This is what the structure factor equation does. The diffracted X-rays are described mathematically using complex numbers because these are able to encapsulate information on both the amplitude and phase. The quantity actually measured is the *intensity* of each diffracted beam, which corresponds physically to the square of the amplitude and is given by the complex number multiplied by its complex conjugate. The structure factor equation shows that every atom in the unit cell contributes to every diffracted beam. A complete set of diffracted intensities must therefore be measured to obtain information about all of the atoms. It does not make sense to measure only a specific region of a diffraction pattern, as might be done for a spectroscopic measurement such as IR or Raman.

6.6   Summary of Key Points

- X-ray diffraction comprises scattering of incident X-rays from electrons in the sample, followed by interference of the scattered X-rays to produce the diffraction pattern.
- X-rays can be represented by complex numbers, which encapsulate information on both the amplitude and phase. Interference can be described by summation on the Argand diagram. The exponential form of a complex number, $A \exp\{i\Phi\}$, provides a concise notation, where A is the amplitude and Φ is the phase.
- For a single crystal, each unit cell scatters X-rays in an identical manner and the diffraction pattern is thereby produced from a periodic distribution of identical scattering objects. For the effectively infinite number of unit cells within a single crystal, this gives an "all or nothing" situation where discrete diffracted beams are seen only along directions where there is complete constructive interference between X-rays scattered from all unit cells.
- Bragg's description provides an intuitive picture of the diffraction angles: $\lambda = 2d \sin \theta$, where d is the d-spacing for a specified set of Bragg planes with Miller indices hkl and 2θ is the angle between the incident and diffracted beams.

- The orientation of a set of Bragg planes can be described concisely by the associated reciprocal lattice vector, which bisects the incident and diffracted beams and has length equal to the reciprocal of the *d*-spacing of the Bragg planes. Each diffracted beam can be associated with a point in the reciprocal lattice, specified by the Miller indices *hkl*.
- The amplitudes and relative phases of X-rays scattered from one unit cell in a particular direction can be calculated using the structure factor equation.
- The structure factor equation contains the atomic scattering factor, *f*, which is a function that encapsulates the scattering behaviour of all of the electron density associated with an atom so that the atom can be represented as a single point.
- The structure factor equation shows that every atom in the unit cell contributes to every diffracted beam. A complete set of diffracted intensities must therefore be measured to obtain information about all of the atoms.

6.7   Case Study: Ropivacaine Hydrochloride Form 2

Ropivacaine (Figure 6.15) is a local anaesthetic, marketed as an injectable solution of the hydrochloride salt of the *S*-enantiomer. Crystalline form 2 of ropivacaine hydrochloride adopts space group $P2_12_12_1$ with unit-cell parameters $a = 10.998$ Å, $b = 11.757$ Å, $c = 27.700$ Å, $\alpha = 90°$, $\beta = 90°$, $\gamma = 90°$.[1] The crystal structure has two ropivacaine cations and two chloride anions in the asymmetric unit $(Z' = 2)$ (Figure 6.16). We will use this structure to give an explicit illustration of some of the concepts in this chapter. The focus here is on diffraction geometry. The example will be extended at the end of Chapter 7 to consider also intensities and symmetry in the diffraction pattern.

The geometry of the diffraction pattern results from the translational symmetry of the crystal structure. We will consider the *ab* plane of the ropivacaine hydrochloride structure with the incident beam

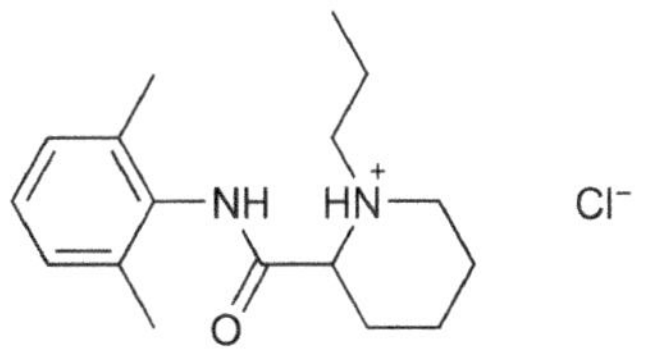

Figure 6.15   Ropivacaine hydrochloride, $C_{17}H_{27}N_2O^+ \cdot Cl^-$.

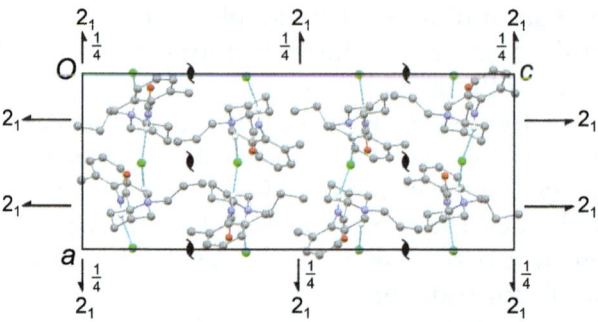

Figure 6.16 Unit-cell contents of ropivacaine hydrochloride projected along the *b* axis. H atoms are shown only on N atoms, with $N^+-H\cdots Cl^-$ hydrogen bonds indicated by dashed lines.

(CuKα, $\lambda = 1.54$ Å) taken to lie in that plane (*i.e.* limiting the discussion to a 2-D scenario). The directions where diffracted beams emerge can be deduced using Bragg's description. The periodicity of the crystal structure along the *a* axis is represented by considering the (100) planes. In ropivacaine hydrochloride, the *d*-spacing of the (100) planes is 10.998 Å, which gives $2\theta = 8.03°$ using Bragg's equation (taking $n = 1$). Knowing that "angle of incidence equals angle of reflection", we can draw directions for the incident and diffracted beams suitable to produce the 100 diffracted beam (Figure 6.17(a)). The underlying physical basis is that the scattering from any point within the crystal is identical to all equivalent points related by translational symmetry. Bragg's construction using the (100) planes allows us to calculate a suitable angular relationship for which the X-rays will emerge in phase if they are scattered from points equivalent by translation along the *a* axis. The calculated 2θ angle applies specifically to the situation where the path difference between X-rays scattered from equivalent points in neighbouring unit cells is equal to one X-ray wavelength ($n = 1$).

Now consider the 200 diffracted beam (Figure 6.17(b)). Physically, this is a simple extension of the argument above, but now we are finding the angular relationship where the path difference between X-rays scattered from equivalent points in neighbouring unit cells is equal to *two* X-ray wavelengths. Bragg's equation using $d = 10.998$ Å and $n = 2$ gives $2\theta = 16.10°$. The process can be continued for 300 ($2\theta = 24.25°$), 400 ($2\theta = 32.53°$), *etc.* As noted in Section 6.3.2, the reflections 200, 300, *etc.* are more commonly presented as reflections from the (200) planes, (300) planes, *etc.*, without including *n* in Bragg's equation. For understanding of the physical process, this is not especially helpful because positions within the (200) planes are

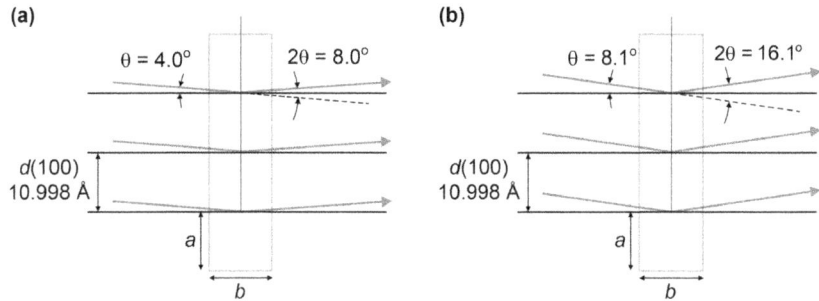

Figure 6.17 Bragg construction for the diffracted beams (a) 100; (b) 200. The dimensions of the unit cell correspond to the *ab* plane for the structure of ropivacaine hydrochloride.

not all related by translational symmetry within the crystal structure, so the significance of "translationally equivalent points scattering in an equivalent way" is not so clear. This reiterates that Bragg's description is a geometrical construction that should not be given too much physical emphasis. To produce each of the diffracted beams 100, 200, 300, *etc.*, it is clear from Figure 6.17 that the incident beam must enter the crystal from different directions. Usually, the incident beam will be fixed in an experimental measurement, so the crystal must be rotated to bring it sequentially into the diffracting position for each diffracted beam.

An alternative way to present the information in Figure 6.17 is to draw the normal to the (100) planes, which corresponds to the direction of **a*** in the reciprocal lattice (Figure 6.18). Integral steps along this line correspond to $1/d(001)$, $1/d(002)$, $1/d(003)$, *etc.*, which defines the reciprocal lattice along **a***. In the orthorhombic crystal system, where all unit-cell angles are constrained to be 90°, it is straightforward to visualise the reciprocal lattice because **a*** is parallel to **a**, **b*** is parallel to **b**, and **c*** is parallel to **c**. The benefit of visualising the reciprocal lattice becomes clear when extending the picture to more than one dimension. Applying the preceding discussion to the (010) planes defines the reciprocal lattice along **b*** (Figure 6.18). For ropivacaine hydrochloride, the spacing of the reciprocal lattice points along **b*** is smaller than along **a*** because the *b* axis of the unit cell is longer than the *a* axis. Now the remainder of the *hk*0 reciprocal lattice points can be added to the diagram and we can immediately deduce the orientations of the Bragg planes for any other *hk*0 reflection. In this case, where the angles between axes are all 90°, it is quite easy to calculate the *d*-spacing of any *hk*0 planes by hand. For 210, as shown in Figure 6.18, $d*(210) = \sqrt{((2a*)^2 + (b*)^2)} = 0.2008 \text{ Å}^{-1}$, so

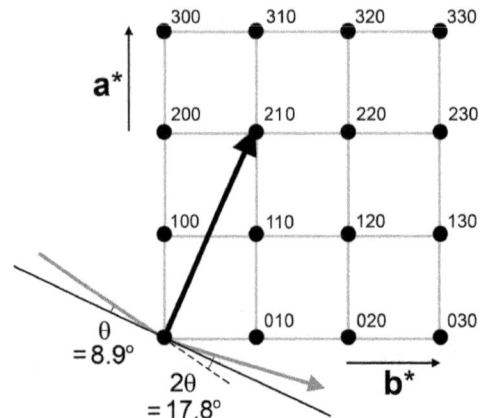

Figure 6.18 Reciprocal lattice construction for ropivacaine hydrochloride. The 210 reciprocal lattice vector is shown, with corresponding orientation of one Bragg plane and the direction of the incident and diffracted beams.

$d(210) = 4.981$ Å, giving $2\theta = 17.8°$ by Bragg's law. It is rarely (if ever) necessary to make such a calculation by hand, but it is helpful to appreciate how this can be achieved.

Reference

1. P.-O. Bergström, A. Fischer, L. Kloo and T. Sebhatu, *J. Pharm. Sci.*, 2006, **95**, 680, [CSD: TIJKIH].

7 Symmetry in Diffraction Patterns

Summary

Symmetry in the atomic positions within the unit cell leads to symmetry in the intensities of the diffracted beams. The intensity-weighted reciprocal lattice is introduced as a tool to visualise a single-crystal X-ray diffraction pattern. The symmetry of the diffraction pattern is the point symmetry of this object. It comprises the point symmetry parts of all symmetry operations in the space group. Space symmetry operations (screws, glides and centring translations) are distinguished from point symmetry operations by additional systematic intensity absences amongst specific groups of diffracted beams. The combination of the point group of a diffraction pattern and any systematic absences reveals the crystal's space group. Under normal scattering conditions, the diffraction pattern always contains an inversion centre, so $I(hkl) = I(\bar{h}\bar{k}\bar{l})$. This inversion centre combines with the true point group of the diffraction pattern to give its Laue group, which is usually what is measured. Under conditions where significant anomalous scattering occurs, the true point group can be measured and it is possible to determine the absolute structure (handedness) of a non-centrosymmetric crystal structure. This allows absolute configuration to be determined for chiral API molecules that crystallise in non-centrosymmetric space groups.

7.1 Introduction

Chapter 6 described the fundamental relationship between a single crystal and its X-ray diffraction pattern. In the context of this book,

Pharmaceutical Crystallography: A Guide to Structure and Analysis
By Andrew Bond
© Andrew Bond 2019
Published by the Royal Society of Chemistry, www.rsc.org

Chapter 6 is a partner to Chapter 2 in that it refers specifically to diffraction from single crystals and it assumes only the crystal's inherent translational symmetry. We have seen in Chapters 3 and 4 that crystals frequently display other types of symmetry, and we must now consider how that affects the diffraction pattern. There is a huge amount of detailed analysis that can be applied to crystal symmetry, most of which will probably never be needed by the average crystallographer. We are most interested in the practical consequences, with a view towards the experimental analysis to be described in Chapters 8–10.

7.2   Symmetry in Diffraction Patterns

Since the geometry of diffraction is related to the translational symmetry of the crystal, the geometry of the diffraction pattern must reflect the metric features of the crystal lattice. If the geometry of the unit cell is constrained by the crystal's symmetry then the relative orientations of the Bragg planes are similarly constrained. For example, in a cubic unit cell, the {100}, {010} and {001} planes must be mutually perpendicular and have identical d-spacings. Although it can be useful to picture the relationships between Bragg planes in this way, it is more fruitful to view diffraction geometry in terms of the reciprocal lattice. Aside from the simplification that comes from representing many sets of Bragg planes by their normal vectors, this view can more easily be extended to represent the diffracted intensities by imagining each reciprocal lattice point scaled by its associated intensity (Figure 7.1). The combined construction, known as the

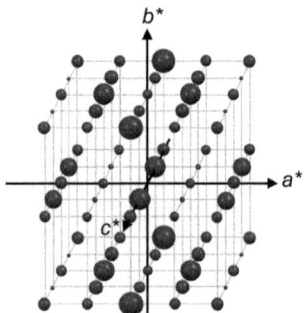

Figure 7.1   Example of an intensity-weighted reciprocal lattice. The volume of each reciprocal lattice point represents the relative intensity of the diffracted beam that it represents.

intensity-weighted reciprocal lattice, allows a coherent discussion of the symmetry of the diffraction pattern because we can treat it as a tangible object and describe its point symmetry.

7.2.1 Symmetry in the Diffracted Intensities

Non-translational symmetry in a crystal structure leads to symmetry in the *intensities* of the diffracted beams. Conceptually, this is easy to understand: if the atoms in the unit cell exhibit symmetry, we should expect the X-rays scattered by the unit cell to be the same in directions related by that symmetry. When we describe the "symmetry of the diffraction pattern", we implicitly refer to the symmetry of the intensity-weighted reciprocal lattice. For example, if the intensities of the diffracted beams hkl and $h\bar{k}l$ are identical, we say that the diffraction pattern has mirror symmetry. This means that the intensity-weighted reciprocal lattice shows mirror symmetry perpendicular to **b*** (Figure 7.2) and we say specifically that the point symmetry of the diffraction pattern is m.

Instinctively, we should expect the point group of the diffraction pattern to be related to the space group of the crystal structure. The relationship can be shown rigorously by considering how the GEPs of the space group contribute to the structure factor equation.

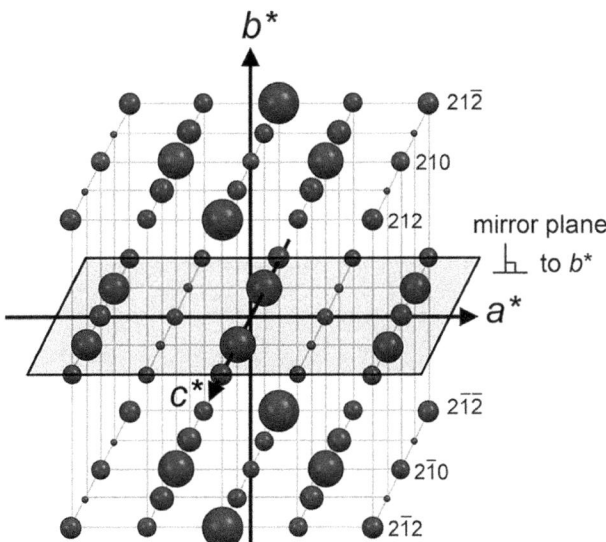

Figure 7.2 Mirror symmetry perpendicular to **b*** in the intensity-weighted reciprocal lattice. A few reflections within the $h1l$ and $h\bar{1}l$ layers are labelled to highlight the relationship.

Table 7.1 Point groups and Laue groups of diffraction patterns.

Crystal system	Point groups	Laue group	Example space groups[a]
Triclinic	1, $\bar{1}$	$\bar{1}$	$P1$, $P\bar{1}$
Monoclinic	2, m, $2/m$	$2/m$	$P2_1$, Cc, $P2_1/c$
Orthorhombic	222, $mm2$, mmm	mmm	$P2_12_12_1$, $Pna2_1$, $Pbca$
Trigonal	3, $\bar{3}$	$\bar{3}$	$R3$, $R\bar{3}$
	32, $3m$, $\bar{3}m$	$\bar{3}m$	$P3_121$, $R3c$, $R\bar{3}c$
Tetragonal	4, $\bar{4}$, $4/m$	$4/m$	$P4_1$, $I\bar{4}$, $I4_1/a$
	422, $\bar{4}2m$, $4mm$, $4/mmm$	$4/mmm$	$P4_12_12$, $P\bar{4}2_1c$, $P4nc$, $I4_1/acd$
Hexagonal	6, $\bar{6}$, $6/m$	$6/m$	$P6_3$, $P\bar{6}$, $P6_3/m$
	622, $\bar{6}2m$, $6mm$, $6/mmm$	$6/mmm$	$P6_122$, $P\bar{6}2c$, $P6_3mc$, $P6_3/mmc$
Cubic	23, $m\bar{3}$	$m\bar{3}$	$P2_13$, $Pa\bar{3}$
	432, $\bar{4}3m$, $m\bar{3}m$	$m\bar{3}m$	$P4_132$, $P\bar{4}3m$, $Ia\bar{3}d$

[a]These examples are the most frequently observed space groups for each point group type, according to statistics extracted from the Cambridge Structural Database (CSD). The triclinic, monoclinic and orthorhombic groups are by far the most common for typical pharmaceutical crystals. Some of the higher-symmetry space groups listed are extremely rare for molecular crystal structures.

We will simply summarise the results in Table 7.1. A straightforward way to deduce the point group from the space group is just to consider the symmetry operations of the space group, discarding the translational components of any space symmetry operations. This means that screw operations should be viewed as the corresponding pure rotations and glide operations should be viewed as mirrors. For example, space group $P2_1$ gives point group 2, space group $P2_1/c$ gives point group $2/m$ and space group $Pbca$ gives point group mmm. To reiterate: the point group refers to the symmetry of the diffraction pattern, which is conveniently viewed as the point group of the intensity-weighted reciprocal lattice.

Under normal experimental circumstances, a diffraction pattern *always* has an inversion centre. This means that the intensity of a diffracted beam *hkl* is identical to its inversion-related counterpart $\bar{h}\bar{k}\bar{l}$, whether the crystal structure contains inversion centres or not. In the language of complex numbers, $F(hkl)$ and $F(\bar{h}\bar{k}\bar{l})$ are complex conjugates. This can be verified by considering the structure factor equation and the resulting Argand diagram (Figure 7.3), where *hkl* and $\bar{h}\bar{k}\bar{l}$ appear as vectors with the same length but opposite phase angle. The consequence is that $I(hkl) = I(\bar{h}\bar{k}\bar{l})$. This result is called

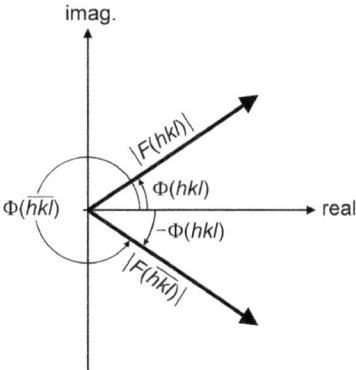

Figure 7.3 Friedel's law expressed on the Argand diagram. The lengths $|F(hkl)|$ and $|F(\bar{h}\bar{k}\bar{l})|$ are equal. Phases $\Phi(hkl)$ and $\Phi(\bar{h}\bar{k}\bar{l})$ are defined in the anticlockwise direction from the real axis. The angle in the clockwise direction from the real axis is $-\Phi(hkl)$. This shows $\Phi(\bar{h}\bar{k}\bar{l}) = -\Phi(hkl)$.

Friedel's law and the reflections hkl and $\bar{h}\bar{k}\bar{l}$ are called *Friedel pairs* (or *Bijvoet pairs*). The consequence of Friedel's law is that the point group usually measured for a diffraction pattern is the point group associated with the space group of the crystal structure, *plus an inversion centre*. The resulting point group is called the *Laue group* (Table 7.1). It is helpful to keep separate the true point group of the diffraction pattern (derived directly from the space group) and the Laue group (measured under normal experimental circumstances, where $I(hkl) = I(\bar{h}\bar{k}\bar{l})$). For example, for space group $P2_1$ the true point group of the diffraction pattern is 2, but the Laue group is $2/m$. For centrosymmetric space groups, the point group and the Laue group are the same.

7.2.2 Systematic Absences

A further important aspect of symmetry is the influence of space symmetry elements (screws and glides) on the diffraction pattern. As before, this can be shown formally by partitioning the atoms into symmetry-related groups and considering the effect on the structure factor equation. This exercise quickly becomes tedious, however, so we will simply state the important results. It has already been said that the effect of a screw axis or glide plane on the point symmetry of the diffraction pattern is identical to the corresponding rotation axis or mirror plane. It is possible to distinguish screw axes and glide planes, however, because they also give rise to characteristic sets of reflections having zero intensity, referred to as *systematic absences*.

Table 7.2 Systematic absences for common space symmetry operations. Only operations with order 2 are listed, as seen in the most common monoclinic and orthorhombic space groups.

Symmetry element	Reflections affected	Condition for absence
Axis parallel to or plane perpendicular to the *a* axis		
2_1 screw axis	$h00$	h odd
b-glide plane	$0kl$	k odd
c-glide plane	$0kl$	l odd
n-glide plane	$0kl$	$k+l$ odd
Axis parallel to or plane perpendicular to the *b* axis		
2_1 screw axis	$0k0$	k odd
a-glide plane	$h0l$	h odd
c-glide plane	$h0l$	l odd
n-glide plane	$h0l$	$h+l$ odd
Axis parallel to or plane perpendicular to the *c* axis		
2_1 screw axis	$00l$	l odd
a-glide plane	$hk0$	h odd
b-glide plane	$hk0$	k odd
n-glide plane	$hk0$	$h+k$ odd

Classes of reflection having zero intensity for common space symmetry operations are summarised in Table 7.2.

The origin of systematic absences can be viewed by considering the physical situation. For example, consider a 2_1 screw parallel to the *b* axis. Table 7.2 shows that this leads to zero intensity for every $0k0$ reflection with odd *k*. Thus, the spacing of the observed $0k0$ diffracted beams appears to be doubled. In the real crystal structure, this means that the periodicity along the corresponding direction must appear to be halved. We know that the 2_1 screw axis parallel to the *b* axis gives pairs of equivalent coordinates x,y,z and $-x, \frac{1}{2}+y, -z$. When we consider the $0k0$ reflections in the structure factor equation, the *x* and *z* coordinates are irrelevant for the sum $hx + ky + lz$, because *h* and *l* are zero. Only the *y* coordinates are relevant, and these appear in pairs: *y* and $\frac{1}{2}+y$. In other words, the atoms appear to be spaced regularly along the *b* axis with separation $\frac{1}{2}b$ (Figure 7.4). The argument only applies to the $0k0$ reflections because any other reflection also has some contribution from the *x* or *z* coordinates. Considering only the *y* coordinates in this way is often described as "projecting the structure onto the *b* axis". A similar argument can be constructed for a glide, where the periodicity along the direction of the glide translation appears to be halved when the structure is projected onto the glide plane (Figure 7.5).

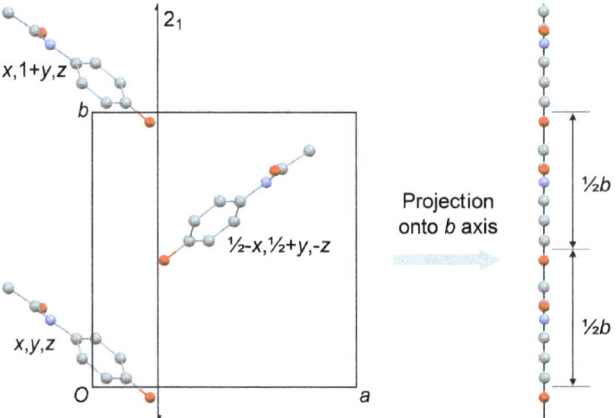

Figure 7.4 Paracetamol molecules related by a 2_1 screw axis parallel to the *b* axis in the crystal structure (left). Projection onto the *b* axis (*i.e.* discarding the *x* and *z* coordinates) produces a 1-D periodic pattern with translational repeat equal to $\frac{1}{2}b$ (right). The result is that the 0*k*0 diffracted beams appear to be spaced twice as far apart.

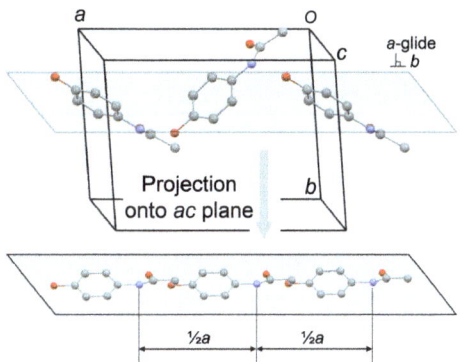

Figure 7.5 Paracetamol molecules related by an *a*-glide in the crystal structure (top). Projection onto the *ac* plane (*i.e.* discarding the *y* coordinates, bottom) produces a 2-D periodic pattern with translational repeat equal to $\frac{1}{2}a$ in the direction of the glide translation. The result is that the *h*0*l* diffracted beams appear to be spaced twice as far apart in the direction of the glide translation.

7.2.3 Centred Unit Cells

Systematic absences also arise for centred unit cells (Section 4.4). As an example, consider a *C*-centred cell, which has equivalent positions x,y,z and $\frac{1}{2}+x,\frac{1}{2}+y,z$ (Figure 7.6). This results in systematic absences for *hkl* when $h+k$ is odd. Unlike screw axes or glide planes, this condition does not refer to any particular class of

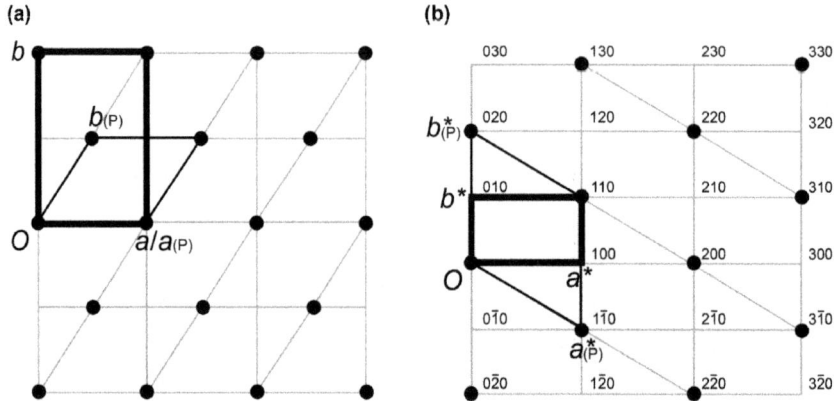

Figure 7.6 Example of systematic absences arising for a centred unit cell. (a) *ab* plane of the real lattice, showing a C-centred cell (thick lines) and corresponding primitive cell (thin lines). (b) *hk*0 plane of the reciprocal lattice, showing the corresponding reciprocal unit cells for the C-centred and primitive cells. The *hkl* labels correspond to the C-centred cell. The reciprocal lattice points for the primitive cell are all present but points with $h+k$ odd are absent for the C-centred cell.

Table 7.3 Systematic absences that arise for centred unit cells in the monoclinic and orthorhombic crystal systems.

Centring type	Condition for absence
P	—
A	$k+l$ odd
B	$h+l$ odd
C	$h+k$ odd
I	$h+k+l$ odd
F	Absent unless *hkl* are all odd or all even

reflections (*i.e.* any particular projection of the structure); it means that half of *all* diffracted beams are absent. The physical reason should be understandable from Chapter 2. We said that the relationship between a crystal structure and its lattice is unique. If we choose to define a centred unit cell that encompasses more than one lattice point, the diffraction pattern cannot change; we are just describing it using a different unit cell. A larger unit cell in real space means that the corresponding reciprocal lattice points will be closer together. The "extra" reciprocal lattice points added for the larger unit cell cannot be associated with any diffracted intensity so the associated diffracted beams must be absent. Table 7.3 summarises the systematic absences that arise for centred unit cells.

7.2.4 Determining the Space Group from the Diffraction Pattern

A crucial part of determining a crystal structure is to choose an appropriate space group. The information that enables this is contained in the intensity-weighted reciprocal lattice. If the point group of the diffraction pattern is established by comparing the intensities, this narrows down the possible space groups as indicated in Table 7.1. Note that the metric features of the crystal lattice are secondary in this assessment: the point group is determined from the *intensities*. The metric symmetry might appear to be higher, *e.g.* a triclinic crystal structure could have $a = b = c$, $\alpha = \beta = \gamma = 90°$, but it is the intensities that matter. Space groups that produce the same point group for the diffraction pattern can generally be distinguished by their systematic absences. For example, if point group $2/m$ is established for the diffraction pattern, there might be either 2-fold rotations or 2_1 screws parallel to the b axis; the 2_1 screw is distinguished by systematic absences in $0k0$ (Figure 7.7(a)). Similarly, there might be either mirrors or glides perpendicular to the b axis; the glide would be distinguished by systematic absences in $h0l$ (Figure 7.7(b)). The combination of the point group and the systematic absences identifies the space group uniquely in most cases. There are a few cases where the assignment remains ambiguous, which must be resolved by trying to describe the structure in the various space groups and deciding which is best. These cases generally arise in higher-symmetry space groups, which are less common for typical small-molecule APIs.

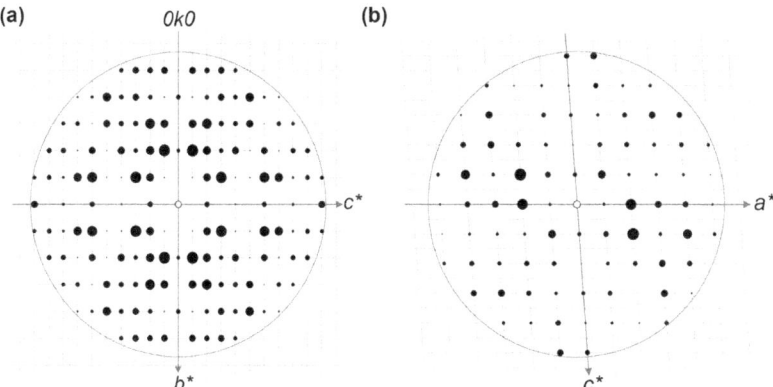

Figure 7.7 Sections through an intensity-weighted reciprocal lattice with point group $2/m$. (a) ($0kl$) plane showing systematic absences along the $0k0$ line (vertical) when k is odd, consistent with a 2_1 screw axis parallel to b; (b) ($h0l$) plane showing systematic absences when l is odd, consistent with a c-glide perpendicular to b. Space group $P2_1/c$ is indicated.

7.2.5 Intensity Statistics

The preceding section glosses over a vital practical point: it is the combination of the *true* point group of the diffraction pattern and any systematic absences that identify the space group, but it is the *Laue group* that is usually measured. If Friedel's law introduces an inversion centre in the diffraction pattern, how can we know whether the space group is actually centrosymmetric? An indication can be obtained by making a statistical analysis of the intensities. The basis is that the atoms in a centrosymmetric crystal structure exist in pairs at x,y,z and $-x,-y,-z$, which leads to a different distribution of intensities compared to a non-centrosymmetric structure (Figure 7.8). Basically, there is a higher probability of obtaining weak reflections in the centrosymmetric case. Intensity statistics can suggest whether a structure is centrosymmetric or not, even when Friedel's law applies. There are circumstances that might be misleading (*e.g.* a non-centrosymmetric structure where atoms making the largest contribution to the scattering just happen to approximate a centrosymmetric arrangement), so statistics are always indicative rather than conclusive.

7.3 Centrosymmetric and Non-centrosymmetric Structures

For molecular crystals in general, centrosymmetric structures are far more prevalent than non-centrosymmetric ones. However, pharmaceutically relevant molecules are frequently chiral, and samples containing only one enantiomer of a chiral compound must crystallise in a non-centrosymmetric space group. For this reason,

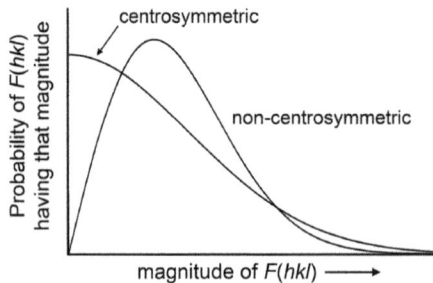

Figure 7.8 Expected distribution of intensities for centrosymmetric and non-centrosymmetric structures. Comparison of experimental measurements to these different distributions can give an indication of whether a crystal structure is centrosymmetric or not.

non-centrosymmetric crystal structures are comparatively more common in the pharmaceutical sciences and it is important to understand some key differences compared to centrosymmetric structures. This is a different matter from detecting an inversion centre using X-ray diffraction (Section 7.2.5); there are some important differences in the nature of the diffraction pattern for centrosymmetric and non-centrosymmetric structures.

The most substantial difference is that the structure factors of centrosymmetric structures are always real numbers. On the Argand diagram, this means that the resultant vectors point one way or the other along the horizontal axis. This can be understood by considering how the structure factor equation is built up on the Argand diagram (Figure 7.9). Each atom in the unit cell contributes a vector and the resultant structure factor is the head-to-tail summation of these vectors. If the atoms exist in centrosymmetric pairs, the summation will include pairs of vectors contributed by equivalent atoms at x,y,z and $-x,-y,-z$. The two vectors in each pair have identical length but opposite phase angle, so any vector that points away from the real axis has a matching vector that takes the resultant back onto the real axis. The result is that any structure factor for a centrosymmetric crystal structure has either $\Phi = 0$ if the structure factor is a positive real number or $\Phi = 180°$ (π radians) if the structure factor is a negative real number. No such restriction exists for a non-centrosymmetric structure, where Φ can take any value in the range 0–360° (0–2π radians). It is interesting to consider a physical picture of what this means: when the X-rays scattered from all atoms in the unit cell of a centrosymmetric structure are summed, the resultant diffracted beam (in *any* direction) is either fully in phase with a

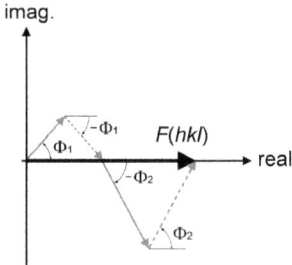

Figure 7.9 Structure factor constructed on the Argand diagram for a centrosymmetric structure. The solid and dashed vectors represent pairs of equivalent atoms with coordinates x,y,z and $-x,-y,-z$. The resultant vector must lie on the real axis. The diagram shows a positive $F(hkl)$, corresponding to phase 0. The resultant $F(hkl)$ could also be negative, corresponding to phase 180° (π radians).

diffracted beam that would be scattered from the origin, or fully out of phase. There is no intermediate possibility. This is just another example of the way that symmetry in a crystal structure manifests itself in the diffraction pattern.

7.3.1 Absolute Structure and Anomalous Scattering

Undoubtedly one of the most valuable applications of single-crystal X-ray diffraction for pharmaceutical compounds is determination of absolute structure. As mentioned in Section 4.6.1, inverting a non-centrosymmetric crystal structure through the origin changes its absolute structure, which means changing its "handedness". The discussion of Friedel's law in Section 7.2.1 suggests that it should not be possible to determine absolute structure. Because the diffraction pattern is always centrosymmetric, the diffraction pattern of a non-centrosymmetric structure should be the same whether it is "right-handed" or "left-handed". Indeed, according to the description of diffraction that we have given so far, we would not be able to determine absolute structure. Without making it explicit, Chapter 6 described only *elastic* scattering of X-rays, where the incident X-rays are scattered without any transfer of energy to the crystal. Under these circumstances, Friedel's law applies. The information required to determine absolute structure comes from another physical process: at appropriate X-ray wavelengths (energies), atoms can absorb X-rays with the consequence that the absorbed X-rays do not contribute to the scattering we have described. The scattering becomes inelastic (meaning that there is some transfer of energy from the X-rays to the atoms in the crystal) and is said to be "anomalous" (Figure 7.10).

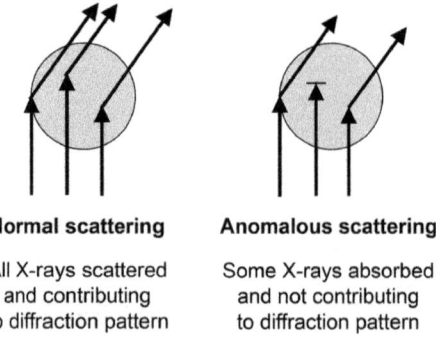

Normal scattering

All X-rays scattered
and contributing
to diffraction pattern

Anomalous scattering

Some X-rays absorbed
and not contributing
to diffraction pattern

Figure 7.10   Schematic diagram showing "normal" (elastic) and "anomalous" (inelastic) scattering.

The term "anomalous scattering" is misleading because it gives an impression of some new type of scattering process. The scattering process relevant to X-ray diffraction is not changed at all; it is just that some of the incident X-rays are no longer scattered in this way. Instead, they are absorbed by atoms, to an extent that can be quantified for each atom type at a specified X-ray wavelength. To describe anomalous scattering in the structure factor equation, an additional term, which we will label $i\Delta f$, is added to the atomic scattering factor. It has been mentioned previously (Section 6.4.4) that the atomic scattering factor already encapsulates numerous effects; inclusion of the $i\Delta f$ term allows the function also to describe the consequences of atom-dependent absorption. Multiplication by i means that the Δf contribution is rotated on the Argand diagram by 90° in the positive (anticlockwise) sense relative to the phase calculated from $2\pi(hx + ky + lz)$ (Figure 7.11). There is nothing "imaginary" about the process and it does not mean that X-rays are scattered with π radians added to their phase. It is simply the way to produce the atom's resultant contribution on the Argand diagram. The normal atomic scattering factor must also be modified, which is just a matter of adding a small (usually negative) number to the value of f. The value of Δf and the associated correction to f are generally much smaller than the actual value of f for each atom type and they can be considered to be independent of θ (in contrast to f itself; Section 6.4.4).

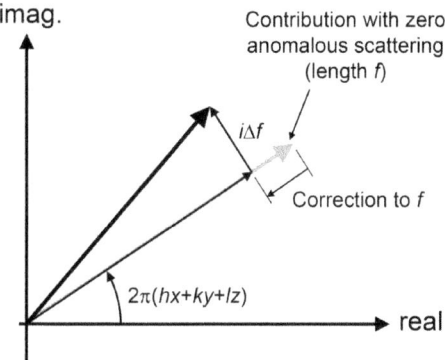

Figure 7.11 Representing anomalous scattering for one atom on the Argand diagram. The normal scattering factor is corrected and an additional term $i\Delta f$ is added, forming a right angle to the normal contribution.

With the introduction of the $i\Delta f$ term into the atomic scattering factor, the structure factor equation can be split into two terms:

$$F(hkl) = \sum_{n=1}^{N} (f_n + i\Delta f_n) \exp\{2\pi i(hx_n + ky_n + lz_n)\}$$

$$F(hkl) = \sum_{n=1}^{N} f_n\exp\{2\pi i(hx_n + ky_n + lz_n)\} + i\sum_{n=1}^{N} \Delta f_n\exp\{2\pi i(hx_n + ky_n + lz_n)\}$$

where f_n is considered to have been corrected as described to account for those X-rays that do not contribute to the scattering. Both terms are summed over all atoms in the unit cell, although the second term may have significant (non-zero) Δf values for only a few atom types.

The important consequence of significant anomalous scattering is that Friedel's law no longer applies. This can be seen on the Argand diagram in Figure 7.12. We saw in Section 7.2.1 that the normal structure factor has the same magnitude but opposite phase for *hkl* and *h̄k̄l̄*. Likewise, the Δf term in the modified structure factor equation must have the same magnitude and opposite phase for *hkl*

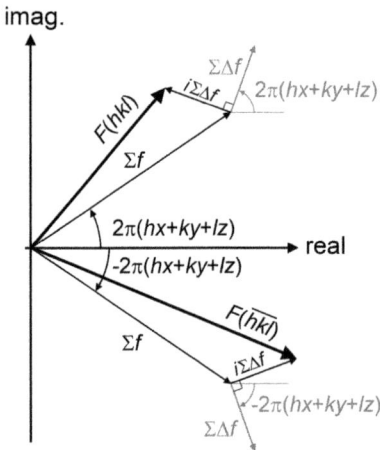

Figure 7.12 Breakdown of Friedel's law for a non-centrosymmetric structure with significant anomalous scattering. The top half of the diagram refers to diffracted beam *hkl* while the bottom half refers to *h̄k̄l̄*. The vectors $\sum f$ indicate the sum of the (modified) normal scattering contributions for all atoms in the unit cell. The grey vectors $\sum \Delta f$ indicate the sum of the Δf terms for all atoms in the unit cell. Multiplication by *i* turns these contributions 90° in a clockwise direction in both cases, which leads to different resultant vectors *F(hkl)* and *F(h̄k̄l̄)*.

and $\bar{h}\bar{k}\bar{l}$. However, the multiplication by i acts to rotate this contribution on the Argand diagram by $90°$ *always in a positive sense*, so that the bottom half of Figure 7.12 is no longer a mirror image of the top half. The resultant structure factors have different phases and different magnitudes, and therefore different intensities. Thus, the diffraction pattern loses the centre of symmetry that is imposed under normal (elastic) scattering conditions and the measured intensity-weighted reciprocal lattice reveals the *true* point group of the diffraction pattern rather than the Laue group. Under standard laboratory conditions, the effect is always subtle. Whether or not it can be detected depends on the magnitude of the differences between $I(hkl)$ and $I(\bar{h}\bar{k}\bar{l})$ compared to the experimental errors in the intensity measurements. Some practical aspects of absolute structure determination are mentioned in Chapters 12–14.

7.3.2 Anomalous Scattering in Centrosymmetric Structures

Given that anomalous scattering enables absolute structure to be determined for non-centrosymmetric crystal structures, it is natural for descriptions of anomalous scattering to focus on non-centrosymmetric structures. Of course, anomalous scattering also operates for centrosymmetric structures, but the practical consequences are not so significant. It was shown in Section 7.3 that the structure factors for a centrosymmetric structure are always real numbers. The construction shown in Figure 7.3, where Friedel pairs are related to each other on the Argand diagram by reflection across the real axis, is therefore simplified because the vectors for hkl and $\bar{h}\bar{k}\bar{l}$ are superimposed on the real axis (Figure 7.13). Adding the contribution from anomalous scattering introduces a vector pointing along the imaginary axis, but it is the same for hkl and $\bar{h}\bar{k}\bar{l}$. Thus, anomalous

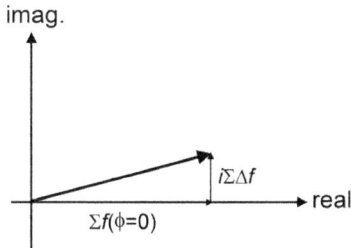

Figure 7.13 Influence of anomalous scattering for a centrosymmetric structure. The diagram is identical for $F(hkl)$ and $F(\bar{h}\bar{k}\bar{l})$, so Friedel's law still applies.

scattering changes the resulting intensities (and therefore must be accounted for when calculating structure factors) but it does not result in *different* intensities for hkl and $\bar{h}\bar{k}\bar{l}$. For a centrosymmetric structure, the point group of the diffraction pattern is always the Laue group, regardless of whether or not there is significant anomalous scattering.

7.4 Summary of Key Points

- Non-translational symmetry in a crystal structure leads to symmetry in the intensities of the diffracted beams. The symmetry of the diffraction pattern refers to the point symmetry of the intensity-weighted reciprocal lattice.
- The point group of the diffraction pattern can be deduced from the symmetry operations of the space group, discarding the translational components of any space symmetry operations.
- Under normal scattering conditions, a diffraction pattern always has an inversion centre, as described by Friedel's law: $I(hkl) = I(\bar{h}\bar{k}\bar{l})$. The consequence is that the point group of a diffraction pattern is the point group associated with the space group of the structure, plus an inversion centre. The resulting point group is called the Laue group.
- Space symmetry operations (screws, glides, unit-cell centring) can be recognised because they produce characteristic sets of reflections having zero intensity, referred to as systematic absences. The combination of the point group of the diffraction pattern and the systematic absences identifies the space group of the crystal structure.
- At an appropriate wavelength of the incident X-rays, atoms absorb X-rays so that they do not contribute to the diffraction pattern. This is described as anomalous scattering, and the result is that Friedel's law does not apply. The measured point group of the intensity-weighted reciprocal lattice then reveals the true point group of the diffraction pattern, rather than the Laue group.
- Under conditions where it is possible to measure the differences between intensities introduced by anomalous scattering, it is possible to determine the absolute structure (handedness) of a non-centrosymmetric crystal structure. This enables absolute configuration to be determined for chiral molecules.

7.5 Case Study: Ropivacaine Hydrochloride Form 2 (Continued)

Continuing the Case Study at the end of Chapter 6, we can now consider the intensities and symmetry of the diffraction pattern for form 2 of ropivacaine hydrochloride. The picture of the reciprocal lattice constructed in Chapter 6 can be augmented by representing each reciprocal lattice point with a volume scaled to the corresponding diffracted intensity. The result is shown in Figure 7.14. Of primary interest is the *symmetry* of the resulting intensity-weighted reciprocal lattice.

The space group of the structure is $P2_12_12_1$, which means that it has 2_1 screw axes running parallel to each of the unit-cell axes. The true point group of the resulting diffraction pattern is 222, which can be deduced simply by discarding the translational components of the screw axes. This means that the intensity-weighted reciprocal lattice shows 2-fold rotation symmetry around each of its axes. The following relationships are established between the diffracted intensities:

$$I(hkl) = I(h\bar{k}\bar{l}) = I(\bar{h}k\bar{l}) = I(\bar{h}\bar{k}l)$$

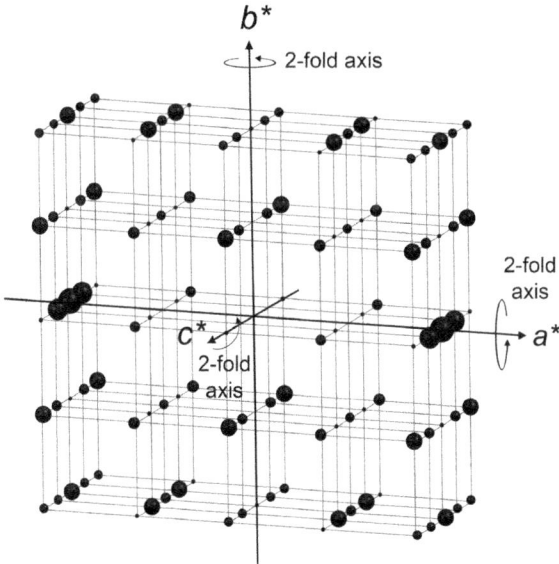

Figure 7.14 Intensity-weighted reciprocal lattice for ropivacaine hydrochloride form 2. The intensities are calculated from the known crystal structure using the structure factor equation.

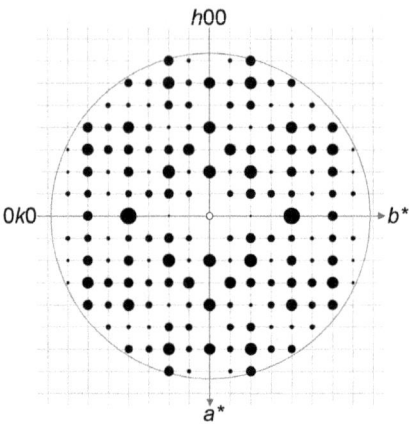

Figure 7.15   *hk*0 plane calculated for ropivacaine hydrochloride form 2, showing
systematic absences for *h*00 and 0*k*0, indicative of 2_1 screw axes along
both the *a* and *b* axes.

Under experimental conditions where there is no significant
anomalous scattering, Friedel's law will also impose an inversion
centre on the diffraction pattern so the relationships become:

$$I(hkl) = I(hk\bar{l}) = I(h\bar{k}l) = I(\bar{h}kl) = I(\bar{h}\bar{k}l) = I(\bar{h}kl) = I(h\bar{k}\bar{l}) = I(hk\bar{l})$$

The point symmetry of the intensity-weighted reciprocal lattice is
now *mmm*, which is the Laue group for the orthorhombic crystal
system. Under these conditions, only $\frac{1}{8}$ of the diffracted intensities
(an "octant") are unique; the remainder of the intensity-weighted
reciprocal lattice can be constructed by applying the symmetry operations of the Laue group.

The fact that the crystal structure contains 2_1 screw axes is revealed
by systematic absences in the diffracted intensities. This can be
illustrated by taking appropriate 2-D slices through the intensity-
weighted reciprocal lattice. For example, the *hk*0 plane calculated for
ropivacaine hydrochloride is shown in Figure 7.15. The characteristic
sign of a 2_1 screw axis parallel to one of the unit-cell axes is the pattern
of alternating zero and non-zero intensities along the corresponding
reciprocal lattice axis. Specifically, the 2_1 axis parallel to the *a* axis is
revealed by the *h*00 reflections, which are absent for odd *h*. The 2_1
axes parallel to the *b* and *c* axes are revealed by similar patterns in the
0*k*0 and 00*l* reflections, respectively. The combination of the point
group and these systematic absences reveals space group $P2_12_12_1$.

8 Single-crystal X-ray Diffraction (Part 1)

Summary

This chapter describes instrumentation for single-crystal X-ray diffraction then focusses on the geometrical aspects of single-crystal analysis. A typical measurement proceeds by establishing the geometry of the diffraction pattern then using that information to devise a strategy to collect the required set of intensities. The diffraction pattern is measured as a set of images while the crystal is rotated to bring it sequentially into suitable orientations to produce the various diffracted beams. Beams that hit the detector produce spots on the diffraction images, which appear and disappear as the crystal rotates. The measured diffraction angles are converted to reciprocal lattice points, then a suitable set of lattice vectors is chosen so that all measured beams are assigned integral *hkl* indices. Once the diffraction pattern has been indexed in this way, the crystal can be aligned to measure any diffracted beam or to identify crystal faces to describe the morphology.

8.1 Introduction

An overview of the process for measuring a single-crystal X-ray diffraction pattern is illustrated in Figure 8.1. The crystal lattice is determined by measuring the geometry of the diffraction pattern. A unit cell is chosen and each diffracted beam is assigned *hkl* indices referring to the specified unit cell. The intensities of all diffracted beams are then measured and the output is a list of *hkl* values with associated intensities. The next two chapters describe the practical

Pharmaceutical Crystallography: A Guide to Structure and Analysis
By Andrew Bond
© Andrew Bond 2019
Published by the Royal Society of Chemistry, www.rsc.org

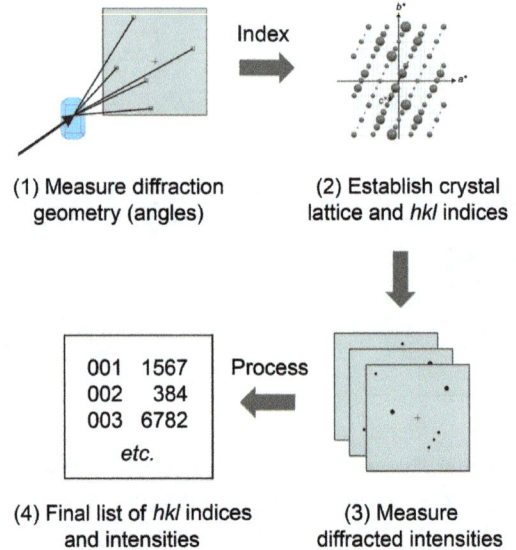

(1) Measure diffraction   **(2) Establish crystal**
geometry (angles)   **lattice and *hkl* indices**

(4) Final list of *hkl* indices   **(3) Measure**
and intensities   **diffracted intensities**

Figure 8.1 Overview of the process of a typical single-crystal X-ray diffraction analysis.

aspects of this process. This chapter gives an introduction to the instrumentation then describes the geometrical aspects. Chapter 9 focusses on measurement and analysis of the diffracted intensities. The division is artificial—we must always measure both the geometry and intensities of the diffraction pattern to determine a crystal structure—but there is quite a lot of material to cover, so it is convenient to divide it in this way. We cannot deal explicitly with any particular type of instrument, or specific features that may be implemented in a particular software package, so the description is as generic as possible. The fundamentals of the process are identical on all modern instruments, so the aim is to provide the understanding to support specific instrument training.

8.2   Overview of a Single-crystal X-ray Diffractometer

A single-crystal X-ray diffractometer (Figure 8.2) has four basic components: (1) a source to provide the incident X-ray beam; (2) a motorised *goniometer* to orient the crystal with respect to the incident beam; (3) a detector to measure the diffracted X-rays and (4) a control system with associated software and user interface. Modern

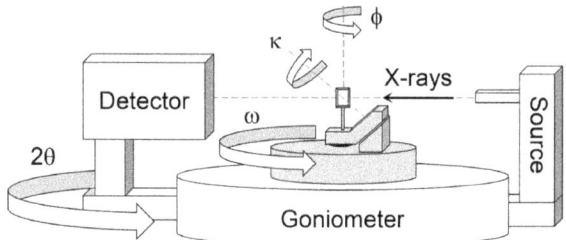

Figure 8.2 Schematic representation of a single-crystal X-ray diffractometer with "kappa geometry". All components are linked to a control system with an associated user interface.

instruments present these components in a highly-integrated and user-friendly package that enables anyone to collect single-crystal X-ray data with minimal training. Indeed, technical developments in instrumentation mean that the barrier to using a single-crystal X-ray instrument is no higher than for any other piece of analytical equipment.

8.2.1 The X-ray Source

In a standard laboratory setting, the conventional X-ray source is the "sealed tube", in which a stream of electrons emitted from a filament is accelerated through an evacuated tube towards a metal target. Modern instruments commonly use *microfocus* sources, which produce greater X-ray flux over a smaller area while operating at far lower power than conventional tubes. The most common targets are Cu and Mo, which emit characteristic X-rays with wavelengths close to 1.54 and 0.71 Å, respectively. Since shorter wavelength corresponds to higher energy, Cu radiation has lower energy and interacts more strongly with a chemical sample (*i.e.* fewer X-rays pass straight through). This generally makes Cu radiation the first choice to study pharmaceutical crystals, since they are often composed of lighter elements (C, H, N, O, *etc.*) that scatter X-rays relatively weakly. Cu radiation is also advantageous for absolute structure determination because it enhances the effects of anomalous scattering. As we have seen in Section 6.3, the geometry of diffraction is influenced by the wavelength of the incident X-rays, and the longer wavelength of Cu radiation means that the diffraction pattern is spread over a wider angular range compared to Mo radiation. We have also seen in Chapter 6 that there is a minimum value of the *d*-spacing between Bragg planes that can be observed in a diffraction experiment, which is equal to $\frac{1}{2}\lambda$. The minimum value of d that can be measured using Cu and Mo radiation is 0.77 and 0.36 Å, respectively. Both are sufficient to determine a crystal structure to "atomic resolution".

8.2.2   The Goniometer

The *goniometer* orients the crystal and detector with respect to the incident X-ray beam. Typically, there are three axes of movement, as shown in Figure 8.2: (1) the ω (omega) axis rotates around the vertical axis of the diffractometer; (2) the κ (kappa) axis swings the crystal within an inclined plane that moves with the ω circle; (3) the ϕ (phi) axis rotates around the crystal's local mount axis. The goniometer must be able to orient the crystal with respect to the incident beam to enable a suitable orientation for diffraction to occur and in such a way that the diffracted beam will hit the detector. Recalling Bragg's description of diffraction geometry, the incident beam, diffracted beam and normal to the relevant Bragg planes lie in a common plane. This means that the crystal must be oriented so that the associated reciprocal lattice vector lies close to the horizontal plane of the instrument for the diffracted beam to hit the detector face (Figure 8.3).

8.2.3   The Detector

Detectors for single-crystal X-ray diffraction are now almost exclusively 2-D area detectors. Numerous technologies exist, but they all comprise a 2-D array of pixels that map and count incoming X-rays (Figure 8.4). The position at which a diffracted beam hits the detector is determined by identifying pixels that register a significant number of X-ray counts above any background level. Consistent with our general approach, we can imagine the task of the detector to be partitioned into two aspects: it must be able to measure the geometry of the diffraction pattern and the intensities of the diffracted beams.

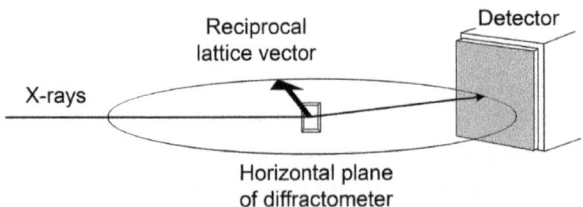

Figure 8.3   Crystal in diffracting position with the reciprocal lattice vector (= bisector of incident and diffracted beams) aligned close to the horizontal plane of the diffractometer.

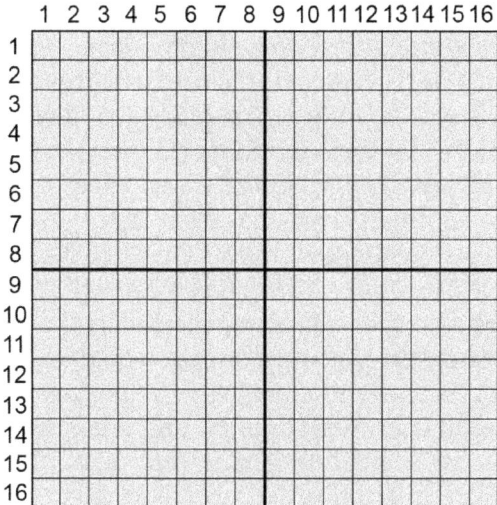

Figure 8.4 Schematic representation of a 2-D area detector. The detector comprises an array of pixels, each of which counts incoming X-ray photons. Real detectors contain many more than 16×16 pixels!

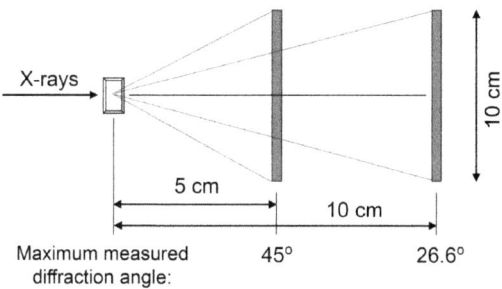

Figure 8.5 Angular range accessible on a detector face with dimension 10 cm at distances of 5 and 10 cm from the crystal. Bringing the detector closer to the crystal allows more of the diffraction pattern to be measured on each image.

The detector moves around the 2θ axis of the diffractometer, aligned so that the incident beam strikes the specified centre of the detector when $2\theta = 0$. The combination of the detector's 2θ position and the location of the pixels on the detector face allow the diffraction angle to be measured.

Out of the horizontal plane, the measurement range depends on the height of the detector and its distance from the crystal. A typical detector with a square active area of 10×10 cm gives a maximum diffraction angle of $45°$ when the detector is 5 cm from the crystal (Figure 8.5). Moving the detector closer to the crystal has the advantage that it covers a greater angular range so it can measure more of the diffraction pattern on a single image. However, there is a disadvantage: moving the detector closer reduces the *spatial resolution* of the diffraction image. Imagine two divergent diffracted beams with a small angle between them (Figure 8.6). If the detector is far back, these beams strike two different pixels with one or more blank pixels between them. Thus it is clear that they are two separate diffracted beams. As the detector moves closer to the crystal, however, the gap between the beams on the detector face will decrease and they will end up striking adjacent pixels or eventually the same pixel. Now it is impossible to detect that they are two separate beams—the two beams cannot be *resolved*. Practically, with the pixel sizes available on modern detectors, and the typical angular separation between diffracted beams for a small-molecule pharmaceutical crystal, it is possible to measure diffraction patterns effectively with a crystal-to-detector distance around 5 cm or less. However, the user should be aware of the idea of angular resolution: if a crystal has a long unit-cell axis, the angular separation between the diffracted beams associated with this axis will be small and it may be necessary to move the detector further away from the crystal to achieve adequate separation of the diffracted beams. Often, the diffractometer control software will suggest when this is likely to be necessary.

Measuring the intensities of the diffracted beams requires the detector to count incoming X-rays in a linear fashion, *e.g.* a relative intensity of "10 X-rays" must provide a relative count of "10" in the detector's electronic output. We can rest assured that the detector manufacturers take care of this. One thing that the user must be aware of, however, is the detector's *dynamic range*, which refers to the maximum number of X-rays that a given pixel can count before it

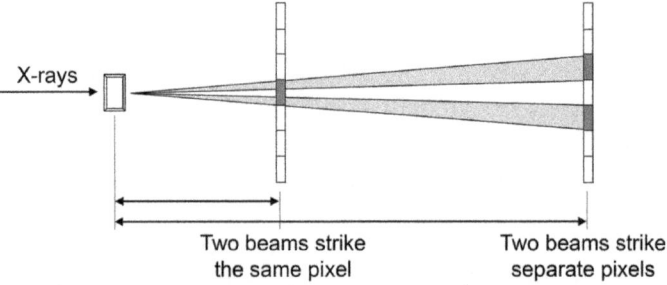

Figure 8.6   Illustration of spatial resolution. At shorter crystal-to-detector distances, separate beams may strike the same pixel on the detector face so they cannot be distinguished. Moving the detector further from the crystal may allow the beams to be resolved.

needs to be reset. When measuring, it must be ensured that the strongest diffracted beams do not exceed the maximum number of counts for any pixel. If they do, the detector will simply stop counting at those pixels and provide intensity values that are systematically under-measured. This is referred to as *overloading* or *topping* of the detector and it is one of the more serious mistakes that a user can make during a single-crystal measurement. We will return to this in Chapter 9.

8.2.4 The Control and Data Processing Software

The control and data processing software manages all aspects of data collection, including movement of the crystal and detector, opening and closing of the X-ray shutter, and read-out and processing of the diffraction images. Modern software systems present a user-friendly interface, which is generally highly automated so that it is possible to proceed through an entire analysis with little or no user intervention. This is one of the main features that has increased the accessibility of the single-crystal technique, but it brings dangers related to understanding and recognising key decision points. The remainder of this chapter, and Chapter 9, is largely concerned with retaining user understanding within this increasingly automated environment.

8.3 Setting Up the Analysis

Crystallisation of APIs and associated solid forms is an experimental skill that is best learned by experience. We assume here that suitable crystals are available for analysis. So what constitutes a "suitable" crystal? Basically, it should be a single crystal and give a diffraction pattern that is sufficiently strong to be measured to a reasonable resolution on the available instrumentation. We have seen some factors that determine how strongly a crystal diffracts: (1) larger crystals scatter more strongly than smaller crystals; (2) atoms with more electrons scatter X-rays more strongly, so crystals containing "heavy" elements will produce more intense diffraction than a crystal with the same volume containing only "light" elements; (3) Cu radiation interacts more strongly with the sample than Mo radiation. For a given chemical compound, we must live with the atom types that we have. It may be possible to make some derivative of an API to introduce heavier atoms (for example, a hydrochloride or hydrobromide salt), but pharmaceutical crystals frequently contain only light atoms that scatter X-rays relatively weakly. In a given laboratory, it may be possible to choose between Cu and Mo radiation, but often it isn't. Likewise, the power of the available instruments is limited. Thus, the major variable that can be controlled is usually the crystal itself.

8.3.1 Choosing a Crystal

The best possible crystal for an X-ray diffraction measurement would be spherical. The advantage of a sphere is that the path length for an X-ray entering and leaving the crystal is identical in any direction. This is important because X-rays are subject to some degree of absorption as they pass through the crystal. If the crystal habit is highly anisotropic, there will be some directions where the X-rays have a much shorter path through the crystal and some that have a much longer path (Figure 8.7). This leads to *anisotropic absorption*, which can be a problem because it attenuates the intensities of the diffracted beams differently, distorting the measured set of relative intensities. We will see in Chapter 9 how this can be corrected for, but like any experiment, it is better to minimise systematic errors than to correct for them. Of course, crystals are not naturally spherical but the aim should be to find one with a shape as regular as possible. For needles and plates, it may be desirable to cut the crystal to obtain a more regular shape. Cutting can be difficult, however, and reducing the volume of the crystal will decrease its overall scattering power. Sometimes, we just have to accept the crystal shape and size that we have.

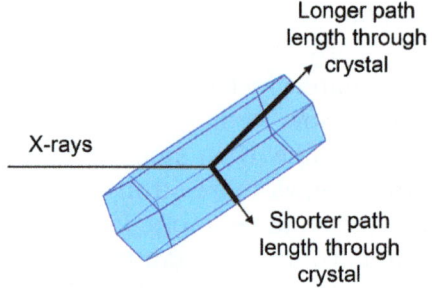

Figure 8.7   Anisotropic absorption for a non-spherical crystal. Different path lengths lead to different degrees of attenuation of the various diffracted intensities. A correction must be made during the data processing to obtain reliable relative intensities.

It is important that the volume of the crystal within the incident X-ray beam is consistent in all crystal orientations. Otherwise, the measured set of intensities will be distorted by the fact that different beams are measured with different exposed crystal volumes. This requires either that the crystal dimensions should not be larger than the diameter of the incident beam, or that the crystal should be so large that it "fills" the incident beam in all orientations. The former situation is preferable because a larger crystal increases absorption. In practice, the requirement for constant illuminated volume is partially relaxed by some final corrections made to the intensities (Section 9.2.5). As stated above, however, it is better to minimise systematic errors than to correct for them, so better results should be obtained by selecting a regular-shaped crystal that is wholly accommodated within the incident beam. Where this is not possible, acceptable results might still be obtained regardless of the crystal size and shape.

8.3.2 Mounting the Crystal

The crystal is mounted on a *goniometer head*, which attaches to the diffractometer's goniometer. Often, the goniometer head can remain attached to the goniometer and crystal mounting can be achieved with a detachable magnetic mount (Figure 8.8). The crystal must be fixed in the geometrical centre of the goniometer, so its centre remains stationary as the crystal is rotated. A traditional way to achieve this is to glue the crystal onto the end of a glass fibre, mounted on a metal pin. The glass fibre should be thin to minimise its contribution

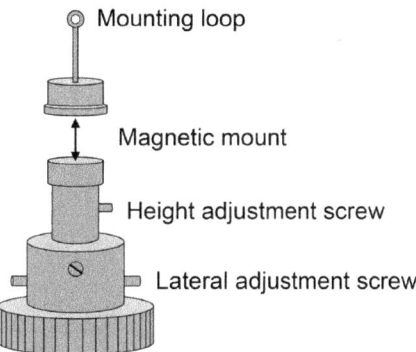

Figure 8.8 Schematic diagram of a goniometer head with detachable magnetic mount and mounting loop. Two lateral adjustment screws provide movement along perpendicular directions.

to background X-ray scattering, but sufficiently rigid to support the crystal's weight. It is probably now more common to use specifically designed crystal mounts, made from X-ray transparent materials that give negligible background scattering. Frequently, these have a loop with diameter similar to that of the incident X-ray beam (*ca.* 100 μm for a typical microfocus source) and the crystal is held within the loop by glue or viscous oil. Mounting crystals using oil rather than glue has several advantages: (1) placing crystals in a drop of oil makes them easier to manipulate on a microscope slide; (2) air- or moisture-sensitive crystals are protected from the atmosphere during mounting; (3) attachment using oil is reversible, so the crystal mount can be used many times. Commonly, single-crystal measurements are carried out under a stream of cold N_2 gas, which causes the viscous oil to freeze and "clamp" the crystal firmly to the mount. For room-temperature data collections, it may still be necessary to use glue. "Superglue" (cyanoacrylate) is convenient because it is less viscous and faster drying, but it can sometimes dissolve organic crystals. An alternative is two-component epoxy glue.

8.3.3   Centring the Crystal

Once mounted, the crystal must be aligned with the geometrical centre of the diffractometer. Crystal centring is achieved by screw adjustments on the goniometer head: one for the crystal height, and two for lateral movement of the crystal (Figure 8.8). The adjustment is made by viewing the crystal through a video microscope, usually aided by a grid or crosshairs to keep track of the crystal's position (Figure 8.9). Practically, the diffractometer should have a defined

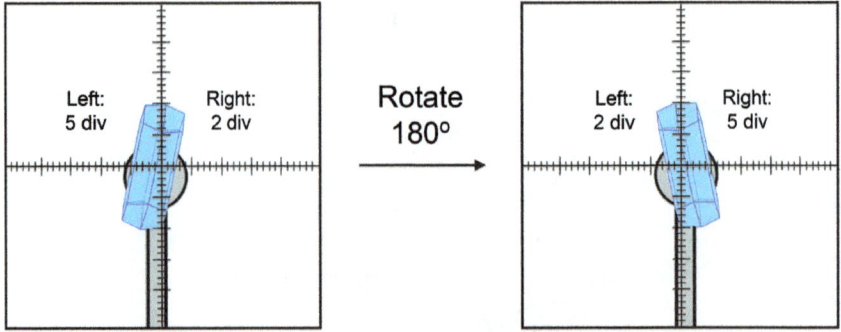

Figure 8.9   Centring of the crystal. The labels refer to where the crystal crosses the horizontal axis. The crystal should be moved 1.5 divisions towards the centre (leaving 3.5 divisions either side of the vertical line of the crosshair).

position to allow centring, which orients one of the lateral adjustment axes parallel to the view direction of the video microscope, with the other lateral axis and the height axis perpendicular to the view direction. The aim is to position the crystal so that a 2-fold rotation around any of the three adjustment axes will leave its centre in the same position. Since the video microscope will invariably have an associated crosshair, it is tempting to align the crystal immediately to the centre of the crosshair. This will work only if the crosshair is properly aligned with the instrument centre! Crystal centring should ensure that the crystal's centre remains stationary during movement of the goniometer axes. Incorrect centring will cause problems with geometrical measurements and could distort the measured intensities if rotation varies the volume of the crystal within the incident beam.

8.4 Measuring the Geometry of Diffraction

A typical single-crystal analysis proceeds by establishing the geometry of the diffraction pattern, then the intensities. An alternative is to measure the whole diffraction pattern before any analysis is made; this option is discussed in Section 9.3. In that case, the subsequent data analysis must still follow these same basic steps, so it remains important to understand the following sections. Typically, the geometry is established by measuring some fraction of the diffraction pattern, then the established geometrical information is used to devise a strategy to collect the complete set of intensities in a controlled manner.

8.4.1 Measuring a First Section of the Diffraction Pattern

The initial aim is to measure a sufficient fraction of the diffraction pattern to identify the crystal lattice. The measured diffraction angles will be converted to reciprocal lattice points as described in Chapter 6, so the first task is to measure a sufficient number of diffraction angles. This can be achieved by rotating the crystal around the ϕ or ω axis, taking a sequence of diffraction images (Figure 8.10). We have seen in Chapter 6 that the crystal must be placed in a specific orientation with respect to the incident beam for a particular diffracted beam to be produced so the crystal is rotated to realise an appropriate diffracting position for each diffracted beam. As the crystal rotates, diffracted beams are produced sequentially in various

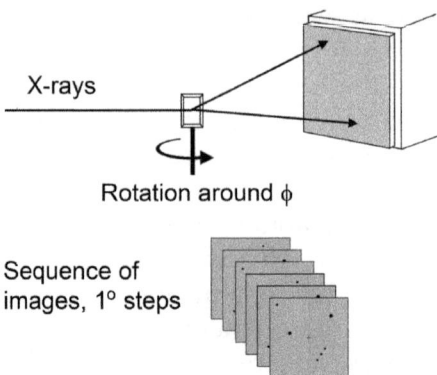

X-rays

Rotation around φ

Sequence of
images, 1° steps

Figure 8.10   Schematic representation of an initial scan around φ. Diffraction
images are read from the detector in steps of 1° (for example).
A sufficient angular range must be measured to enable indexing.

directions and those hitting the detector face are measured. If the
incident beam is constantly on and the detector measures continuously in real time, spots are seen to appear and disappear on the
diffraction image. This confirms the nature of the diffraction process;
we do not simultaneously see all diffracted beams and we do not see
spots travel across the detector face while the crystal rotates, as they
would if the diffracted beams were constantly present. The crystal's
mosaicity (Section 6.3.5) means that there is a small angular range
over which the crystal will be in an appropriate diffracting position,
so diffraction spots remain on the image over a small range of the
rotation. Spots at lower diffraction angle stay longer than spots at
higher diffraction angle. This is most easily explained by picturing the
reciprocal lattice: as the crystal rotates at a constant angular rate, the
reciprocal lattice points further from the origin (corresponding
to higher-angle diffracted beams) must be moving faster than those
closer to the origin. Thus, reciprocal lattice points at higher angle
move more rapidly into and out of the diffracting position.

Practically, the continuous rotation must be broken down into a
sequence of images, like a movie is broken down into a sequence of
frames. For example, a 180° rotation of the crystal could be broken
down into 180 images, each covering 1° of the rotation. If the detector
image can be read out rapidly, the crystal can rotate continuously and
the incident beam can stay on constantly, as described above. This is
referred to as a *shutterless* mechanism. The frames are separated by
the detector reading and storing them at regular intervals through the
continuous rotation. If the detector takes longer to read out images,
the rotation must be made step-wise with the incident beam turned

on and off by a shutter mechanism. The shutter is opened, the detector measures while the crystal is rotated through a specified angle, the shutter is closed while the detector image is read, then the detector is reset and the measurement continues. Whatever the practicalities, we end up with a sequence of diffraction images covering an angular region of the diffraction pattern.

The width of each image in the sequence determines the angular resolution of the measurement. If we take a "wide" image, where the crystal is rotated by several degrees, a diffracted beam may appear on the detector somewhere near the beginning of the rotation, pass through its maximum as the rotation continues, then disappear (Figure 8.11(a)). The resulting image will show the diffracted beam as a spot, but accurate information on the angular centre of the beam cannot be obtained—we can only say that the centre is somewhere within the rotation interval. Taking more images with a narrower width obviously allows us to be more specific about the angle at which a diffracted beam actually appears and disappears (Figure 8.11(b)). In principle, we should take images that are as narrow as possible to measure accurately the geometry of the diffraction pattern. In practice, a suitable image width depends on the width of the diffracted beams. Very narrow beams must be sampled with narrow images but there is little point taking a huge number of slices through very wide diffracted beams. We have already said that the angular width of the diffracted beams depends on the width of the incident beam, which is characteristic of the instrument used, and on the crystal mosaicity, which varies from crystal to crystal. Hence, a suitable image width

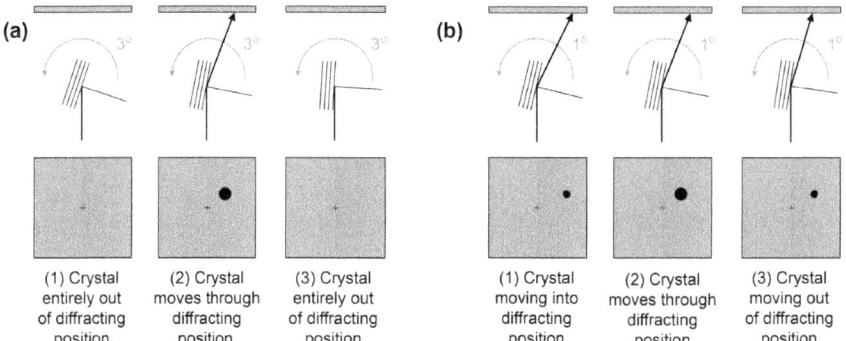

(a)			(b)		
(1) Crystal entirely out of diffracting position	(2) Crystal moves through diffracting position	(3) Crystal entirely out of diffracting position	(1) Crystal moving into diffracting position	(2) Crystal moves through diffracting position	(3) Crystal moving out of diffracting position

Figure 8.11   Wide frames (a) *vs.* narrow frames (b). In (a), the diffracted beam is seen only on the middle image, and its centre can only be said to be somewhere within the specified 3° rotation. In (b), the crystal is seen to move in to, through and out of the diffracting position on successive images. The centre of the diffracted beam can be localised to the 1° rotation recorded on the middle image.

varies from instrument to instrument and crystal to crystal. At this initial stage of an actual analysis (prior to determining the unit cell), nothing is known about the crystal so we should just choose a standard image width appropriate for the instrument. For a typical laboratory instrument, this is likely to be around $1°$.

For a crystal in an arbitrary orientation, a rotation of $180°$ around the ϕ axis should produce a sufficient number of diffracted beams to enable the diffraction geometry to be established. Some diffractometer control software may implement a more sophisticated strategy, where several segments are measured from different regions of the diffraction pattern. A strategy like this is often referred to as a *matrix strategy*, and the collected images as *matrix images*, because they are designed to enable determination of the crystal's *orientation matrix* (see Section 8.4.2). The angular sections to be measured must generally be wider for Cu radiation than for Mo radiation because Cu spreads the diffraction pattern over a wider angular range.

8.4.2 Indexing

To convert the measured diffraction images into a list of diffracted beams, each image must be scanned to identify where diffracted intensity has been observed on the detector. This step is commonly called *harvesting*. The available information in each measured image is actually just a 2-D array of numbers corresponding to the intensity counted at each pixel. When this is represented graphically, diffracted beams are clearly visible to the human eye as spots. For the computer, however, it is necessary to define some threshold value to decide whether a given pixel has registered diffracted intensity or not. Every pixel in an image will count *some* intensity due to background scattering and (probably) electronic noise. The threshold specifies the minimum measured intensity to be registered as a diffracted beam. The control software will often have an algorithm to assess a suitable threshold for harvesting or the user can provide a value. At this stage, we are only interested in the *positions* of the diffracted beams (not their intensities), so the aim is to specify a threshold that captures as many diffracted beams as possible without assigning too many false beams. When the images have been scanned, the diffracted beams are converted to reciprocal lattice vectors (according to Section 6.3.3), which can be viewed in a *reciprocal lattice viewer* (Figure 8.12). This is a vital stage because it gives the first comprehensible representation of the diffraction pattern. We know that the diffraction angles converted to reciprocal lattice vectors should yield a regular lattice.

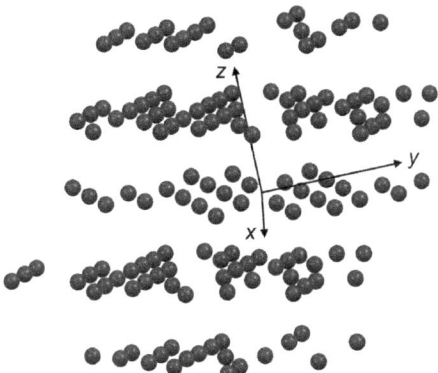

Figure 8.12 Diffraction geometry illustrated in a reciprocal lattice viewer. The reciprocal lattice points are obtained from measured diffraction angles and displayed relative to a generic (Cartesian) set of axes referring to the instrument. Indexing means choosing a more appropriate set of axes so that all measured points lie at integral values of the reciprocal lattice vectors.

Now we can see whether this is the case. Probably only a fraction of the diffraction pattern is available at this stage, but it needs to be sufficient to make this judgement. If the derived reciprocal lattice does not look regular, the crystal may not be single, or it may be that the threshold for harvesting the measured images is set too low so that spurious peaks have been included. Usually, a reciprocal lattice viewer allows points to be filtered by intensity, so the latter can be investigated by gradually eliminating points associated with the lowest intensities.

To plot the derived set of reciprocal lattice vectors, it is necessary to use some reference axes. At this stage, the geometry has been measured only in relation to a generic (Cartesian) set of axes on the instrument, so the reciprocal lattice vectors are initially plotted relative to those. The next step is to define suitable basis vectors for the reciprocal lattice so that all of the measured points are assigned integral *hkl* indices. This is called *indexing* the diffraction pattern, and it is here that the dimensions of the crystal lattice are determined. At the heart of the process is the *orientation matrix*, which links the reciprocal lattice axes to the instrument axes. Like many of the more technical aspects, the orientation matrix is likely to be hidden behind the scenes of the diffractometer control software so the average user may never know of its existence. However, the concept of measuring the diffraction pattern on the instrument then defining at the indexing stage how the reciprocal lattice is aligned relative to the instrument must be understood to appreciate the geometrical aspects of data collection. The orientation matrix is the quantitative link in this process. When it is known, any crystal orientation can be "dialled up" by specifying the appropriate values of ω, κ and ϕ, and we know exactly the direction in which each diffracted beam will emerge. Since the matrix captures the crystal's orientation on the diffractometer, it is specific to each data collection and becomes redundant once a crystal is removed from the instrument.

A simple method for indexing is to sort the measured reciprocal lattice vectors into increasing length order then select the three shortest non-collinear vectors as potential basis vectors. It can then be established how many of the measured reciprocal lattice vectors have integral *hkl* indices (within some experimental tolerance) referred to those vectors. If the percentage is lower than some specified value (say 90%), a new set of three basis vectors can be chosen and the process can be repeated until a suitable fit is obtained. If the measured points form a genuine lattice, an exhaustive search should eventually yield a suitable set of basis vectors that will fit all observed diffracted beams. Practically, robust indexing algorithms are implemented in all diffractometer control software. If the indexing algorithms fail, it is because of the crystal or the way that it has been mounted or centred, so there is definitely some reason to investigate. For example, the crystal might not be single, which may mean that it is not worth further investigation. It could also be that an insufficient number of diffracted beams have been measured, either because the "matrix strategy" was inappropriate, or the crystal did not diffract with sufficient intensity under the applied measurement conditions, or the threshold for identifying diffraction spots on each detector image was set too high. Small unit cells might require more of the diffraction pattern to be measured because there are not so many diffracted beams in a given angular range. The reciprocal lattice viewer should provide a clue to these potential problems. Ultimately, if the diffraction pattern cannot be indexed, it is not possible to make much progress—it is advisable where possible just to try a new crystal.

8.4.3 Selecting the Bravais Lattice

The established basis vectors for the reciprocal lattice can be transformed to basis vectors for the real crystal lattice, which can then be examined to identify the most appropriate Bravais lattice. This is usually achieved by establishing the reduced unit cell (Section 2.2) then applying a standard set of transformation matrices to yield unit cells that should be consistent with each potential Bravais lattice type. The process works by seeing how well the results conform to the required constraints on the unit-cell parameters. For example, applying the appropriate standard matrix to convert the reduced unit cell to a cubic P unit cell is expected to give a result with $a=b=c$ and $\alpha=\beta=\gamma=90°$. If it does, cubic P is an appropriate choice of Bravais lattice. If it does not, cubic P is not an appropriate choice. In practice, all 14 potential Bravais lattices are often presented in a table, with some quantitative assessment of how closely the lattice conforms to each type (an example is given in the Case Study at the end of the chapter). There will always be 100% agreement between the measured lattice parameters and the triclinic P lattice, because there are no geometrical constraints in that case. The other 13 possibilities will deviate to some extent from 100% agreement due to experimental uncertainties, but the aim is to identify any higher-symmetry lattice type that appears to provide an appropriate fit to the measured lattice parameters. The diffractometer control software will usually suggest the best fit on the basis of some figure-of-merit. The user can override this choice if there is some reason to do so, but the software will generally make a sensible choice on the basis of the information that is currently available.

8.4.4 How Do We Know If the Indexing Is Correct?

Once the Bravais lattice has been selected, the geometrical parameters are refined by a least-squares process to optimise the agreement between the observed diffraction geometry and the established lattice parameters (and crystal orientation). Any unit-cell angles constrained by the symmetry of the Bravais lattice will not be refined. Numerous instrument parameters, such as the crystal-to-detector distance, detector centre, crystal centre, *etc.*, can be optimised, but the user should keep a close eye on such parameters to ensure that they retain physical sense. For example, it is unlikely that the detector distance, detector centre, *etc.* require adjustments any more than fractions of a millimetre, so if these parameters should drift significantly, there is probably some instability in the least-squares process that should be investigated. The least-squares process refines up to nine primary parameters (the nine elements of the orientation matrix), which can be thought of as the six metric parameters of the crystal lattice (some of which might be constrained by the specified symmetry) plus three parameters that specify how the lattice is aligned with respect to the instrument. If we also refine some instrument parameters, we may get up to around 15 parameters at this stage. For a robust least-squares process (Section 12.2.1), it is desirable to have many more observations than parameters—a rough guide of 10 observations per parameter indicates that at least 150 observed diffracted beams are desirable to yield a robust initial refinement of the geometrical parameters. Later, these parameters will be refined against the entire diffraction pattern, but it is desirable that the initial estimate is as good as possible.

At this stage, how should the result be assessed? Various numerical indicators might be presented by the control software. For example, histograms of the deviation of the reciprocal lattice points from integral *hkl* values or the average deviation of the measured diffraction angles from the calculated diffraction angles. All such measures can be informative. However, the best way to assess whether the indexing is correct is to *look at the diffraction images*. The control software now has enough information to predict exactly where diffracted intensity should be seen on the detector in any crystal orientation, so it is possible to overlay calculated positions for the diffraction spots on each measured image (Figure 8.13). The user should look at how well these positions agree with what is observed. Every observed diffraction spot should have a corresponding predicted spot. If there are numerous observed spots that are not predicted, the indexing is not correct. The converse is not true: there may be numerous predicted

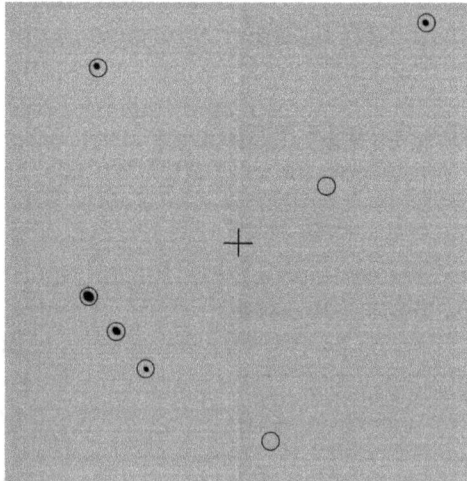

Figure 8.13 Predicted positions of diffracted beams (open circles) overlaid on a measured diffraction image. For correct indexing, all observed spots should be predicted (as in this case), but it is not necessary for all predicted spots to be observed.

spots that do not have a corresponding observed spot. Remember that we are currently assessing the diffraction *geometry*, and don't yet know anything about the diffracted intensities. It could be that a predicted diffracted beam is a systematic absence or just happens by coincidence to have negligible or zero intensity. If there is some general systematic pattern of missing spots in many images, this may be indicative of a problem. For example, if every second predicted spot along some direction is not observed, it may be that one of the unit cell parameters is actually half as long as currently specified.

As described in Section 8.4.1, the control software must know the angular beam width to make an accurate prediction of what each diffraction image should look like. This is usually specified as the *mosaicity* parameter. The value can be controlled or refined to provide the best agreement between the observed and calculated diffraction images. A mosaicity value that is too large will lead to many more predicted diffraction spots than are observed on a given image. A value that is too small will mean that observed diffracted beams may not be predicted. It should be kept in mind that it might be necessary to fine-tune the mosaicity estimated by the control software when assessing the agreement between observed and predicted diffraction images.

8.5 Measuring Crystal Morphology

For pharmaceutical crystals, it is often useful to describe the morphology, as discussed in Chapter 5. The practical procedure is referred to as *face indexing* because it establishes Miller indices for the visible crystal faces. We are able to do this once we have measured the geometry of the diffraction pattern because we can now align the crystal in any (mechanically accessible) orientation relative to the diffractometer's video microscope. Practically, the procedure often involves recording a video using the diffractometer's microscope while the crystal is rotated around the ϕ axis. An analysis tool can then be used to mark the crystal faces and measure the perpendicular distance of each face from the crystal's centre (Figure 8.14). This information can be used to reconstruct the crystal habit as described in Chapter 5. The combination of high-resolution video microscopes and rapid initial determination of a crystal's lattice parameters make it easy to establish the morphology of a single crystal, and it may be valuable to do this for every crystal analysed.

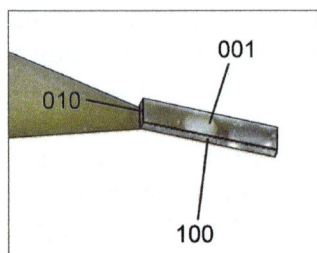

Figure 8.14 Example of face indexing for a crystal mounted on a loop. The face normals are added by the diffractometer control software on the basis of the established crystal lattice. The habit is constructed by specifying the observed distances of each labelled face from the crystal's centre.

8.6   Summary of Key Points

- Instrumentation for single-crystal X-ray diffraction comprises an X-ray source, a goniometer to orient the crystal relative to the incident beam and a detector that records diffracted intensity. Most modern instruments are equipped with 2-D area detectors.
- Laboratory instruments generally use either Cu $(\lambda \approx 1.54$ Å$)$ or Mo $(\lambda \approx 0.71$ Å$)$ radiation. Cu radiation is preferable for pharmaceutical crystals because it interacts more strongly with the sample. The longer wavelength means that the diffraction pattern is spread over a larger angular range so it takes longer to measure the pattern.
- A typical single-crystal analysis proceeds by establishing the geometry of the diffraction pattern, then the intensities. The established geometrical information is used to devise a strategy to measure the complete set of intensities in a controlled manner.
- Measured diffraction angles are converted to reciprocal lattice points using the geometrical construction described in Chapter 6. When a sufficient fraction of the reciprocal lattice has been defined, an appropriate set of basis vectors can be established for the reciprocal lattice so that all measured reciprocal lattice points are assigned *hkl* indices. This indexing establishes the crystal lattice and the orientation of the crystal on the diffractometer.
- The best way to assess whether the indexing is correct is to look at the agreement between the observed diffraction images and the positions expected for diffraction spots with the established crystal lattice and orientation. The exact appearance of each image will depend on the width of the diffracted beams, which is quantified by the mosaicity parameter.

- Once the crystal lattice and orientation are established, crystal faces can be identified and the morphology of the crystal can be described.

8.7 Case Study: Aspirin Form I

Aspirin is polymorphic and displays close structural similarity between forms I and II.[1] This case study follows a single-crystal X-ray analysis to determine the structure of form I.[2] The geometrical aspects are described in this chapter then the intensity measurements and data processing are discussed in Chapter 9. It is assumed that a suitable single crystal has been mounted and properly centred on the diffractometer, which uses Cu radiation. The first step is to collect a sufficient fraction of the diffraction pattern to establish the orientation matrix. The applied method aligns the crystal along the vertical axis of the instrument and rotates $180°$ around the ϕ axis in $1°$ steps with the detector placed 50 mm back from the crystal, centred on $2\theta = 0$. Many other strategies are possible; the approach taken will probably depend on the specific instrument and options within the control software. With typical detector dimensions 10×10 cm, the edge of the detector in the horizontal plane reaches $2\theta = 45°$ in the position described, which corresponds to $d \approx 2.0$ Å for Cu radiation. Eventually, it will be necessary to measure to much higher 2θ angles, but this angular range should be sufficient for initial determination of the lattice parameters. Harvesting the initial data frames using a relatively conservative threshold (*i.e.* requiring a high intensity relative to the background) yields 248 measured reflections. Applying a conservative threshold at this stage eliminates false reflections that are actually noise, possibly at the expense of some weaker reflections that are genuine. Again, a reasonable threshold will depend on the detector type and strength of diffraction for the specific crystal and it may be necessary to experiment with the threshold to produce a suitable set of reflections for indexing.

It is vital at this stage to look at the constructed reciprocal lattice (Figure 8.15). Each of the 248 measured reflections is converted to a point in reciprocal space (using the construction described in Chapter 6) and it should be checked that the result looks like a genuine lattice. Figure 8.15 shows a regular array of points as we would expect. The next step is to deduce an appropriate set of reciprocal lattice axes so that all measured points fall at integral multiples of each reciprocal lattice direction, *i.e.* find $\mathbf{a}^*$, $\mathbf{b}^*$ and $\mathbf{c}^*$ so

(a)　　　　　　　　　　　　　　　　(b)

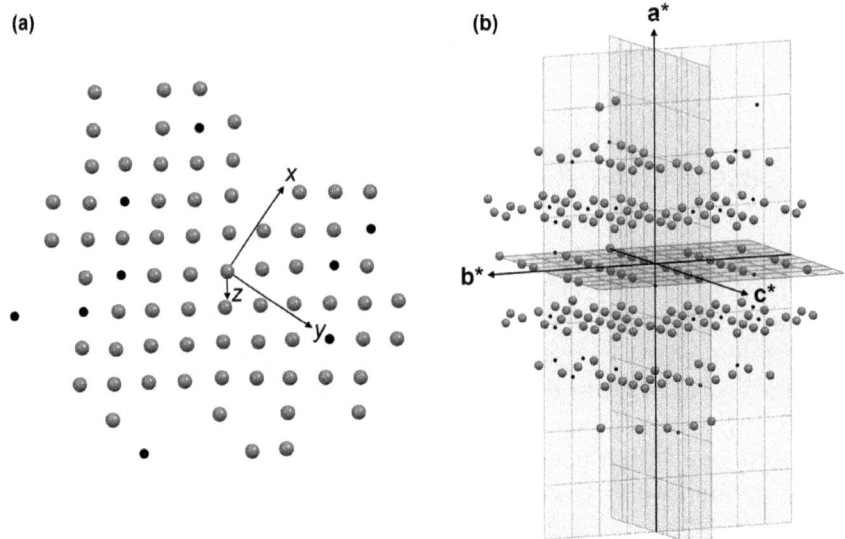

Figure 8.15　Reciprocal lattice view for aspirin form I. (a) Prior to indexing, the measured reciprocal lattice points are described against the instrument axes, labelled *x, y, z*. (b) After indexing, the grid indicates the reciprocal lattice axes for the applied unit cell. It can be seen in (b) that all reciprocal lattice points fall on the grid, corresponding to integral multiples of the reciprocal lattice vectors. In both diagrams, the smaller black dots indicate relatively weak reflections.

that all measured points are described by $h\mathbf{a}^* + k\mathbf{b}^* + l\mathbf{c}^*$ with integral *hkl*. Practically, this just means running an algorithm for indexing. The result obtained for aspirin is as follows:

Bravais Lattice	FOM	a (Å)	b (Å)	c (Å)	α (°)	β (°)	γ (°)
Orthorhombic *F*	**0.01**	**6.59**	**23.60**	**23.62**	**80.21**	**106.13**	**106.17**
Orthorhombic *I*	**0.02**	**6.59**	**15.21**	**18.06**	**89.97**	**68.68**	**90.02**
Orthorhombic *C*	0.79	15.21	16.82	6.59	90.08	90.02	89.96
Orthorhombic *P*	**0.06**	**6.59**	**11.34**	**11.34**	**84.23**	**89.93**	**89.95**
Monoclinic *C*	0.86	16.82	15.21	6.59	89.98	90.08	90.04
Monoclinic *P*	0.77	11.34	6.59	11.34	90.07	95.77	89.95
Triclinic *P*	1.00	6.59	11.34	11.34	84.23	89.93	89.95

The table presents the reduced unit cell for the established lattice parameters plus derived unit cells referring to each Bravais lattice type up to orthorhombic, obtained by applying known transformation matrices to the reduced cell (Section 8.4.3). The figure-of-merit (FOM) indicates how well the transformed parameters conform to the expected metric constraints. Bold rows in the table highlight those that are not reasonable. The lattice is consistent with monoclinic *P* and also orthorhombic *C* (or monoclinic *C* with $\beta \approx 90°$). At this stage, we cannot say which of these possible unit cells is "correct". All we can say is that the lattice is consistent with the metric constraints for triclinic *P*, monoclinic *P* or orthorhombic *C* unit cells.

We will decide finally which unit cell to use after the diffracted intensities are measured (Chapter 9). For now, we may as well choose the reduced cell. All we are doing is choosing a valid set of lattice vectors to establish the orientation matrix and gain control over the next stages of the analysis. Choosing the reduced cell is sufficient to decide whether the indexing looks to be a fair interpretation of the diffraction pattern. The reciprocal lattice viewer now shows the established reciprocal lattice axes, and it can be seen in Figure 8.15(b) that the vast majority of the 248 measured reciprocal lattice points fall on a grid marking integral multiples of the reciprocal lattice vectors. We should also look at the measured diffraction images (Figure 8.16), which show that the observed diffracted beams match expected positions. It may be necessary to adjust the crystal's mosaicity (controlling the width expected for the diffracted beams) to produce the best match for the diffraction images. Figure 8.16 shows several predicted positions where diffraction spots are not seen, which may correspond to systematic absences or just coincidentally weak reflections, but there are no observed reflections that are not predicted.

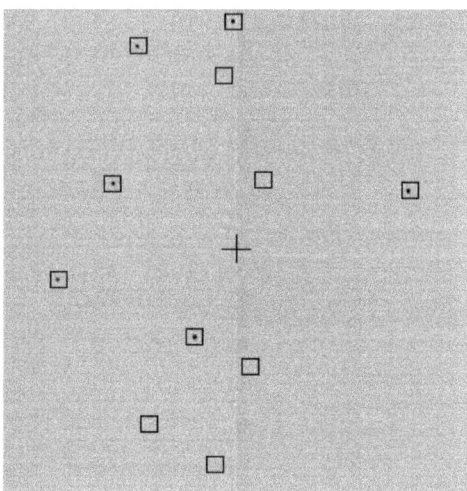

Figure 8.16   Diffraction image taken from the initial images collected for aspirin. The squares show the positions predicted for diffraction spots if the chosen unit cell is correct. All observed spots are predicted.

Finally, there are a few numerical indicators to quantify the checks described above. The indexing algorithm will indicate how many of the measured diffracted beams are fitted by the established lattice parameters. This assessment must be made against some threshold for how far the measured diffracted beam position can deviate from expectations. The user should be aware that subtle changes in the threshold may give a very different quantitative assessment. A more detailed indication might be presented as a histogram showing the number of diffracted beams lying within particular ranges from integral *hkl* values. This effectively quantifies the visual fit described using the reciprocal lattice viewer. To quantify the visual fit to the diffraction images, the deviations between observed and predicted diffracted beam positions in the horizontal and vertical directions on the 2-D detector face are usually given. Some angular measure is also required for the deviation between the observed and predicted maximum of each diffracted beam as the crystal rotates. This will be influenced by the angular width of each diffraction image, as described in Section 8.4.1. It is deliberate here to avoid any firm statement of threshold values to indicate correct or incorrect indexing. The numerical indicators can be helpful, but an informed user should probably place more confidence in the visual comparisons.

To summarise: at this stage of the analysis, we have established the lattice parameters (translational symmetry) of aspirin form I and how the crystal is aligned on the diffractometer (*i.e.* the orientation matrix). The orientation of the crystal can now be controlled to bring any reflection into the diffracting position in such a way that it can be measured on the detector. A strategy can now be devised to measure the intensities of all diffracted beams; this will be discussed in Chapter 9.

References

1. A. D. Bond, R. Boese and G. R. Desiraju, *Angew. Chem., Int. Ed.*, 2007, **46**, 618.
2. P. J. Wheatley, *J. Chem. Soc.*, 1964, 6036, [CSD: ACSALA].

9 Single-crystal X-ray Diffraction (Part 2)

Summary

After establishing the lattice parameters and orientation of the crystal on the diffractometer, the diffracted intensities can be measured in a controlled manner. Typically, a strategy to measure the required intensities is calculated with the user specifying parameters such as the maximum diffraction angle, redundancy, image width and measurement time per image. The fraction of the diffraction pattern to be measured depends on its point symmetry, which is commonly inferred from the metric features of the lattice but cannot be known conclusively until all of the diffracted intensities have been analysed. A sensible approach is to make the most conservative assumptions to ensure that sufficient data are measured. An alternative is simply to measure the complete diffraction pattern, if sufficient time is available. After data collection, the diffraction images are integrated, and a multi-scan correction is applied to minimise the effects of systematic errors such as anisotropic absorption or poor crystal centring. The final outcome of a single-crystal analysis is a statement of the unit cell and a list of intensities with *hkl* indices referring to that unit cell.

9.1 Introduction

The geometrical aspects of the single-crystal measurement are established in Chapter 8. The control software can now calculate goniometer angles to place the crystal in the required orientation to observe any diffracted beam on the detector. With this information, the diffracted intensities can be measured in a controlled manner,

Pharmaceutical Crystallography: A Guide to Structure and Analysis
By Andrew Bond
© Andrew Bond 2019
Published by the Royal Society of Chemistry, www.rsc.org

which means choosing which intensities to measure and devising a specific strategy to measure them. We could also just measure everything and analyse the pattern later; this approach is considered in Section 9.3. The benefits of taking a controlled approach come partly in the previous stage. If the steps in Chapter 8 have been followed, we have a good idea of the crystal quality and whether it is actually worthwhile to collect the entire diffraction pattern. We might also have checked whether the unit-cell parameters correspond to some known crystal structure that does not need to be measured again. Then there is a matter of efficiency: if the geometry of the crystal has been established, we know exactly where to look to measure the diffracted intensities that we need. This chapter describes the steps associated with collecting and processing the diffracted intensities.

9.2   Measuring the Diffracted Intensities

When the geometry and orientation of the crystal are established, the control software can consider a list of possible movements that can be made by the goniometer and construct an appropriate subset to achieve the desired intensity measurements. This is a unique calculation each time, even when measuring a crystal with a known structure, because it depends on the orientation of the crystal on the diffractometer. The user must specify a few overall targets for the strategy then decide the image width and measurement time for each image. Various aspects of the process are considered below.

9.2.1   How Much of the Diffraction Pattern Should Be Measured?

To establish an efficient strategy, the key factor is the expected point symmetry of the diffraction pattern. If the point group is known, it is known which intensities are related to others by symmetry, so it is only necessary to measure the unique set. For example, we saw in Section 7.2 that the point group for a centrosymmetric monoclinic crystal is $2/m$. This means that $I(hkl) = I(h\bar{k}l) = I(\bar{h}k\bar{l}) = I(\bar{h}\bar{k}\bar{l})$, and we only have to measure the intensity for one diffracted beam out of each group of four. In this context, the complete diffraction pattern is usually referred to as a "full sphere", so it is necessary to measure only a "quadrant" of the sphere (Figure 9.1). If the crystal structure is non-centrosymmetric, the true point group would be either 2 or m, and it would be necessary to measure the appropriate "hemisphere".

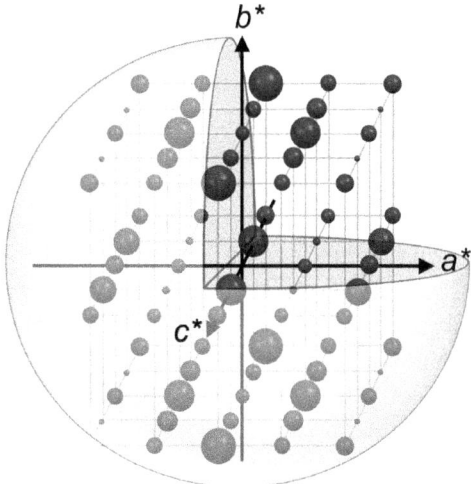

Figure 9.1 Schematic view of the intensity-weighted reciprocal lattice, indicating a quadrant of the "full sphere" (cut out) required to be measured for point group 2/*m*.

We should consider what is known about the point group at this stage of the analysis. The only information so far is an indication of the Bravais lattice, obtained by measuring the geometry of the diffraction pattern. However, we know from Chapter 4 that the metric features of the lattice do not necessarily indicate the true symmetry of the crystal structure. Valid judgements can be made about *higher* symmetry—for example, the space group cannot be cubic if the lattice is not consistent with a cubic Bravais lattice—but lower symmetry cannot be excluded on the basis of geometry alone. Nonetheless, most diffractometer control software will assume the point group of the diffraction pattern from the chosen Bravais lattice and devise the measurement strategy on that basis. Why? Because experience shows that the Bravais lattice *is* probably a good indicator of the crystal's symmetry. The approach will work in the majority of cases and it is usually implemented because it is more efficient. However, this is one point where the user should understand the decision that is made. The geometrical basis for choosing the point group is inherently suspect and users should be on the lookout for cases where the assumed point symmetry turns out to be incorrect. The consequence of making an incorrect assumption is that some required intensities may not be measured. If this is discovered later, it will be necessary to repeat the analysis—assuming that crystals are still available.

We should also consider the issue of centrosymmetry. Is it necessary to measure both $I(hkl)$ and $I(\bar{h}\bar{k}\bar{l})$? It is not known at this stage whether the space group is centrosymmetric or not. We know that an enantiomerically pure sample of a chiral molecule must crystallise in a non-centrosymmetric space group, so we can be certain of non-centrosymmetry in that case. But *any* sample could potentially adopt a non-centrosymmetric crystal structure so surely we must assume that the structure is non-centrosymmetric and devise a strategy accordingly? Actually, this is rather like the previous decision. Experience shows that the crystal structure of a non-chiral molecule (or a racemic mixture of enantiomers) is more likely to be centrosymmetric, so centrosymmetry might be assumed as the default because it reduces the number of intensities to be measured. Again, the user should be aware if centrosymmetry has been assumed by the control software and either override the choice or recognise that there could be a need to collect more data if the structure is subsequently found to be non-centrosymmetric.

9.2.2 Resolution, Completeness and Redundancy

When a crystal structure is determined, every measured diffracted beam contributes to the resulting image of the electron density. This will be considered in more detail when looking at solving and refining structures in Chapters 11 and 12. Broadly speaking, the beams at low-angle (large *d*-spacing) define the gross features of the structure while the beams at higher-angle (smaller *d*-spacing) add progressively finer details. The *resolution* of the final image of the electron density improves progressively as higher-angle reflections are included (Figure 9.2). Atomic positions are identified from the centres of peaks in the electron density, so the quality of the final crystal structure

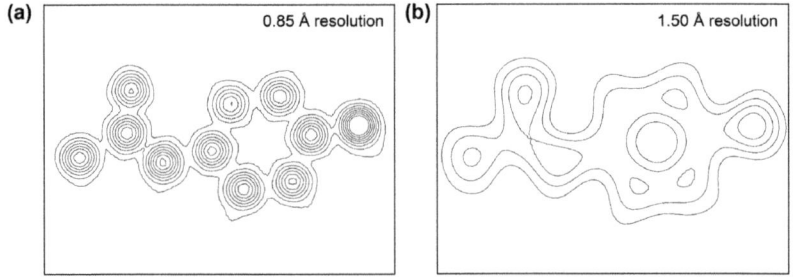

Figure 9.2   Images of the electron density for the non-H atoms of a paracetamol molecule within the crystal structure of form I. Contours are drawn at intervals of 1 e $Å^3$. At 0.85 Å resolution, the atoms are well defined and the positions of their centres are obvious. At 1.50 Å resolution, this is clearly not the case. Chapter 11 will describe how these images are produced.

depends crucially on how clearly such features are defined. The question is what resolution is required for the electron density to give a sufficient representation of the atomic positions? In the current context, what is the largest diffraction angle that must be measured to produce an acceptable crystal structure?

The standard guideline for a small-molecule crystal structure is to collect data to a *d*-spacing that is less than the shortest interatomic distance. The relevant distance in a pharmaceutical crystal is *ca.* 1 Å for bonds involving H, so the standard guideline is to collect data to 0.85 Å resolution. The corresponding maximum 2θ angle is *ca.* 130° for Cu radiation, or *ca.* 50° for Mo radiation. Data could be measured to higher angle, but this is not necessary for a standard structure determination. Restricting the measured data to lower resolution (*i.e.* larger minimum *d*-spacing) will start to compromise the image of the electron density (as in Figure 9.2) and thereby reduce the accuracy of the final atomic coordinates. The only reason to do this is out of necessity. We saw in Chapter 6 that diffracted beams are always less intense at higher diffraction angles because the atomic scattering factors decrease with increasing diffraction angle. A crystal that diffracts only weakly could have measurable intensities only to a limited resolution. There is no point measuring out to 0.85 Å if the intensities at that resolution are too weak to be measured. Data sets extending to "limited resolution" can still be valuable if they allow the structure to be solved. The key consideration is what we are hoping to obtain from the measurement. Do we just need a coarse image of the molecular structure, or are we interested in some subtle differences between particular bond lengths or angles? This will be discussed in more detail in subsequent chapters. At this stage, we note that we should always aim to collect data to at least 0.85 Å resolution, that we might be able do better than this (but we generally don't need to), and that there can still be value in collecting data to lower resolution if that is all the crystal permits.

The term *completeness* describes the fraction of data that is measured to the maximum diffraction angle. For example, we say that "data are measured with 98% completeness to 0.85 Å resolution". For a standard single-crystal analysis, there is no good reason for the completeness ever to be far from 100%. It could be that a limited number of reflections cannot be measured due to mechanical constraints, but this is likely to affect very few reflections on an area-detector instrument. Completeness generally refers to the assumed point symmetry of the diffraction pattern rather than the "full sphere". For point group 2/*m*, for example, 100% completeness refers to measurement of all data within a quadrant of the sphere. The term *redundancy* (or *multiplicity*) indicates the number of independent measurements of equivalent reflections. This equivalence can refer to two things: (1) reflections that are equivalent under the point symmetry of the diffraction pattern and (2) the same reflection measured more than once with the crystal in different orientations. As for any experimental measurement, redundancy is advantageous because averaging should help to reduce the effects of statistical error. With an area-detector instrument, the redundancy will always exceed 1.0 because any strategy that brings all unique reflections onto the detector face in some set of images will inevitably bring other reflections onto each image. High redundancy is valuable for application of a multi-scan correction (Section 9.2.5) so the aim should be to achieve as high a redundancy as possible within the time available.

9.2.3   Image Width and Image Measurement Time

When a strategy has been established to measure all required intensities, it is necessary to choose an appropriate image width. For example, the strategy might include a 180° sweep around the ω axis with the crystal placed in some particular orientation. Should this involve 180 images of 1° width, 90 images of 2° width, *etc.*? The geometrical aspects of this were discussed in Section 8.4.1. Although the lattice parameters have (usually) already been determined, they will be refined against the full set of measured data at the end of the analysis, so it is still relevant to consider the geometrical aspects. When discussing image width, there are two limiting scenarios: (1) the image width is greater than the angular width of the diffracted beams so each reflection falls completely onto one image (as in Figure 8.11(a)) and (2) the image width is less than the angular width of the diffracted beams so each beam is distributed across several consecutive images (Figure 8.11(b)). These scenarios are called "thick slice" and "thin slice", respectively. For measurement of the diffracted intensities, thick-slice images have the potential advantage that they count all intensity for a diffracted beam on one image, which maximises the signal-to-noise ratio. If the intensity is distributed across several thin-slice images, there will be proportionately greater noise on each image. Conversely, thin-slice images are required to build up an accurate picture of the diffracted beam *profile*, which is used by many integration procedures (Section 9.2.4) to count weaker intensities. A suitable image width may therefore be linked to the algorithm that will be used later to integrate the images. The control software will probably suggest a suitable image width in the course of setting up the measurement strategy. Such a suggestion will require suitable prior assessment of the lattice parameters and crystal mosaicity and the user should keep a sceptical eye on these to ensure they are reasonable. A typical image width on a laboratory instrument is likely to be in the range 0.5–2°.

The measurement time per image should be sufficient to ensure that intensity is observed for the weakest reflections at the highest measured diffraction angles. The important feature is signal-to-noise, often referred to as "*I upon sigma*" ($I/\sigma(I)$), where $\sigma(I)$ is the uncertainty associated with each measured intensity. For detectors with low inherent noise, the image time can be increased essentially without limit. For detectors with higher inherent noise, there will come a limit where increasing the measurement time will not improve the signal-to-noise ratio. If it is necessary to approach this limit, the diffraction pattern is probably too weak to be measured to the desired resolution with the available instrument. Since the intensities decrease with increasing diffraction angle, a measurement time sufficient for the weakest reflections at higher angle is likely to exceed the requirements for the stronger reflections at lower angle. It is vital that the intensities at low angle do not exceed the maximum value that the detector can measure. This was mentioned in Section 8.2.3: overloading (or topping) the detector will lead to severe systematic under-measurement of the affected strong intensities. There is no way to recover from this after the measurements have been made—the data are likely to be useless.

One way to handle the measurement of strong low-angle reflections and weak high-angle reflections is to apply different measurement times for low-angle and high-angle images. Often, to cover all of the diffraction pattern it is necessary to measure several sets of diffraction images with the detector at different 2θ positions, especially with Cu radiation where the diffraction pattern is spread across a larger angular range (Figure 9.3). When several detector positions are required, the higher-angle images can be measured for longer without overloading, and the integration software can subsequently normalise the images to account for the different measurement times. The rate at which the diffracted intensity decreases with diffraction angle might depend on the crystal and/or the measurement temperature. In some cases, it might be necessary to measure higher-angle images for substantially longer than low-angle images. An alternative is simply to measure each image for the same time, whilst actively checking for overloading and responding if it occurs. For a shuttered mechanism,

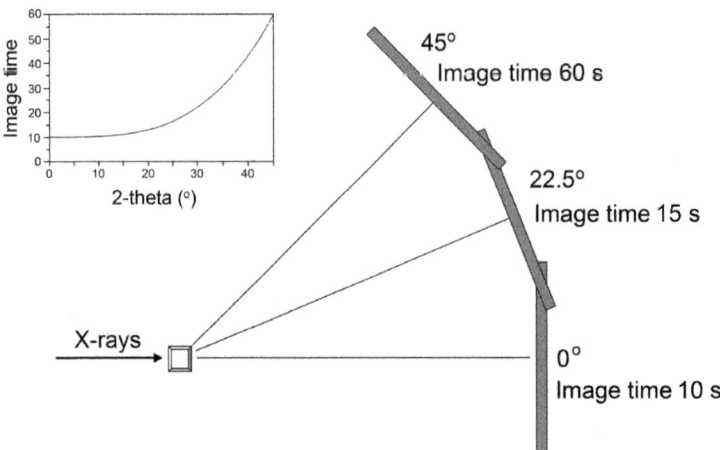

Figure 9.3 Schematic illustration of image measurement time varying as a function of diffraction angle. The inset plot indicates the chosen time profile up to $2\theta = 45°$. The longest image time should (ideally) be sufficient to observe diffracted intensity to 0.85 Å, while the shortest should not overload the detector for the strongest reflections.

where the process proceeds in steps, each image can be checked as it is read. If any pixel is found to have reached its maximum value, the crystal can be moved back to the starting point of the image rotation and the image can be re-measured for a shorter time. The same cannot be applied for a shutterless mechanism because the crystal rotates continuously as the images are read. In that case, it is possible to measure an additional set of low-angle images rapidly to provide a (non-overloaded) assessment of the strongest intensities (a *"fast scan"*), which can be substituted into the final dataset if any reflections are found to be overloaded in the main images. This is of course just a slightly different approach to measuring low-angle and high-angle images for different times. Obviously, a crystal that over-loads the detector at the shortest possible measurement time diffracts too strongly to be measured effectively on the instrument. It must be cut to reduce its volume, or the incident beam must be attenuated.

In practice, choosing appropriate image measurement times is an imprecise art. The control software might make a suggestion on the basis of the images measured in the initial assessment of the diffraction geometry (the matrix images). Such an assessment will be improved if the matrix images extend out to the highest diffraction angles to be measured so that it can be based on actual information rather than extrapolation. With experience of a given instrument, the user should develop a feel for how reliable any automatic suggestion might be. Likewise, the requirements for the measured intensities might depend on the integration software to be used (Section 9.2.4). Integration software can be highly effective at assessing weak reflections, even if they look very weak to the human eye. This is another area where the user should develop a feel for the performance of a particular instrument, including its data-processing software.

There is always the practical consideration of time available: it may not be possible to measure a crystal for several days to reach the highest diffraction angles. Often it is necessary to choose the best possible strategy within a fixed time window. If there is a choice between collecting more data (high redundancy) with shorter individual image times or fewer data with longer image times, redundancy is preferable because it benefits subsequent corrections to be applied (Section 9.2.5). If it seems that data simply cannot be observed to 0.85 Å resolution, it might be preferable to limit the analysis to slightly lower resolution (*e.g.* 1 Å resolution) and to make the best possible measurement to that resolution in the time available. If it becomes crucial to re-examine the structure to higher resolution, subsequent effort should probably be directed towards preparing a larger crystal that will scatter more strongly.

When all aspects of the data-collection strategy have been decided, the diffractometer is left to collect a sequence of diffraction images, which could take anything from minutes to days. It is always advisable to watch the arrival of the first few images of a data collection. If it becomes clear that the first few images have no observable data, or if every image is overloaded, the data collection should be stopped to make the necessary adjustments. When returning to the instrument at the end of the data collection, it should be checked (as a minimum) that the final images still look as expected. Intensity should be visible to about the same level as the start of the data collection (*i.e.* the crystal has not degraded) and the observed reflection positions should still correspond to the predicted positions (*i.e.* the crystal has not moved on its mount). Any problems of this sort are likely to reveal themselves at the data-processing stage, but it is nonetheless helpful for the user to maintain a focus on the primary diffraction images.

9.2.4 Integration

When data collection has completed, the next step is to count the intensities from the measured images; this is known as *integration*. The basic task is to assign each pixel in each measured image to a specific reflection *hkl*, then to sum the intensities of the pixels associated with each reflection. In practice, the process is not quite this straightforward because it is difficult to assess which pixels actually contribute to each diffracted beam, particularly for weak reflections where the measured intensity might be only slightly higher than the surrounding background. Ultimately, integration is carried out by the control software and there is little for the user to do. However, it is helpful to have an overview of the process to recognise when things might be going wrong and to appreciate which decision points might help or hinder the integration.

A key concept mentioned in Section 9.2.3 is the diffracted beam profile, which refers to the shape of the intensity distribution for each beam. Usually, this is viewed in reciprocal space (Figure 9.4). In the actual experiment, diffracted beams hit the detector face generally at some oblique angle thereby defining elliptical spots whose shape will vary depending on their position on the detector face. If the diffracted beam is spread over several images (as for thin-slice images), the shape of the spot on the detector face will probably also vary from image to image. Converting to a 3-D representation in reciprocal space eliminates these experimental distortions to produce comparable profiles for each measured beam. It is not necessary to describe the details; the important point is that most integration algorithms work by constructing 3-D intensity profiles that should be broadly comparable for all diffracted beams. The profile information can then be used effectively to establish which pixels should contribute to weaker diffracted beams.

Clearly, the diffraction geometry must be established before integration to provide a good initial estimate for the position of each

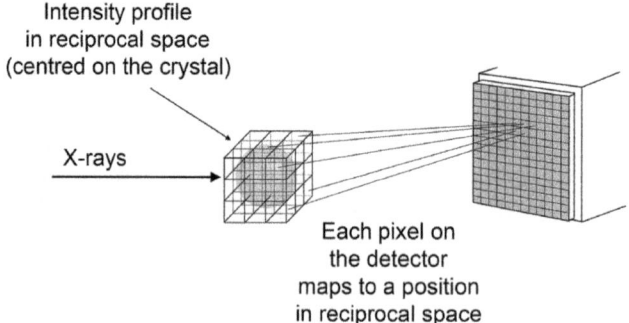

Intensity profile
in reciprocal space
(centred on the crystal)

X-rays

Each pixel on
the detector
maps to a position
in reciprocal space

Figure 9.4   Diffraction images in real space converted to 3-D profiles in reciprocal space. The diffraction angle formed at each pixel on the detector can be converted to the corresponding position in reciprocal space. The intensity profiles in reciprocal space should be broadly comparable for all diffracted beams, so they can be scaled to model weaker reflections.

diffracted beam. The geometry can be monitored and optimised during the integration by checking that diffracted intensity is actually being found at the positions calculated from the current geometrical information. At the end, the lattice parameters are refined against the positions established for all of the diffracted beams to obtain the best possible final assessment of the diffraction geometry. It is necessary here to specify the Bravais lattice type for the unit cell in order to apply any constraints on the lattice parameters. The Bravais lattice is *still* not certain at this stage (the discussion in Section 9.2.1 still applies), so the user should be aware of any constraints that may be applied in this final geometry refinement and be prepared to go back to the integration if the assumed Bravais lattice is found to be incorrect.

9.2.5 The Multi-scan Correction

After integration, a crucial final step is to correct the raw intensities to account for systematic (non-random) errors. A generally unavoidable source of systematic error is anisotropic absorption, where the incident and diffracted beams are attenuated to different degrees in different directions due to a non-spherical crystal habit (see Figure 8.7). With earlier point-detector instruments, specific measurements were made to correct for this, and the term *absorption correction* has persisted. In fact, the corrections applied on a modern diffractometer account for any systematic error (*e.g.* variation of the irradiated volume due to poor crystal centring, crystals larger than the beam, *etc.*), so they are better described by the term *multi-scan correction*. The correction is a vital part of the data processing and must be applied in every case, whether or not anisotropic absorption or some other systematic error is thought to be significant.

The basic aim of the multi-scan correction is to minimise the differences between multiple measurements of intensities that should be equivalent, including different measurements of the same diffracted beam and diffracted beams expected to be equivalent according to the point symmetry of the diffraction pattern. The corrections are implemented as functions that depend on the incident and diffracted beam directions and the algorithms fit parameters to provide the best agreement. It is here that high data redundancy becomes essential to provide a sufficient number of intensities to compare, so data-collection strategies should always aim to maximise redundancy. It is also necessary to specify the point group of the diffraction pattern, which *still* is not known conclusively because it requires a comparison of the final list of intensities, and possibly a decision on the optimal space group to describe the structure once it has been established. Hence, the user should be aware of the point group applied for the multi-scan correction and again be prepared to go back to this stage if a different point group becomes clear after structure determination. This applies particularly to centrosymmetric *vs.* non-centrosymmetric point groups. If the aim is to determine absolute structure by considering the differences between $I(hkl)$ and $I(\bar{h}\bar{k}\bar{l})$ (Section 12.4), it is not helpful to apply a multi-scan correction that aims to make $I(hkl)$ and $I(\bar{h}\bar{k}\bar{l})$ equivalent!

9.3   Measure First, Analyse Later

Repeatedly in Chapters 8 and 9, we have hinted at an alternative approach to measure a single-crystal X-ray diffraction pattern. In the process that has been described so far, information is gradually accumulated by establishing the geometry of the diffraction pattern then using that to deduce a strategy for controlled measurement of the required diffracted intensities. This is firmly linked to the theory discussed up to this point. In a practical sense, the approach is based on a desire for controlled efficiency during the measurement of the intensities. On older instruments, where intensities were measured one-at-a-time by a point X-ray counter, collecting a full sphere of data rather than a hemisphere would double the data collection time, potentially adding *days*. On modern instruments, the difference is likely to be no more than a few hours. Under these circumstances, a viable alternative is simply to measure the entire diffraction pattern and to analyse it later (Figure 9.5). The strategy to collect a full sphere of data can be programmed without needing to know anything about the crystal. The required movements of the goniometer circles can be made to ensure that the whole of reciprocal space is measured. Without an initial assessment of the width of the diffracted beams, it is necessary to choose some standard image width and also to make some initial assessment of the image time to ensure that data are observed to the target resolution. Then the diffractometer can be left to collect the entire diffraction pattern.

To *interpret* the measured diffraction pattern, the process basically returns to what we have described: the geometry must be established, then the images integrated. It may be advantageous to have the whole reciprocal lattice available when attempting to index the diffraction pattern, rather than some fraction obtained from a limited set of

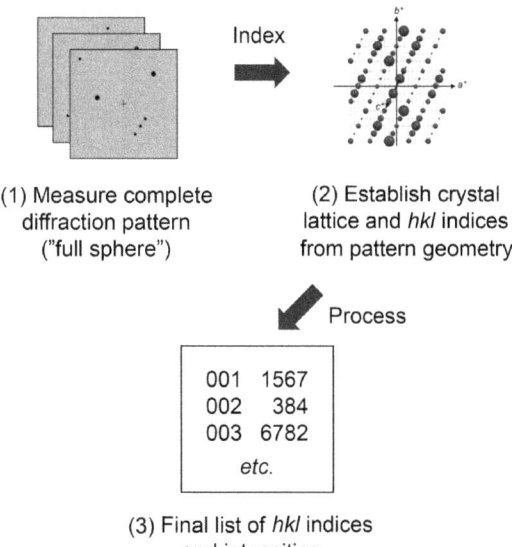

(1) Measure complete
diffraction pattern
("full sphere")

Index

(2) Establish crystal
lattice and *hkl* indices
from pattern geometry

Process

001	1567
002	384
003	6782
etc.	

(3) Final list of *hkl* indices
and intensities

Figure 9.5 Overview of the "measure first, analyse later" approach to single-crystal X-ray diffraction. Compare to Figure 8.1.

matrix images. The major advantage of measuring the full sphere, however, is that no assumption has been made about the point group of the diffraction pattern before measuring it. The point group must eventually be specified for an appropriate unit-cell refinement and multi-scan correction, but all data are available if it is later necessary to choose a lower symmetry. Ultimately, the strategy of "measure first, analyse later" is safest, and it seems like the most sensible approach if sufficient time is available. It is arguably less satisfying because the user is less involved with the measurement, but the involvement and requirement for understanding come during the data analysis.

9.4 What's the Outcome of a Single-crystal Data Collection?

The *primary* data obtained from a single-crystal analysis is the set of measured diffraction images (Figure 9.6). These should sample an appropriate section of the reciprocal lattice or a complete sphere if the "measure first" approach has been followed. After indexing and integration, the result is a list of measured intensities with assigned *hkl* values referring to the specified unit cell. It is vital to appreciate that the unit cell and associated intensities are an *interpretation* of the diffraction pattern. Hopefully the interpretation has been made

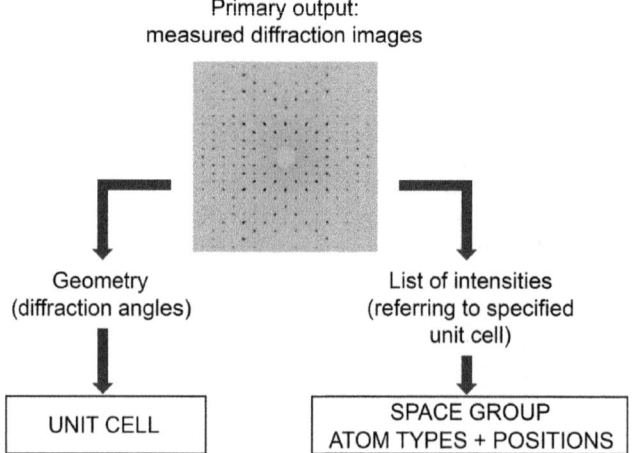

Figure 9.6   Schematic representation of the outcome of a single-crystal X-ray analysis. The unit cell and list of measured intensities are an interpretation of the measured diffraction images.

carefully, but if things should go wrong at a later stage the investigation should return to the measured diffraction images. Is the specified unit cell actually an appropriate interpretation of the measured diffraction pattern? Has the integration counted all intensities, or might it have missed some due to an incorrect choice of unit cell? Do the diffracted beams have a reasonable angular width or are they unreasonably wide? Have any of the images been "overloaded"? Are there indications of twinning or disorder (see Chapter 13)?

Assuming that a satisfactory interpretation of the diffraction images has been made, there are some key indicators in the final set of intensities that provide information on the data quality and choice of point group. The most prominent is the *merging R-factor*, or *internal R-factor*, which is usually given the symbol R_{int}. This is a measure of the agreement between the intensities that are expected to be equivalent according to the point group, defined as follows:

$$R_{\text{int}} = \frac{\sum \left\{ \left| I(hkl)_{\text{obs}} - I(hkl)_{\text{average}} \right| \right\}}{\sum I(hkl)_{\text{obs}}}$$

where both summations include all diffracted beams for which more than one symmetry equivalent is available. The value is given as a decimal in the range 0–1, or as a percentage. A smaller value of R_{int} indicates better internal agreement amongst the measured intensities. For a perfect case, the differences between equivalent reflections would be zero, so $I(hkl)_{\text{obs}} = I(hkl)_{\text{average}}$ for all diffracted beams and R_{int} would be zero. Of course, the value of R_{int} depends on which reflections are considered to be equivalent (which data are "merged"), so R_{int} provides an indication of whether the chosen point group is correct. An example is given in the Case Study at the end of the chapter.

The first step in analysing a measured set of intensities is usually to confirm the Bravais lattice and Laue group of the diffraction pattern. We basically return to the discussion in Section 8.4.3, but now in possession of the measured intensities. All potential Laue groups consistent with the geometry of the lattice can be considered, now with their corresponding R_{int} values. The correct one usually has the highest symmetry with a reasonable R_{int} value. A "baseline measure" for R_{int} is provided by its value for the data merged according to Laue group $\bar{1}$. This merges only data that have been measured more than once, and the Friedel pairs, $I(hkl)$ and $I(\bar{h}\bar{k}\bar{l})$. The R_{int} value for a correct higher symmetry Laue group should be comparable to the baseline value, noting that R_{int} will always increase to some extent when a greater number of data are merged. It is also informative to look at R_{int} as a function of diffraction angle. The value of R_{int} will increase for data at higher diffraction angle (lower d-spacing) because the intensities become weaker and therefore more prone to experimental uncertainty, but the increase should be smooth rather than erratic.

With the Laue group established, it is possible to investigate the space group of the crystal structure from the intensity statistics and pattern of systematic absences. Automated algorithms exist for this purpose. As described in Section 7.2.5, this usually identifies the space group uniquely, or at least narrows it down to a small number of alternatives. Another approach, which is probably now more common, is to determine the crystal structure using the space group with the lowest symmetry consistent with the Laue group then to detect any additional symmetry from the established structure. This can provide a few alternative descriptions of the structure in different space groups that can be considered further during structure refinement. More details are given in Chapter 11.

9.5 Summary of Key Points

- If the lattice parameters and orientation of the crystal on the diffractometer are established, a strategy can be devised to measure all diffracted intensities.

- An efficient strategy will measure only the unique set of intensities according to the point symmetry of the diffraction pattern, but this cannot be known conclusively until after the intensities are measured. Assumptions are commonly made on the basis of the symmetry indicated by the lattice parameters and the user must be cautious to ensure that sufficient data are collected. The safest approach is to measure the entire diffraction pattern, if sufficient time is available.
- The standard guideline for a small-molecule pharmaceutical crystal is to measure intensities to 0.85 Å resolution. For crystals that diffract less strongly, data sets extending to limited resolution can still be valuable if they allow the structure to be solved but they may not allow detailed comparison of specific bond distances or angles.
- The measurement time per image should be sufficient to ensure that intensity is observed for the weakest reflections at the highest measured diffraction angles, but care must be taken to avoid overloading the detector at lower diffraction angles. It is often helpful to measure images at higher angle for longer than those at lower angle. For a shutterless mechanism, it is usually necessary to collect additional rapid low-angle images from which the intensities of very strong reflections can be assessed, in case they overload the main diffraction images.
- Integration counts the intensities from the measured images. The procedure often employs profile-fitting methods that produce more reliable estimates of weaker intensities. Optimised unit-cell parameters are obtained by a final least-squares refinement against the positions of all measured diffracted beams.
- The multi-scan correction minimises the effects of systematic errors. The applied point group must again be treated with caution because this still may not be known conclusively until the structure is finally solved. High redundancy is always beneficial for the multi-scan.
- The merging R-factor (R_{int}) provides a measure of the agreement between the intensities that are expected to be equivalent according to the specified point group. A smaller value of R_{int} indicates better internal agreement amongst the measured intensities.
- The primary data obtained from a single-crystal analysis is the set of measured diffraction images. After indexing and integration, the derived result is a list of measured intensities with assigned *hkl* values referring to the specified unit cell.

9.6   Case Study: Aspirin Form I (Continued)

Continuing the Case Study on aspirin form I, the next step is to devise a strategy to collect the intensities. The lattice parameters are consistent with either a triclinic P, monoclinic P or orthorhombic C unit cell. By default, the diffractometer control software suggests orthorhombic C because it is the highest symmetry consistent with the lattice. If the orthorhombic crystal system is assumed and the structure is centrosymmetric, the point group of the diffraction pattern will be *mmm* and it is necessary to measure one eighth of all diffracted beams. If the crystal system is actually monoclinic, the point group will be $2/m$ for a centrosymmetric structure and it is necessary to measure one quarter of all diffracted beams. In the event that the crystal system is triclinic, it will be necessary to collect at least half of all diffracted beams. Which should we choose? Some further indications come from experience: space-group statistics show that monoclinic P and triclinic P are dramatically more common than orthorhombic C for molecular crystals. In fact, space groups $P2_1/c$ (*ca.* 35%) and $P\bar{1}$ (*ca.* 25%) together account for *ca.* 60% of all published molecular crystal structures, while the most common C-centred orthorhombic space group accounts for <0.3%. Hence, there is a reason to be cautious about orthorhombic C. The most sensible strategy would probably be to collect a hemisphere of data, since this will be sufficient for either triclinic P (assuming a centrosymmetric structure) or monoclinic P.

   The unit cell and point group of the diffraction pattern must be specified to calculate the strategy. We could choose the reduced cell ($a = 6.59$ Å, $b = 11.34$ Å, $c = 11.34$ Å, $\alpha = 84.23°$, $\beta = 89.93°$, $\gamma = 89.95°$) and specify point group $\bar{1}$, or the monoclinic P cell ($a = 11.34$ Å, $b = 6.59$ Å, $c = 11.34$ Å, $\alpha = 90°$, $\beta = 95.77°$, $\gamma = 90°$) and specify point group 2. Both approaches will collect a hemisphere of data. The *hkl* values assigned to each diffracted beam will depend on the specified unit cell, but the indices can be transformed later if a different unit cell is chosen. For this discussion, we proceed with the monoclinic P unit cell and collect a hemisphere of data according to point group 2. This will measure all unique $I(hkl)$ and $I(\bar{h}k\bar{l})$ and assume that $I(h\bar{k}l)$ and $I(\bar{h}k\bar{l})$ are equivalent due to 2-fold symmetry. The details of the calculated strategy and measurement time per image will depend on the available instrumentation. The important point is that data should be measured to (at least) the standard 0.85 Å resolution, with sufficient measurement time to observe intensities out to that resolution and as much redundancy as time will permit.

After data collection, the next step is integration. At this stage, we specify the monoclinic *P* unit cell, which we suspect is most likely to be correct. At the end of the integration, the unit-cell parameters are refined by a least-squares procedure against 9040 measured reflection positions. The refined parameters are $a = 11.3105(4)$ Å, $b = 6.5683(2)$ Å, $c = 11.3178(4)$ Å, $\alpha = 90°$, $\beta = 95.8017(12)°$, $\gamma = 90°$. The figures in brackets are standard uncertainties, which express the experimental errors; more details on uncertainties are discussed in Chapter 14. Since the monoclinic *P* cell was specified, the values of α and γ are constrained to 90° in the refined unit cell, with no associated error. The integrated data are finally passed to the multi-scan correction where it is necessary to specify the point group of the diffraction pattern. We proceed on the assumption that the point group is $2/m$, which imposes the conditions $I(hkl) = I(h\bar{k}l) = I(\bar{h}\bar{k}\bar{l}) = I(\bar{h}k\bar{l})$. If we later discover that the space group is not consistent with point group $2/m$, the multi-scan correction must be re-applied with the correct point group. The final output is 13709 corrected intensities referring to the optimised unit cell.

With the full list of diffracted intensities, the symmetry of the diffraction pattern can be analysed in detail. We already know that the lattice parameters are consistent with either a monoclinic *P* or orthorhombic *C* unit cell, but now we can assess whether the intensities are consistent with the corresponding Laue groups. The table below shows the possible monoclinic and orthorhombic unit cells with associated R_{int} values:

Bravais Lattice	a (Å)	b (Å)	c (Å)	α (°)	β (°)	γ (°)	R_{int}
Orthorhombic *C*	15.170	16.790	6.568	90	90	89.96	0.566 [2067]
Monoclinic *P*	11.311	6.568	11.318	90	95.80	90	0.010 [1319]

The R_{int} value for the monoclinic *P* cell is based on averaging intensities according to Laue group $2/m$ (giving 1319 groups of equivalent reflections) while the orthorhombic *C* unit cell applies Laue group *mmm* (giving 2067 groups of equivalent reflections). Clearly, R_{int} for the orthorhombic unit cell is very much larger than for the monoclinic cell, which proves finally that orthorhombic *C* is not appropriate. To stress, the geometrical features of the lattice are consistent with the orthorhombic *C* unit cell, but the unreasonably high R_{int} value shows that the intensities are not consistent with Laue group *mmm*.

As described in Section 7.2.5, an analysis of the distribution of the intensities can now give an indication of whether the space group is centrosymmetric. This analysis may be presented graphically or summarised by some numerical indicators; in this case, the data are found to be consistent with a centrosymmetric structure. Finally, it is necessary to examine systematic absence conditions. This could be shown graphically by plotting appropriate sections of the intensity-weighted reciprocal lattice (as in Section 7.2.2) but is more frequently shown within analysis software as a summary table. For example:

Exceptions to systematic absence conditions:				
	2_1	a	c	n
N	12	763	757	758
$N\,[I > 3\sigma(I)]$	0	3	362	361

For each axis direction, the table shows the number of measured reflections (N) that should be absent for each space symmetry element and indicates how many of them are considered to be observed (defined as having intensity greater than 3 times the associated standard deviation). For a 2_1 screw axis parallel to the b axis, twelve $0k0$ reflections are measured with k odd, of which none are considered to be observed. Hence, we can conclude that the 2_1 screw axis is present. Similarly, there are 763 measured reflections that should be absent for an a-glide perpendicular to the b axis, of which only 3 are observed. Hence, an a-glide is likely to be present. By contrast, there are 757 measured reflections that should be absent for a potential c-glide perpendicular to the b-axis, of which 362 are observed. Hence, space group $P2_1/a$ is indicated. The standard setting of this space group type is $P2_1/c$, so the a and c axes of the unit cell can be exchanged within the processing software (almost certainly automatically) to give a final unit cell of $a = 11.3178(4)$ Å, $b = 6.5683(2)$ Å, $c = 11.3105(4)$ Å, $\alpha = 90°$, $\beta = 95.8017(12)°$, $\gamma = 90°$. The *hkl* values in the intensity list would be updated accordingly, and these can now be taken forward to structure solution, as described in Chapter 11.

10 Powder X-ray Diffraction

Summary

Powder X-ray diffraction is an essential partner to single-crystal X-ray diffraction because it is applied to a bulk solid rather than a selected single crystal. Powder diffraction can be viewed as a collection of simultaneous single-crystal measurements from a large number of crystals in all orientations. Averaging of the sample means that the diffracted beams are distributed around cones with half-angle 2θ, which appear as rings when measured on a 2-D detector. If the sample orientation is properly averaged, the diffracted intensity is spread uniformly around the rings so it is necessary only to scan along a radius to produce a representative plot of diffracted intensity *vs.* 2θ. Powder diffraction can be measured in transmission geometry, where the incident beam passes through the sample, or in reflection geometry, where the incident beam appears to be reflected from the surface of the sample. The relationship between a crystal structure and its powder diffraction pattern is described and some applications of powder diffraction for pharmaceutical solids are considered.

10.1 Introduction

Powder X-ray diffraction is an essential partner to single-crystal X-ray diffraction, because powder diffraction is carried out on a bulk solid rather than a selected single crystal. In this chapter, powder diffraction is described as a collection of simultaneous single-crystal measurements from a large number of crystals in all orientations. The 3-D diffraction pattern from a single crystal is averaged into a pattern that has effectively only one spatial dimension, which is the Bragg angle.

Pharmaceutical Crystallography: A Guide to Structure and Analysis
By Andrew Bond
© Andrew Bond 2019
Published by the Royal Society of Chemistry, www.rsc.org

If seeking to determine a crystal structure, this means that a powder diffraction pattern contains significantly less information than a single-crystal diffraction pattern. However, the powder pattern contains additional information about the bulk sample. For a pharmaceutical solid, it can assess phase purity (whether the crystals in the sample all have the same crystal structure), preferred orientation (related to the crystal habit and preferred alignment of crystals in the sample), crystallite size and strain, and whether a solid is crystalline at all. We start by describing powder diffraction with reference to what we now understand about single-crystal diffraction, then describe how to measure a powder pattern and the information that can be obtained. In the pharmaceutical sciences, the most common abbreviation for the powder technique is probably XRPD, meaning "X-ray powder diffraction". The alternative PXRD refers to "powder X-ray diffraction". The two abbreviations are obviously equivalent and interchangeable.

10.2   Relating Powder Diffraction to Single-crystal Diffraction

Powder diffraction can be viewed as a collection of simultaneous single-crystal diffraction measurements, with a large number of crystals randomly oriented with respect to the incident beam. All of the discussion from Chapters 6 and 7 applies, but we have to consider how the diffraction pattern is influenced by the nature of the powder sample. We saw in Section 6.3.2 that a crystal must be in an appropriate orientation relative to the incident beam for a diffracted beam to emerge. It was also mentioned how a crystal in a diffracting position could be rotated $360°$ around the incident beam, staying in the diffracting position but changing the direction in which the diffracted beam emerges. Thus, the diffracted beam could lie anywhere on the surface of a cone, with the crystal at its apex and with half-angle 2θ, where θ refers to Bragg's description (Figure 10.1). For a sufficiently large, randomly-oriented powder sample it can be assumed that all possible crystal orientations are represented, so the diffracted intensity actually does define these cones. If the pattern is measured using a 2-D detector perpendicular to a fixed incident beam, the result is a set of concentric circles referred to as *powder rings*. If the number of crystals in the powder is sufficiently large, and if their orientations are truly random, the intensity around each cone will be distributed uniformly. In that case, the relative intensities of the powder rings can be measured just by scanning along a radius of the rings. This approach is

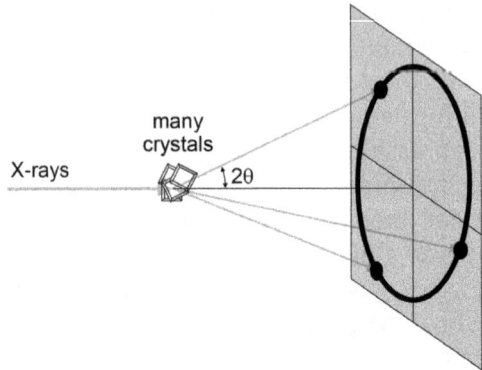

Figure 10.1 Many crystals in random orientations produce concentric powder rings on the face of a 2-D detector. Bragg's law applies, but the diffracted beam from each crystal could lie anywhere on the surface of a cone with half-angle 2θ. If the crystal orientation is properly averaged, the intensity is distributed uniformly around each ring so it is necessary only to measure along the radius.

applied in most laboratory powder diffractometers. It may become problematic if the intensity is not distributed uniformly around each ring, as for a sample exhibiting *preferred orientation* (Section 10.2.4).

10.2.1 Peak Position

On account of the spatial averaging of the diffraction pattern, the only geometrical information remaining for each diffracted beam is its Bragg angle. Accordingly, the position of a peak in a powder diffraction pattern refers to its 2θ value in degrees. A plot of intensity *vs.* 2θ is overwhelmingly most common in the chemical and pharmaceutical literature. Bragg's equation shows that the 2θ value depends on the wavelength of the incident X-rays, so it is necessary to specify the applied X-ray wavelength when presenting PXRD data in this way.

10.2.2 Peak Intensity

The intensity of a diffracted beam contributing to a PXRD pattern is inherently the same as the corresponding intensity in the single-crystal diffraction pattern, which can be calculated using the structure factor equation. For a powder sample, however, only the Bragg angle θ is relevant to determine the peak position so all diffracted beams with the same θ value overlap. The resulting peak intensity is the sum of the intensities for all overlapping peaks. The "inherent" reason for peak overlap is crystal symmetry, as discussed in Chapter 7. For example, if the diffraction pattern has Laue group $2/m$,

$d(hkl) = d(h\bar{k}l) = d(\bar{h}\bar{k}l) = d(\bar{h}k\bar{l})$, so a general peak in the PXRD pattern will actually comprise an overlay of four symmetry-related diffracted beams. In that case, the peak is said to have *multiplicity* equal to 4. Some values of *hkl* may lie on symmetry elements within the Laue group (*e.g.* 0*k*0 or *h*0*l* for 2/*m*), so their multiplicity is reduced accordingly. An important case occurs for Friedel pairs, *hkl* and $\bar{h}\bar{k}\bar{l}$, which *always* overlap in a PXRD pattern. It is therefore impossible to determine absolute structure using PXRD.

In addition to symmetry-related overlap, many peaks in a PXRD pattern overlap simply by coincidence, meaning that the size and shape of the unit cell happens to result in similar 2θ values for several diffracted beams. The extent of coincidental overlap increases substantially at higher 2θ. For a typical small-molecule pharmaceutical crystal, it is common to see isolated peaks up to about $2\theta \approx 30°$ (CuKα radiation), but the PXRD pattern above this angle generally comprises a mass of overlapping peaks (Figure 10.2).

10.2.3 Peak Shape

Bragg's equation indicates a peak position for each diffracted beam, which in principle corresponds to an infinitely sharp line (a "delta function"). Diffracted beams in real PXRD patterns have a finite width, which can be considered to arise from two distinct sources: (1) contributions from the instrument and (2) contributions from the sample.

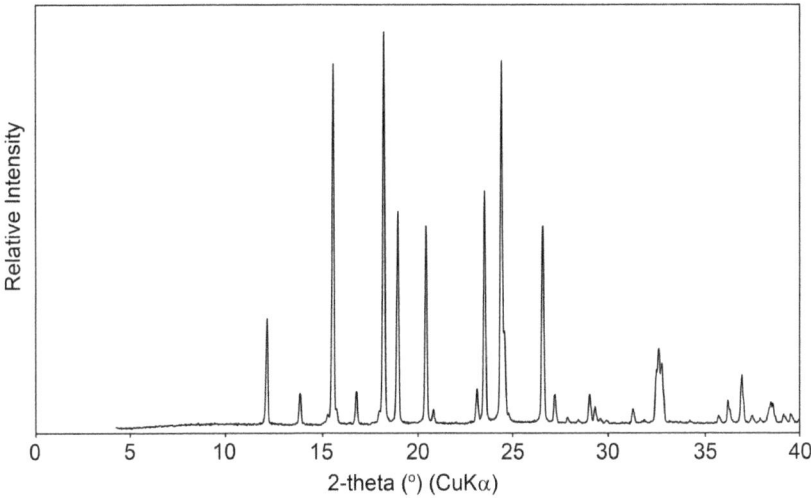

Figure 10.2 Measured PXRD pattern (CuKα radiation) of the monoclinic polymorph of paracetamol (as in the Case Study of Chapter 4), presented as intensity *vs.* 2θ.

The instrument contribution reflects the nature of the incident beam, for example whether it is parallel or diverging, plus the measurement geometry and other features of the beam path. The overall effect on the width of PXRD peaks is referred to as the *instrument profile function*. It can be measured for an instrument by taking a standard sample known to give a negligible contribution to the peak shape. Contributions from the sample can have numerous origins. One is a consequence of crystallite size, where smaller crystallites produce broader peaks (Figure 10.3(a)). The important property is not necessarily "particle size", but "domain size", referring to the size of "perfect" crystalline regions within the sample. A second sample effect comes from *strain*, which describes local distortions of the crystal structure. Depending on the nature of the distortions and their orientation relative to the incident and diffracted beams, strain can cause shifts in peak position and/or peak broadening (Figure 10.3(b)). Broadening due to crystallite size has the same effect on all diffracted beams, but strain broadening can have different effects on different diffracted beams. Quantitative relationships can be derived between peak width

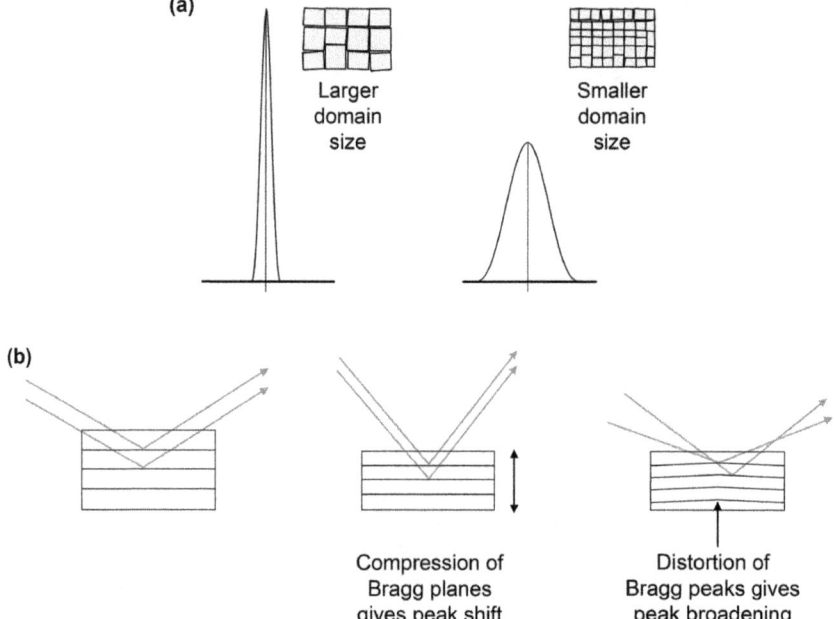

Figure 10.3   Peak broadening due to (a) crystallite size; (b) crystallite strain. Broadening due to crystallite size has the same effect on all diffracted beams. Crystallite strain can take many different forms and can have different effects on different beams.

and crystallite size/strain, which can be applied in the context of "profile analysis".

10.2.4 Preferred Orientation

Simplifying the measurement of powder rings to measurement along their radius assumes that the diffracted intensity is distributed uniformly around the rings. If the crystallite orientations are not uniformly averaged, there will be some preferred directions for diffracted beams to emerge on the surface of each diffraction cone and the relative intensities measured by a particular radial scan may not therefore be representative (Figure 10.4). In this scenario, the sample is said to exhibit *preferred orientation*. There is an obvious reason why this might occur: if the crystallites have an anisotropic habit, they are more likely to lie in certain orientations. For example, plates are more likely to lie with their faces parallel in a flat-plate sample while needles in a capillary are more likely to be aligned with their long axes parallel. For this reason, it is desirable to grind a powder before preparing a PXRD sample to produce crystallites that are as regular as possible. Preferred orientation affects intensities, not peak positions. It can be very difficult to eliminate preferred orientation completely, so peak positions in a PXRD pattern are generally far more reliable than peak intensities.

10.2.5 Simulating a PXRD Pattern

To emphasise the relationship between single-crystal and powder diffraction patterns, it is useful to consider how a PXRD pattern can

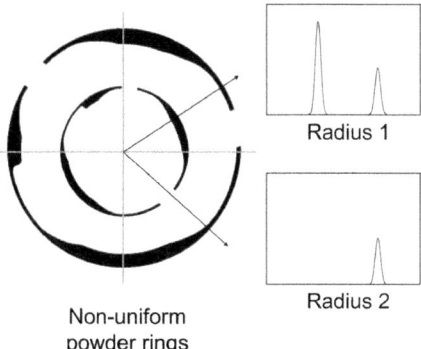

Non-uniform powder rings

Radius 1

Radius 2

Figure 10.4 Schematic non-uniform distribution of intensity around powder rings due to preferred orientation. The two radial scans illustrated produce very different peak intensities.

be simulated from a crystal structure. A practical procedure, summarised in Figure 10.5, is as follows:

(1) Use the lattice parameters to calculate the *d*-spacing for each value of *hkl*. Use Bragg's equation to convert *d* to 2θ for the desired X-ray wavelength (usually CuKα).

(2) Use the structure factor equation to calculate $F(hkl)$ for each value of *hkl* up to a desired maximum diffraction angle. Square each $|F(hkl)|$ to produce $I(hkl)$. Plotting all $I(hkl)$ against 2θ produces an "idealised" PXRD pattern comprising a set of sharp lines. Since only relative intensities are important, the pattern can be scaled to some arbitrary maximum.

(3) Introduce a peak-shape function and overlay it on each of the lines, scaled to the corresponding $I(hkl)$ value. Then sum the total intensity at each 2θ value to produce the full PXRD pattern.

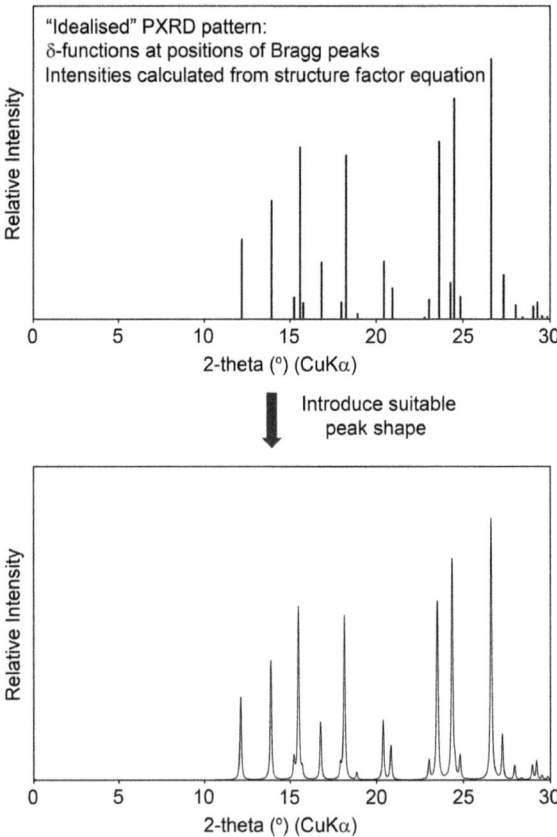

Figure 10.5 The process of simulating a PXRD pattern from a crystal structure.

A suitable peak shape for a simulated PXRD pattern is a "pseudo-Voigt" function, which is a weighted mixture of a Gaussian and a Lorentzian function. By comparison with experimental patterns, the Gaussian function is generally found to be too broad while the Lorentzian function is found to be too narrow, so a mixture of the two is about right. To introduce the peak-shape function, it is the *area* of the peak that represents the intensity: thus, a pseudo-Voigt function is defined with its centre at the required value of 2θ, and with its area scaled to the total intensity calculated in step (2). In an experimental PXRD pattern, the peak shape will actually be determined by instrumental and sample factors, which can be described by various analytical functions. In the simulated pattern, the pseudo-Voigt function is just an appropriate choice to produce a reasonable simulation. The simulation can be extended to take account of preferred orientation, which can be useful when comparing experimental PXRD patterns to those simulated from known crystal structures. In an experimental PXRD pattern, some background intensity must also be present, which again will contain contributions from both the instrument and the sample and will vary with 2θ.

10.3 Powder X-ray Diffraction Measurements

The instrument to measure a PXRD pattern is essentially identical to that described in Chapter 8 for single-crystal X-ray diffraction. However, the nature of the diffraction pattern means that several components can be simplified and/or optimised for the powder measurement. There are several common ways to carry out a PXRD measurement in a standard laboratory, so there is a wider variety of instrumentation. Some generic descriptions are given below.

10.3.1 Transmission (Debye–Scherrer) Geometry

One way to measure a PXRD pattern is directly comparable to a single-crystal instrument, with the powder sample held in a capillary aligned along the ω axis (Figure 10.6). This is referred to as *transmission* (Debye–Scherrer) *geometry* because the incident and measured X-ray beams pass through the sample. The capillary is usually spun around its axis during the measurement, partly to reduce the effects of preferred orientation, and partly to bring more crystallites into the diffracting position at any given diffraction angle. To produce peaks with a reasonable width, transmission geometry requires a parallel or

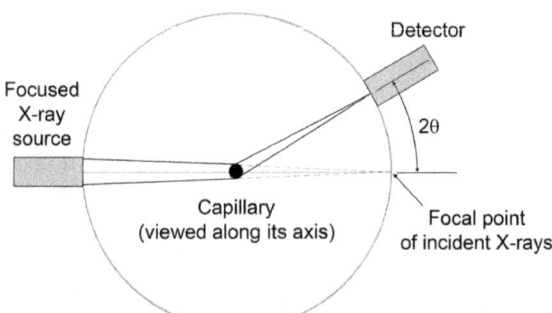

Figure 10.6 Transmission (Debye–Scherrer) geometry. The view is along the capillary axis, which corresponds to ω in the single-crystal description. The diagram corresponds to looking down onto the horizontal plane of the single-crystal instrument in Figure 8.2. The focussed X-ray beam produces peaks of reasonable width at the detector.

focussed incident beam. One of the main practical advantages is that the capillary set-up requires only a small amount of material. Since the X-ray beam must pass through the sample, transmission geometry is less suitable for highly-absorbing materials. This rarely applies to pharmaceutical samples, which typically contain only light elements.

10.3.2 Reflection (Bragg–Brentano) Geometry

Probably a more common set-up for laboratory powder diffraction is *reflection* (Bragg–Brentano) *geometry*, where samples are prepared as flat plates and it appears that the diffracted beams are reflected from the sample surface (Figure 10.7). The main technical advantage of reflection geometry is that it can produce narrow diffracted beams by making use of an apparent focussing effect. If the incident beam is divergent (*i.e.* it gets wider as it moves away from the source), the "reflected" beam must be convergent (*i.e.* it gets narrower as it moves away from the sample). Matching the sample-to-detector distance to the source-to-sample distance produces diffracted beams with width comparable to that produced at the source, without requiring any actual focussing device.

Goniometers for reflection geometry come in two main types. If the incident beam is fixed, the detector moves around 2θ, as for a single-crystal instrument. At the same time, the sample is rotated by an angle θ around the ω axis to maintain the geometry of Bragg's construction and bring more crystallites into the diffracting position (Figure 10.7). This is known as θ–2θ movement. An alternative is to keep the sample horizontal whilst moving the detector *and* source by an angle θ in opposite directions, known as θ–θ movement (Figure 10.8). This

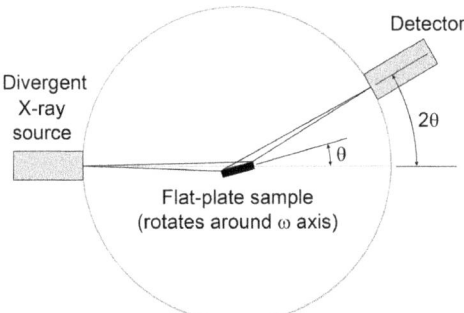

Figure 10.7 Reflection (Bragg–Brentano) geometry with a fixed incident beam. As the detector moves around 2θ, the sample rotates by θ to maintain the geometry of Bragg's construction.

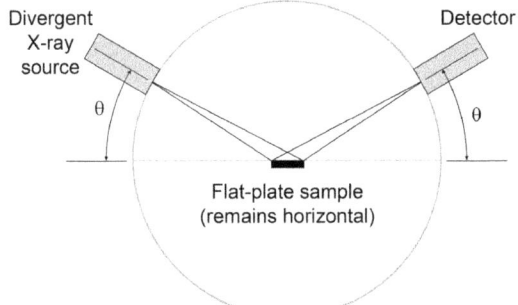

Figure 10.8 Reflection geometry showing θ–θ movement. The sample is fixed in a horizontal orientation while the source and detector rotate by θ in opposite directions, corresponding directly to the typical representation of Bragg's construction.

maintains the same geometry as θ–2θ movement, but requires a smaller angular range of movement for the detector. It is mechanically more demanding to move the source, but this is relatively straightforward in the one dimension required for a PXRD measurement. A practical advantage of θ–θ movement is that the sample remains static and horizontal, which gives greater flexibility in the types of samples that can be analysed. There is no danger that the sample will move or fall off the sample holder and it is easier to accommodate other types of sample environments or automated sample changers.

In reflection geometry, the X-rays interact largely with a surface layer of the sample. The path length of X-rays through the sample is minimised compared to transmission geometry, which is advantageous for highly-absorbing samples. However, a problem can arise with more transparent samples (such as typical pharmaceutical

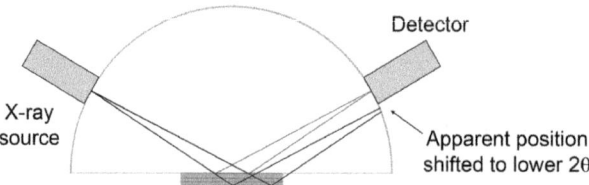

Figure 10.9 Sample transparency in reflection geometry. X-rays penetrate the sample and appear to be scattered from a position below the instrument centre. This leads to an apparent offset in 2θ. The effect is greatly exaggerated in the diagram. In practice, the errors are generally fractions of a degree, increasing with increasing 2θ.

solids) because the incident X-rays penetrate further into the sample and diffraction can appear to arise from a range of positions beneath the surface. The consequence is that the sample appears to be displaced from the centre of the instrument, which leads to variability in the peak positions (Figure 10.9). For highly transparent samples, many X-rays will pass straight through and be reflected from the sample holder, contributing only to the background. This effect can be minimised by so-called "zero-background" sample holders, which are plates cut from a single crystal of silicon, oriented to avoid the diffracting position for any observable reflection of Si.

10.3.3 The X-ray Source

As for single-crystal instruments, the standard laboratory PXRD source is the sealed tube. The most common anode is Cu because its relatively long X-ray wavelength (*ca.* 1.54 Å) spreads the diffraction pattern over a wider angular range, helping to minimise peak overlap. The X-rays produced by a Cu target actually comprise several characteristic wavelengths:

	Wavelength (Å)	Relative intensity
CuKα(1)	1.5406	1
CuKα(2)	1.5444	0.5
CuKβ	1.3922	0.2

For an X-ray diffraction measurement, it is desirable to select just one wavelength to avoid an overlay of diffraction patterns corresponding to each wavelength. If the X-ray wavelengths (and hence energies) are sufficiently far apart, they can be separated by a filter which comprises a metal foil that selectively absorbs one wavelength.

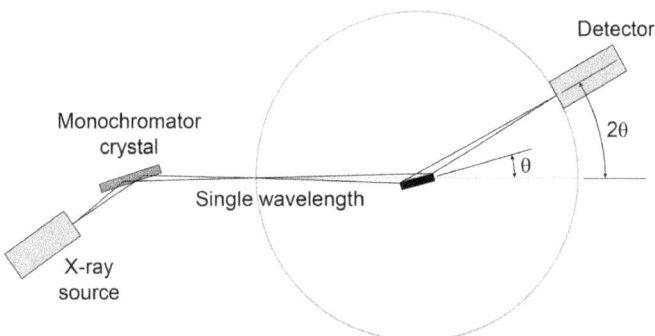

Figure 10.10   Inclusion of a monochromator in the incident beam path in reflection geometry. The monochromator crystal is aligned precisely to satisfy the Bragg condition for just one wavelength.

Ni foil is appropriate to absorb CuKβ, so powder diffractometers using Cu radiation usually have a Ni filter somewhere in the beam path. When the X-ray wavelengths are closer together, as for CuKα(1) and CuKα(2), it is not possible to separate them effectively using a filter. An alternative is to use a *monochromator*, which exploits the selectivity of X-ray diffraction. The incident beam is passed through a single crystal aligned precisely in the diffracting position for one wavelength and the diffracted beam is then used for the PXRD measurement (Figure 10.10). Typical monochromator crystals are Si or quartz (SiO_2). The disadvantage of a monochromator is that it reduces the intensity of the incident beam because it eliminates a significant proportion of the incoming X-rays. Including a mono-chromator probably also makes it impractical to adopt a θ–θ arrangement for reflection geometry, because the set-up is more complex and must maintain a very precise alignment.

For CuKα(1) and CuKα(2), we can consider the consequence of *not* including a monochromator. What will the diffraction pattern look like if both wavelengths are present? For a few representative diffracted beams, the differences in 2θ for peaks produced by the two wavelengths are as follows:

d (Å)	2θ (°), CuKα(1)	2θ (°), CuKα(2)	$\Delta 2\theta$ (°)
8.838	10.000	10.025	0.025
2.252	40.000	40.103	0.103
1.343	70.000	70.198	0.198
1.006	100.000	100.337	0.337
0.850	130.000	130.610	0.610

The difference is negligible at lower 2θ, but greater than 0.5° at very high 2θ. Whether or not these angular differences would be resolved into separate peaks depends on the peak widths. A typical full-width-at-half-maximum (FWHM) for a crystalline API on a laboratory PXRD instrument might be around 0.1° in 2θ. The typical maximum 2θ measured using CuKα radiation is 50° or less because the intensity of diffraction has generally dropped off significantly by this angle (due to the drop off of the atomic scattering factors) and peak overlap has become so severe that the pattern is very difficult to interpret. To this diffraction angle, there is very little difference between patterns measured with monochromated and non-monochromated CuKα radiation (Figure 10.11). Non-monochromated radiation is frequently chosen for the practical advantage of maximising the incident beam intensity (and minimising the cost). The wavelength for non-monochromated CuKα radiation is quoted as an average of the two constituent wavelengths, weighted according to their relative intensity, *i.e.* $(2\times1.5406 + 1\times1.5444)/3 = 1.5418$ Å. This is clearly a simplification; the incident X-rays do not actually have $\lambda = 1.5418$ Å.

10.3.4   The Incident Beam Path

While a single-crystal diffractometer has a "pencil-like" incident beam, a powder diffractometer usually has a line source to illuminate a larger volume of the sample. For transmission geometry using a capillary, the line of the incident beam extends along the capillary axis. In the absence of any other device, the incident line diverges as it moves away from the source, becoming both longer (*axial divergence*) and wider (*angular divergence*). The incident beam path includes various components to control the shape of this divergent beam (Figure 10.12). The axial width can be controlled by a simple mask that truncates the line to the desired length. It is common also to include *Soller slits*, which comprise a row of vertical slits which block X-rays not travelling on appropriate parallel paths. Controlling axial divergence is applicable for both transmission and reflection geometry.

The requirement to control angular divergence of the incident beam depends on the measurement geometry. Transmission geometry requires the beam to be focussed in order to produce peaks with reasonable width in the diffraction pattern. This can be achieved using focussing mirrors, which comprise layers of material that effectively reflect the incident X-rays. Careful shaping of the mirrors can produce parallel or focussed beams from an incoming divergent beam. For a capillary transmission set-up, the aim is to focus the beam on the

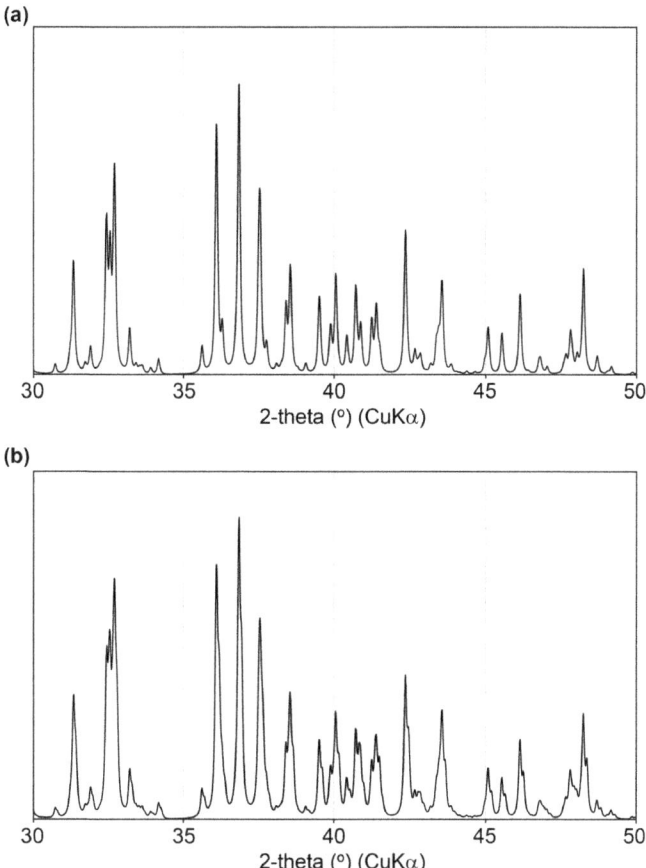

Figure 10.11 PXRD pattern simulated for monoclinic paracetamol in the range $2\theta = 30\text{--}50°$ (pseudo-Voigt peak shape with full-width at half-maximum = 0.1° in 2θ). (a) Using monochromated CuKα radiation ($\lambda = 1.5406$ Å); (b) using non-monochromated CuKα radiation ($\lambda_{ave} = 1.5418$ Å). Shoulders can be seen in (b) on the right-hand side of each peak, becoming more clear as the diffraction angle increases. However, there is little overall difference between the two patterns.

detector rather than the sample (Figure 10.6), because diffraction at the sample does not provide any additional focussing effect.

For reflection geometry, the principal need to control angular divergence is to match the size of the incident beam to the surface of the flat-plate sample (Figure 10.12). It is desirable to illuminate as much of the sample as possible, whilst avoiding background scattering that will add to noise. The angular width of the incident beam is controlled by passing the beam through a narrow slit called the *divergence slit*. A narrow divergence slit illuminates a smaller area, while a wider slit illuminates a larger area. For a given angular width, the illuminated

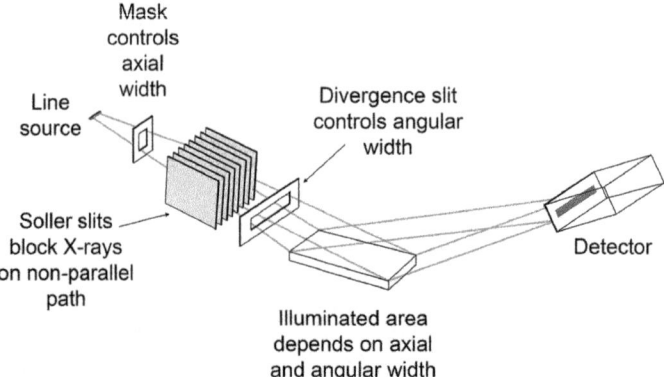

Figure 10.12 Schematic view of a representative incident beam path for reflection geometry, with components to control the beam shape.

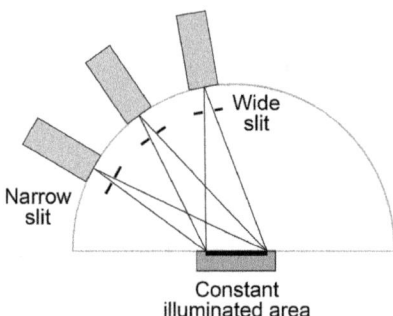

Figure 10.13 Variable divergence slit.

area will be greater at lower angles of incidence than at higher angles. A suitable slit size should therefore cover the available sample area at the smallest diffraction angle to be measured. There is an obvious problem here: since the same incident X-ray beam is distributed over a different sample area as a function of diffraction angle, the measured peak intensities will be systematically affected. This can be addressed using a *variable divergence slit* that changes its width as a function of diffraction angle so as to maintain a constant illuminated area (Figure 10.13). This is practically more demanding (and therefore more expensive) so a fixed-slit arrangement is probably more common in a standard laboratory instrument. Obviously, the relative intensities measured with a variable divergence slit will appear quite different from those measured with a fixed slit, so it is important to specify the experimental set-up when reporting a PXRD pattern.

An additional *anti-scatter slit* is often included in the incident beam path (and sometimes matched in the diffracted beam path). This is the angular equivalent of the Soller slits in that its purpose is to block X-rays on an inappropriate path. The anti-scatter slit should be wider than the beam at the corresponding radius. Its purpose is not to shape the incident beam, but simply to catch any stray X-rays that would otherwise add to the background.

10.3.5 The Detector

Given the 1-D nature of the radial measurement, a powder diffract-ometer is commonly equipped with a 1-D detector comprising a strip of measurement pixels. The angular range covered by a strip of a particular length depends on the sample-to-detector distance (Figure 10.14). For reflection geometry, we have seen that the sample-to-detector distance should match the source-to-sample distance to exploit the apparent focussing effect, but how large should the instrument radius be? On first sight, it may seem desirable for the detector to be very close to the sample to maximise the measured angular range. However, the detector comprises a certain number of pixels with a specific width, so this would mean that each pixel would individually measure a larger angular range. In that case, X-rays covering a large angular range would all be measured on the same pixel, and precise knowledge of the angle of each diffracted beam would be lost, *i.e.* there would be a significant reduction in the *angular resolution*. The same effect was discussed for single-crystal X-ray diffraction in Section 8.2.3. To achieve the best angular resolution, the detector should move away from the sample. This has its disadvantages, however, not least that the instrument size would become excessive. A typical compromise for a laboratory diffractometer is a radius of 240 mm.

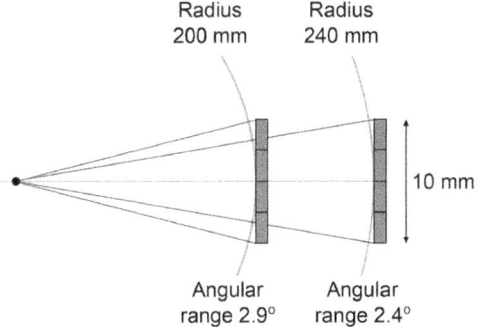

Figure 10.14 1-D strip detector with angular range as a function of distance. If the detector comprises a row of 128 pixels, each pixel would cover an angular range of 0.022° at 200 mm or 0.018° at 240 mm.

10.3.6   Control and Data Processing Software

The control and data processing software for a powder diffractometer can be more straightforward than for a single-crystal instrument. There is no requirement to consider which part of the diffraction pattern to measure or how to measure it: it is simply a matter of specifying the angular range and the measurement time. The latter may vary more than for a single-crystal measurement because the aims for collecting PXRD data can be very different. If the aim is solely to confirm that a particular crystal structure is representative of the bulk, or to establish that the sample is amorphous, then a very rapid measurement may suffice. If, however, the aim is to extract accurate unit-cell parameters or peak profiles, or even to solve or refine a structure from the PXRD data, more care must be taken to obtain accurate peak positions, well-defined peak profiles, representative intensities and adequate signal-to-noise. Some examples are given in Section 10.5.

10.4   The Sample

The sample on a powder diffractometer must be suitably centred to produce accurate peak positions. For transmission geometry, centring means aligning a capillary along the ω axis, so that it spins exactly on the axis. The practicalities are the same as for centring of a single crystal (Section 8.3) except that angular adjustment of the capillary will also be required; goniometer heads with angular adjustments are available for this purpose. For reflection geometry, the sample surface must be flat and aligned at the geometrical centre of the instrument. Typically, samples are loaded in holders with a shallow well and levelled so that the sample surface is aligned with the upper face of the sample holder. The holder is then held by a clip or some other mechanism that aligns its upper face relative to the instrument. The consequences of an incorrect sample height are largely as described earlier for sample transparency (Figure 10.9). If the sample is too low, peak positions will be systematically shifted to lower angle, and *vice versa*. If the sample is not level, inaccuracy of the peak positions will vary with diffraction angle.

For a sample to achieve a suitable "powder average", it must have a sufficient number of crystallites covering all possible orientations. An insufficient number of crystallites may mean that some diffracted beams are not seen because there are no crystallites in a suitable diffracting position. Samples are typically spun during a measurement to help with averaging; a capillary is spun around its axis and a flat-plate is rotated about its normal. If reliable peak intensities are required, it is necessary to eliminate as far as possible any doubts over the existence of preferred orientation. For this purpose, it can be useful to measure a PXRD pattern several times, in as many different ways as possible. With a flat-plate sample, a sample might be ground until it appears uniform, then measured. Then the same sample can be ground further and measured again. If the relative intensities change, the first measurement must have been subject to preferred orientation. In principle, the process could be continued until two successive PXRD patterns show the same relative intensities. Of course, there are usually associated practical difficulties such as whether there is enough sample, whether peak broadening becomes more significant on account of smaller particle size, or whether the solid form might actually change with prolonged grinding. If possible, the same sample might be measured as a flat plate and capillary. Any preferred orientation should be different for the two methods, so comparable intensities in both measurements adds confidence. It is possible to obtain reliable peak intensities, but it generally requires more attention than just loading one flat-plate sample and running a standard measurement.

10.5   Extracting Information from a PXRD Pattern

Extracting information from a PXRD pattern can be viewed as "reverse engineering" of the simulation process in Section 10.2.5. If the crystal structure is known, it should be possible (at least in principle) to fit a PXRD pattern completely by defining a suitable peak profile function and perhaps applying a correction for preferred orientation to the peak intensities. This type of "full profile fitting" is the best way to extract information from a PXRD pattern. If unit-cell parameters are known for a structure but not atomic co-ordinates, the PXRD pattern can still be fitted in this way if peak intensities can be assigned by some other method. One option is Pawley refinement, where the diffracted intensities are treated as additional numerical parameters to be optimised during the profile fitting. If nothing is known about the crystal structure, a PXRD pattern still provides a "fingerprint" that can be used to identify a solid form by visual inspection. Solid-form identification by matching PXRD patterns is probably the most common application of PXRD in the pharmaceutical sciences. We cannot cover all of the possibilities for analysing PXRD data, or to describe any of them in significant detail. The following sections are intended simply to give an overview of the various levels of analysis in the context of the discussion so far.

10.5.1 Solid Form Identification

Ideally, single-crystal diffraction should *always* be accompanied by PXRD to establish that the structure obtained from the selected single crystal is representative of the bulk sample. Often, it is sufficient just to make a visual comparison between the measured pattern and the pattern simulated from the established crystal structure. There may be subtle differences between peak positions, due to sample preparation or perhaps a temperature difference between the determined crystal structure and the measured PXRD data. Probably more substantial differences will exist between peak intensities, due to preferred orientation or details of the measurement method, but the human eye is generally very good at recognising and matching patterns. When working with a specific API, PXRD patterns will often be known for a set of solid forms, either measured locally or published in the literature. In that case, it can be straightforward to say whether a PXRD pattern matches any of those patterns or whether it might represent a different solid form. On a larger scale, databases of experimental and PXRD patterns exist, which can be searched using a variety of pattern-matching algorithms.

10.5.2 Indexing and Unit-cell Refinement

Establishing a unit cell and assigning associated *hkl* indices to peaks in a PXRD pattern is complicated by compression of the diffraction pattern into one dimension, plus the extensive peak overlap. The problem is harder for lower-symmetry crystal systems, which are most prevalent for pharmaceutical materials. Numerous algorithms are available to suggest possible unit cells for a list of peak positions. In practice, it is helpful to try as many of these algorithms as possible in the hope that one of them might indicate the correct unit cell. There are a few issues to consider beyond that. The first is how the list of peak positions will be obtained. It may be easy to deduce a position for a strong isolated peak, but how will peak positions be extracted from a group of overlapping peaks? Of course, peak-fitting procedures exist, but in practice it becomes very difficult to extract more than *ca.* 20 unambiguous peaks from a typical pharmaceutical sample measured on a laboratory PXRD instrument. For the indexing algorithms to be effective, it is usually important to list the positions of *all* peaks up to a given diffraction angle. Since the process works with only a small number of peaks, it is likely that a few missed peaks will have a substantial influence on the results.

Then we must know how to recognise the correct unit cell. Since there will generally be only a small number of clearly-defined peaks, and each one could be subject to a window of experimental error, this can be very difficult. Quantitative measures exist, based on the differences between observed and calculated peak positions, but these can often be comparable for numerous possible (inequivalent) unit cells. A useful indicator is the unit-cell volume, which can suggest whether the result is reasonable for the expected molecular contents. A practical guideline is the "18 $Å^3$ rule", which states that non-H atoms occupy on average *ca.* 18 $Å^3$. Of course, this can be problematic if an unknown solid form turns out to have an unexpected composition, such as a solvate/hydrate with unpredictable stoichiometry. In a truly unknown case, the best way to assess a unit cell established from a PXRD pattern is probably to carry out a Pawley refinement, as described below.

10.5.3 Pawley Refinement

Pawley refinement fits peak positions and profiles, treating the intensities as additional parameters to be freely adjusted. Hence, it enables unit-cell parameters to be refined and peak profiles to be determined without needing to know the full crystal structure. The peak profile must include contributions from the instrument and from the sample, both of which can be represented by a combination of analytical functions intended to model physical effects. The exercise basically follows the path defined in Section 10.2.5, with the intensity of each peak (the *area* defined by the peak profile) also optimised to provide the best fit to the whole pattern. It is necessary to model the background intensity, which will vary across the 2θ range. Typically this is achieved by linear interpolation (*i.e.* defining values at a few points across the range, then joining them up with straight lines) or by more complicated polynomial functions. To make an effective Pawley fit, the measured PXRD pattern must be sufficiently detailed to enable proper fitting of the peak profiles (Figure 10.15). Hence, it is necessary to consider the angular resolution of the measurement (Section 10.3.5) if the PXRD data are intended for profile fitting. The fit can be assessed by several quantitative measures (effectively comparable to those described for single-crystal data in Chapter 12), but the most informative guide is

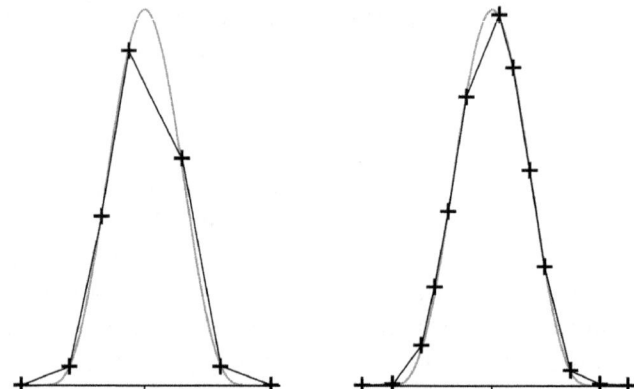

Figure 10.15 Peak profile fitting. The smooth grey line in each case indicates the "true" peak shape. (a) Measuring too few points across the peak will make it difficult to fit a meaningful peak shape. (b) A sufficient number of points across the peak define the peak shape more clearly.

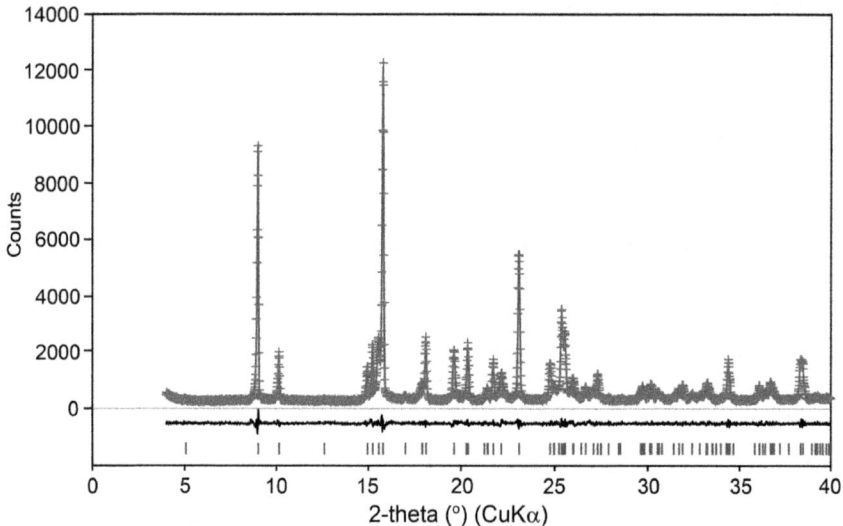

Figure 10.16 A Pawley fit. The measured data points are shown as crosses in the upper curve and the continuous line shows the fitted profile. The lower black line shows the difference between the observed and calculated intensity at each 2θ position. This is relatively flat, indicating a good fit to the data. The bottom "tick marks" indicate the expected peak positions for the established unit cell.

the *difference curve*, which shows the observed intensity minus the calculated intensity across the 2θ range (Figure 10.16). Ideally, the difference curve should be featureless.

10.5.4 Rietveld Refinement

Assuming that the applied unit cell is correct, the Pawley refinement provides an indication of the best possible fit to a measured PXRD pattern. It provides refined unit-cell parameters and appropriate descriptions for the peak profiles and background, given an "optimal" set of intensities. To take this further, the diffracted intensities should be calculated from the crystal structure. It may be necessary at this stage to model preferred orientation to match the calculated intensities to the measured pattern. If this can be achieved, we might then optimise some of the structural parameters (*i.e.* the atomic coordinates) to improve the fit to the PXRD pattern; this is *Rietveld refinement*. For a given structure model, the calculated PXRD profile is constructed basically as described in Section 10.2.5, using the established peak profiles and background function, then the result is compared point-by-point to the measured PXRD pattern, as in Figure 10.16. It is not possible to describe here the details of Rietveld refinement. However, it is worth noting that structure refinement against PXRD data is far more difficult and far less robust than refinement against single-crystal data, principally because of the information that is lost on averaging of the PXRD pattern.

10.6 Summary of Key Points

- Powder diffraction can be viewed as a collection of simultaneous single-crystal measurements from a large number of crystals in all orientations. The 3-D diffraction pattern is averaged into a pattern that has effectively only one spatial dimension, which is the Bragg angle.
- Due to averaging of the crystal orientations, the diffracted intensity from a powder sample is distributed around cones with half-angle 2θ, which appear as concentric rings on a 2-D detector. A representative plot of intensity *vs.* 2θ is measured by scanning along a radius of the rings.
- The position of a peak in a PXRD pattern refers to its 2θ value for the specified X-ray wavelength. All diffracted beams with the same 2θ value overlap in the PXRD pattern.
- The intensity of a peak in a PXRD pattern has contributions from numerous overlapping diffracted beams. The inherent reason for peak overlap is crystal symmetry, but coincidental overlap also becomes significant at higher 2θ.

- The shape of a peak in a PXRD pattern has contributions from the instrument and the sample. Sample contributions include crystallite size and strain. Peak profiles can be modelled using a combination of analytical functions intended to describe these physical effects.
- If the crystallite orientations in the powder sample are not uniformly averaged (*i.e.* the sample exhibits preferred orientation), the diffracted intensity will not be distributed uniformly around each powder ring and the relative intensities measured by a radial scan may not be representative. It is difficult to eliminate preferred orientation entirely, so the intensities in a measured PXRD pattern are generally less reliable than the peak positions.
- A PXRD pattern provides a "fingerprint" of a crystalline solid. Solid-form identification by matching PXRD patterns is probably the most common application of PXRD in the pharmaceutical sciences.

10.7 Case Study: Celecoxib

Celecoxib (Figure 10.17) is a non-steroidal anti-inflammatory drug (NSAID). At the time of writing, the compound has one crystal structure in the Cambridge Structural Database (CSD),[1] but three polymorphic forms are mentioned in the patent literature.[2] The polymorphs are illustrated in the patent by the PXRD patterns shown in Figure 10.18. A PXRD pattern simulated from the crystal structure in the CSD (space group $P\bar{1}$) shows a very good match to form III. There is evidence of some minor preferred orientation in the measured pattern, *e.g.* the pairs of peaks at $2\theta = 10.7, 11.0°$ and $14.9, 16.1°$ both have relative intensities opposite to the expectations from the simulated pattern, but there is no doubt that the measured pattern for form III matches the known crystal structure.

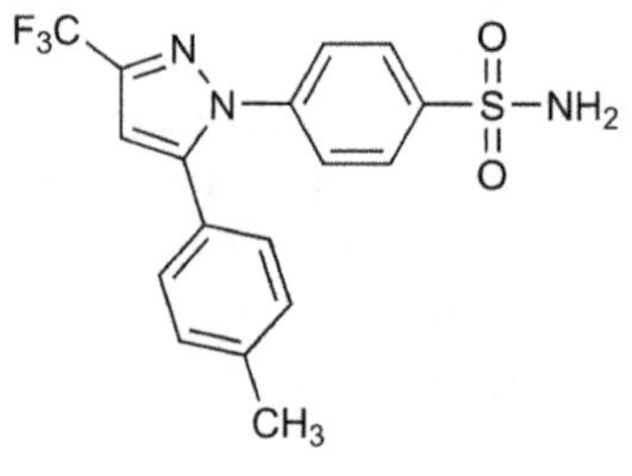

Figure 10.17   Chemical structure of celecoxib, $C_{17}H_{14}F_3N_3O_2S$.

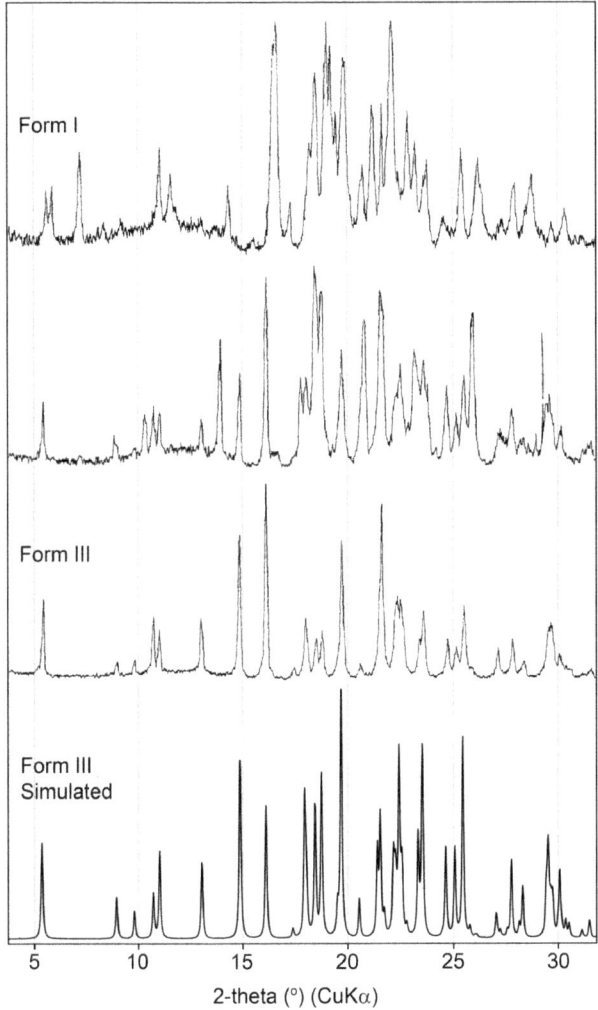

Figure 10.18 PXRD patterns of celecoxib polymorphs reproduced from US7476744B2. The gridlines have been added to clarify the discussion. The bottom pattern is simulated from the crystal structure of form III.[1]

At the top of Figure 10.18, the PXRD pattern of form I is clearly different from form III. The pattern has a noisier baseline and the peaks appear to be broader than those of form III (probably indicating smaller crystallites), but forms I and III are clearly different polymorphs. The central measured pattern in Figure 10.18 is described as a mixture of forms II and III. It is interesting to examine this more closely. The overall shape of the central pattern is comparable to that of form III. Particularly its appearance at low angle (where peaks are generally separated and hence easiest to distinguish) is very similar to form III. The reason to describe the pattern as a mixture is the presence of some additional peaks, specifically at $2\theta \approx 10.3$ and $13.8°$. For mixtures of crystalline forms, the PXRD pattern will be the sum of the patterns for the two forms, weighted according to their ratio in the bulk sample. If we imagine subtracting a suitably scaled pattern of form III from the central pattern, the result will show very few peaks to characterise form II. In the region above *ca.* $20°$, it is difficult to see anything clearly due to the extensive peak overlap. There are differences, but it is hard to conclude much from this region. At lower angle, only the peaks at $2\theta \approx 10.3$ and $13.8°$ will remain for form II after subtraction of the form III peaks. Perhaps there is a hint of a peak at $2\theta \approx 7.2°$, but this actually corresponds to a prominent peak in form I. The apparent shoulder on the peak around $16.6°$ also corresponds to a strong peak in form I, so it seems possible that the central pattern could actually contain some quantity of form I. That aside, the only clear low-angle peaks in the central pattern that do not correspond to form III are at $2\theta \approx 10.3$ and $13.8°$. If the first peak in the form II pattern was genuinely at $2\theta \approx 10.3°$, it would indicate a unit cell significantly smaller than forms I and III. This seems suspicious. An alternative explanation could be that the PXRD pattern of form II contains lines that are coincident with form III. Particularly the line at $2\theta \approx 5.4°$ seems a candidate for both form II and form III on the grounds of the implications for the unit-cell size. In that case (and ignoring the possible minor contamination of form I), might the central pattern actually represent pure form II? It is not unusual for polymorphs to have substantially similar PXRD patterns if the crystal structures are similar. It could be that the PXRD patterns of forms II and III are closely comparable, and the difference is revealed by just a few specific peaks. To clarify this, it would of course be preferable to have crystal structures fully determined, particularly when PXRD data is thought to represent a mixture.

It should be stressed that this discussion considers only the presented PXRD data and is specifically intended to provoke thought. It could well be that the interpretation as a mixture of forms II and III is correct. In the absence of a published crystal structure for form II, however, the PXRD data leave room for doubt and illustrate the care that must be taken when a PXRD pattern does not immediately match a known crystal structure. Incidentally, it is conventional in the patent literature to claim polymorphic forms by listing the positions of a selected number of characteristic peaks in the PXRD pattern. The experimental description in the celecoxib patent contains the following: "*As illustrated in* [Figure 10.18]*, the three forms were easily distinguishable by PXRD. Using a Cu X-ray source (1.54 nm), the characteristic diffractions were observed at 2-theta values of 5.5°, 5.7°, 7.2° and 16.6° for Form I celecoxib and about 10.3°, 13.8° and 17.7° for Form II celecoxib.*" The attentive reader may spot a mistake: the wavelength of CuKα radiation is 1.54 Å, a factor of 10 smaller than the quoted 1.54 nm. Obviously, the claimed peak positions would not be obtained using the quoted wavelength. We must leave it to the lawyers to discuss how that affects the patent.

References

1. R. Vasu Dev, K. Shashi Rekha, K. Vyas, S. B. Mohanti, R. Rajender Kuma and G. Om Reddy, *Acta Crystallogr., Sect. C: Struct. Chem.*, 1999, **C55**, IUC9900161, [CSD: DIBBUL].
2. L. J. Ferro and P. S. Miyake, *U. S. Pat.* US7476744B2 and *Eur. Pat.* ES2236011T3.

11 Solving X-ray Crystal Structures

Summary

Solving a crystal structure means generating an image of the electron density from the measured structure factors. This reverses the process described for generating an X-ray diffraction pattern from a crystal structure. The required mathematics—the Fourier summation—is well defined, but it is necessary to know both the amplitude and phase of each structure factor. The experimental measurement yields the diffracted intensities, from which only the structure factor amplitudes can be extracted. Solving the structure means recovering the lost phase information. For most small-molecule pharmaceutical analyses, structure solution is achieved by direct methods or dual-space methods, both of which are automated and highly effective. If a partial structure is obtained, it can be developed by a Fourier summation using the measured structure factor amplitudes and the phases calculated from the current structure model. The resulting image of the electron density should reveal new atoms that can be added to the model, then the process can be iterated until it converges on the complete structure.

11.1 Introduction

Chapter 6 described how an X-ray diffraction pattern is produced from a crystal structure. Solving a crystal structure means reversing the process to produce the structure from a measured diffraction pattern. We have already seen how to handle the geometry, by constructing the reciprocal lattice from the measured diffraction angles

Pharmaceutical Crystallography: A Guide to Structure and Analysis
By Andrew Bond
© Andrew Bond 2019
Published by the Royal Society of Chemistry, www.rsc.org

then converting the reciprocal lattice to the real lattice. That task is completed during the experimental analysis at the end of the data integration step. The challenge now is to convert the list of measured intensities back to the set of atoms that produced them. Chapter 6 described how the amplitudes and relative phases of the diffracted beams result from interference between X-rays scattered from all atoms in the unit cell. In Section 6.4.3, the structure factor equation was developed to describe this. The mathematics to reverse the process is very similar. It will become clear that "structure solution" amounts to recovering information that cannot be measured, namely the *phases* of the diffracted beams.

11.2   Building the Electron Density From the Structure Factors

In mathematical language, the structure factor equation is a *Fourier transform*. Any Fourier transform has an associated inverse, which in this case describes how the diffracted beams can be recombined to produce the electron density within the unit cell. The required expression is:

$$\rho(xyz) = \frac{1}{V_{\text{cell}}} \sum_{hkl} F(hkl) \exp\{-2\pi i(hx + ky + lz)\}$$

where $\rho(xyz)$ is the value of the electron density at position x, y, z and the summation runs over all of the structure factors, $F(hkl)$. Since structure factors have units of electrons (Section 6.4.3), the $1/V_{\text{cell}}$ multiplier produces units of electrons per unit volume, *i.e. electron density*. The resemblance to the structure factor equation should be clear:

$$F(hkl) = \sum_{n=1}^{N} f_n \exp\{2\pi i(hx_n + ky_n + lz_n)\}$$

We understand that the structure factor equation represents a summation of waves with (relative) phase $2\pi(hx + ky + lz)$ along the direction specified by hkl. This is a direct interpretation of the physical picture of the interference of scattered X-rays. The atomic scattering factor f_n is a scalar multiplier that encapsulates the scattering behaviour of each atom along the specified direction. It is straightforward to picture this summation on an Argand diagram (as in Figure 6.12).

By analogy, $\rho(xyz)$ is constructed from a summation of waves with (relative) phase $-2\pi(hx + ky + lz)$. In this case, however, the structure factors $F(hkl)$ are *complex numbers*. How can we visualise that? We can start by writing the structure factor in terms of its amplitude, $|F(hkl)|$, and phase, $\Phi(hkl)$:

$$\rho(xyz) = \frac{1}{V_{cell}} \sum_{hkl} |F(hkl)| \cdot \exp\{i\Phi(hkl)\} \cdot \exp\{-2\pi i(hx + ky + lz)\}$$

This can be rearranged by noting that multiplication of two exp terms gives a single exp term in which the exponents are summed:

$$\rho(xyz) = \frac{1}{V_{cell}} \sum_{hkl} |F(hkl)| \exp\{i(\Phi(hkl) - 2\pi(hx + ky + lz))\}$$

Now we can visualise what this means: the summation can be viewed on an Argand diagram as head-to-tail summation of a set of vectors, one for each structure factor, with length $|F(hkl)|$ and phase $\{\Phi(hkl) - 2\pi(hx + ky + lz)\}$ (Figure 11.1). The summation runs over all positive and negative values of h, k and l, including 000 (which is just a positive real number). A separate summation is required for each coordinate xyz, producing a resultant value of $\rho(xyz)$. We should expect the result to be a real number equal to the number of electrons per unit volume.

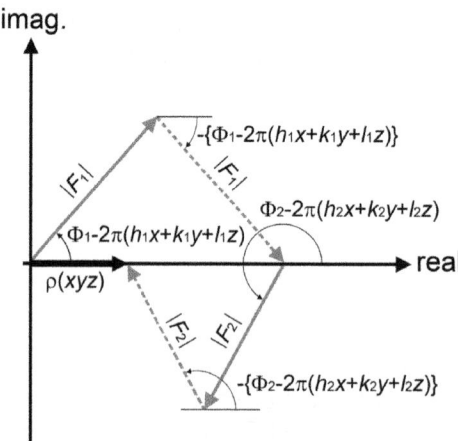

Figure 11.1 Summation of the structure factors on the Argand diagram to produce a value for the electron density, $\rho(xyz)$. The summation applies to the specific coordinate xyz and runs over all structure factors hkl. Only two pairs of structure factors are shown, with the solid and dashed lines representing complex conjugates (hkl and $\bar{hkl}$). The result is a real number.

Now we can move towards a more physical picture by making a few further mathematical manipulations. Firstly, the value calculated for $\rho(xyz)$ can be confirmed to be a real number by noting that the structure factors $F(hkl)$ and $F(\bar{h}\bar{k}\bar{l})$ are complex conjugates, *i.e.* $|F(hkl)| = |F(\bar{h}\bar{k}\bar{l})|$ and $\Phi(hkl) = -\Phi(\bar{h}\bar{k}\bar{l})$. This causes all components along the imaginary axis of the Argand diagram to cancel when summing over all structure factors because $\sin(\Phi) = -\sin(-\Phi)$. Using Euler's notation $(\exp\{i\Phi\} = \cos\Phi + i\sin\Phi)$ and noting that all $i\sin\Phi$ terms cancel in this way; the expression can be written as a sum of cosines:

$$\rho(xyz) = \frac{1}{V_{\text{cell}}} \sum_{hkl} |F(hkl)| \cos\{\Phi(hkl) - 2\pi(hx + ky + lz)\}$$

where the summation is still over all of the structure factors (including 000) at each specific xyz. Since $\cos(\Phi) = \cos(-\Phi)$, this can be written equivalently with $\Phi(hkl)$ and $2\pi(hx + ky + lz)$ exchanged:

$$\rho(xyz) = \frac{1}{V_{\text{cell}}} \sum_{hkl} |F(hkl)| \cos\{2\pi(hx + ky + lz) - \Phi(hkl)\}$$

This is effectively all that we need: it gives a numerical recipe to calculate the value of $\rho(xyz)$ at any position within the unit cell by summing over the set of structure factors. In reference to its mathematical origin, this is called the *Fourier summation* (or *Fourier synthesis*). The resulting 3-D electron density map is commonly called the *Fourier map*.

11.2.1 Visualising the Fourier Summation

The cosine terms that are summed in the Fourier summation do not represent X-rays. They are just cosine functions that add together to produce a value for the electron density at each point within the unit cell. The functions are 3-D (*i.e.* the value depends on the three co-ordinates, *xyz*), with periodicity defined within the volume of the unit cell. Probably more familiar is a 1-D cosine function, which has an oscillating value that varies with one coordinate (*x*) along a line (Figure 11.2(a)). The "wave front" of a 1-D cosine is a point that progresses along the line. By extension, a 2-D cosine function has an oscillating value that varies with two coordinates (*xy*) within a plane (Figure 11.2(b)). For a 2-D cosine, the wave front is a line that progresses across the plane. For a 3-D cosine, the wave front is a plane that progresses through the unit cell (Figure 11.2(c)). In the Fourier

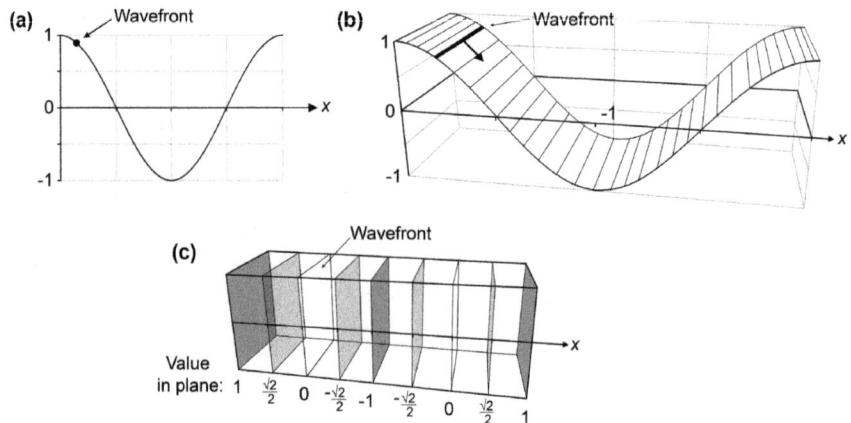

Figure 11.2 Representations of the cosine function: (a) a 1-D cosine takes an oscillating value at each point along a line; (b) a 2-D cosine takes an oscillating value in each line within a plane; (c) a 3-D cosine takes an oscillating value in each plane within a volume.

summation, the direction of each cosine term is the normal to the set of planes denoted by (hkl) (*i.e.* the direction is parallel to the reciprocal lattice vector hkl). The number of repeat cycles within the volume of the unit cell is specified by $2\pi(hx + ky + lz)$. Since the coordinates x, y and z span the range 0 to 1, the products hx, ky and lz span the range 0 to h, 0 to k and 0 to l, respectively, so the sum spans the range 0 to $(h + k + l)$. This means that there are $h + k + l$ planes for which $2\pi(hx + ky + lz)$ will be an integral multiple of 2π, so there are $h + k + l$ repeat cycles of the cosine function per unit cell.

For example, the 001 term can be viewed as a planar wave front oriented parallel to the (001) planes (Figure 11.3(a)). It has one cosine cycle within the unit cell because lz spans the range 0 to 1. If the function has its maximum value at $z = 0$, it has its minimum value at $z = \frac{1}{2}$ and its next maximum at $z = 1$ (equivalent to $z = 0$ by translational symmetry). The 002 term also has its planar wave front oriented in the same way, but with two cosine cycles within the unit cell because lz spans the range 0 to 2 (Figure 11.3(b)). Hence, it has maxima/minima at $z = 0$, $\frac{1}{4}$, $\frac{1}{2}$, $\frac{3}{4}$. The 111 term has its planar wave front perpendicular to the body diagonal of the unit cell, with three cosine cycles per cell because $hx + ky + lz$ spans the range 0 to 3 (Figure 11.3(c)). Visualising the summation of a large set of these 3-D cosine terms is not so easy, and ultimately the calculation can just be considered as a mathematical recipe that gives a value for $\rho(xyz)$ at any given point. Nonetheless, it is useful to have this physical picture of what the Fourier summation represents.

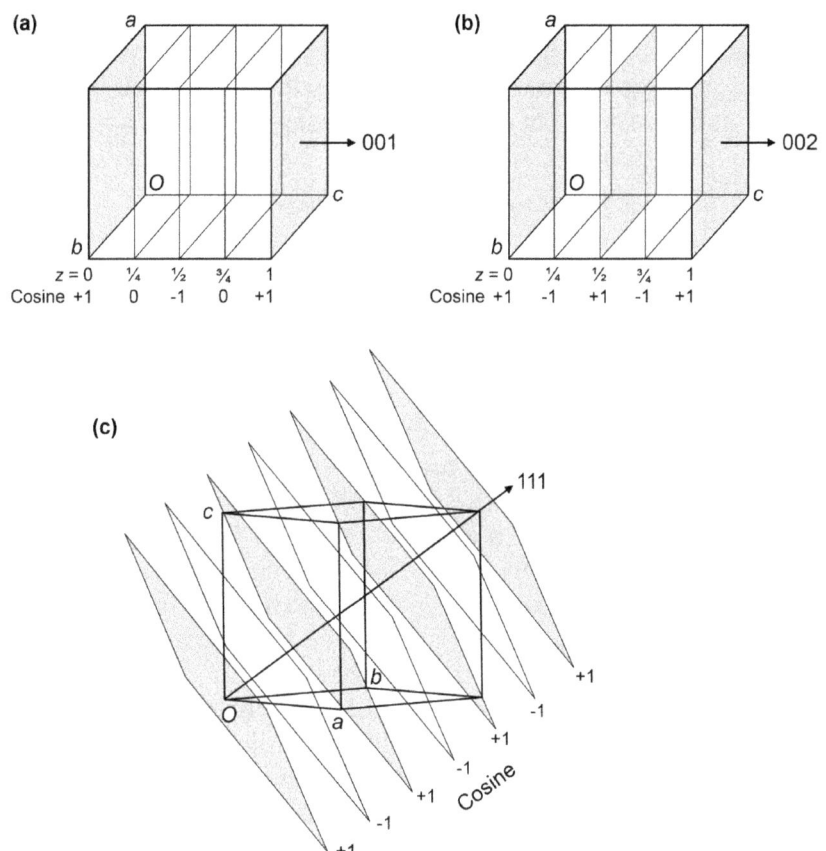

Figure 11.3 3-D cosine terms defined within the unit cell: (a) 001; (b) 002; (c) 111. The value of the cosine within the illustrated planes is shown. In the Fourier summation, this would be multiplied by $|F(hkl)|$.

11.2.2 What If Friedel's Law Doesn't Apply?

The discussion in Section 11.2 asserted that Friedel's law must apply, *i.e.* $F(hkl)$ and $F(\bar{h}\bar{k}\bar{l})$ must be complex conjugates, because the calculated electron density has to be real. But Section 7.3.1 described how Friedel's law does not apply under conditions of significant anomalous scattering, when significant X-ray absorption occurs. In that case, the Fourier summation will produce a complex number. This does not mean that the electron density becomes complex. It means that the mathematical description used to account for both elastic scattering and absorption (Section 7.3.1) produces a complex number for $\rho(xyz)$ when the inverse Fourier transform is applied to the $F(hkl)$ values encompassing both effects. To reveal the true value for $\rho(xyz)$, which must be real, the $F(hkl)$ values can be corrected so that the Fourier summation is applied only to the part of $F(hkl)$ that describes the elastic scattering. As ever, we can rest assured that this is handled by crystallographic software and that any calculated map of $\rho(xyz)$ is appropriate.

11.2.3 Electron Density Maps and Resolution

Compared to the structure factor equation, the Fourier summation has a crucial difference. Structure factors are calculated by summing over the atoms in the unit cell. Since the number of atoms in the unit cell is finite, the calculated value of $F(hkl)$ is exact for a given structure model. When constructing the electron density, however, the summation must be carried out over all structure factors which in principle are infinite in number. In practice, we can only measure a finite number of $F(hkl)$ values, so the obtained $\rho(xyz)$ can only ever be an approximation. The errors that arise in $\rho(xyz)$ from having a finite number of structure factors are called *series termination errors*. The main effect of such errors is to produce "ripples" in the electron density around each atomic position, which become more significant for atoms with a greater number of electrons.

Considering how the Fourier summation builds an image of $\rho(xyz)$, the cosine terms with the fewest cycles per unit cell contribute only broad features to the image. To add finer details requires the cosine terms to oscillate more frequently within the unit cell. These are the terms with larger values of hkl, which correspond to the higher-angle diffracted beams. Hence, the resolution of the image of $\rho(xyz)$ is progressively increased by including structure factors measured at higher diffraction angles. An example of $\rho(xyz)$ for a paracetamol molecule at 0.85 and 1.50 Å resolution was shown in Figure 9.2. As mentioned in Chapter 9, the resolution of a crystal structure is usually described by noting the d-spacing of the highest-angle diffracted beams included in the Fourier summation. Now it should be clearer what this means. A typical resolution for a small-molecule crystal structure is 0.85 Å. Lower-resolution images may still be valuable in cases where the diffraction data are too weak to be measured at higher diffraction angles, but the quality of the 3-D electron-density map and any conclusions drawn from it will be compromised.

11.2.4 Where Are the Atoms?

To describe a crystal structure requires the fractional coordinates of the atoms within the unit cell. It is not possible, however, just to calculate $\rho(xyz)$ at each atomic position because it is not yet known where the atoms are. Instead, it is necessary to reconstruct a complete image of $\rho(xyz)$ within the unit cell with the expectation that it will reveal concentrated regions of electron density that can be identified as atoms. Practically, it is necessary to calculate $\rho(xyz)$ on a grid of specified *xyz* values. A sufficient number of grid points is required to identify the positions of maxima in the electron density, but there is no reason to define an extremely fine grid to describe an image whose resolution is ultimately limited by a restricted number of structure factors. An appropriate grid size will be taken care of by the software used to make the calculation. The positions of atoms are then identified as the centroids of concentrated regions in the electron density (Figure 11.4). The "peak-searching" exercise is carried out automatically by crystallographic software, and the user is usually presented with a graphical representation of the derived peak positions. It is vital to appreciate that this is an *interpretation* of the electron density map, which may conceal informative features of $\rho(xyz)$. For example, a peak position may be derived from a spherical region in $\rho(xyz)$, as expected for a well-defined atom, but it might also be derived from some elongated region that could indicate significant thermal motion or disorder (Figure 11.5). The derived atom position may be the same in the two cases, but the electron density tells a different story. This will be considered further in Chapter 12.

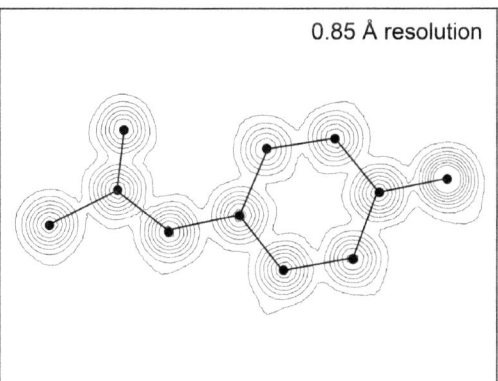

Figure 11.4 Defining the positions of atoms as the centroid of concentrated regions of electron density. A 2-D slice is illustrated through the plane of the paracetamol molecule. This image is identical to Figure 9.2(a), but with the positions of the atoms indicated.

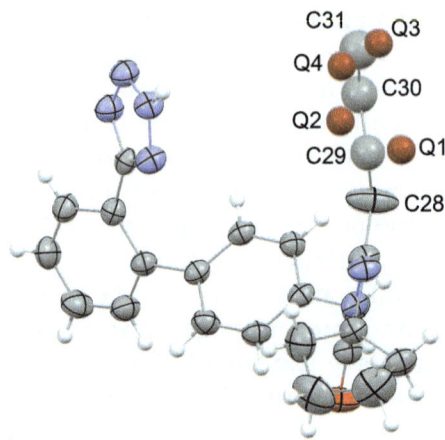

Figure 11.5 Additional peaks in the electron density (labelled Q1–Q4) around the *n*-butyl chain in form B of irbesartan (using data from ref. 1). In this case, the single set of atomic positions are not fully representative of the underlying electron density in this region.

11.3 The Phase Problem

Although the mathematics to calculate $\rho(xyz)$ is well defined, there is a problem. The Fourier summation uses the structure factors, which comprise both an amplitude, $|F(hkl)|$, and a phase, $\Phi(hkl)$. Experimental measurement of a diffraction pattern provides only the *intensities* of the diffracted beams. The value of $|F(hkl)|$ can be obtained as the square root of the measured intensity (after suitable corrections and scaling), but $\Phi(hkl)$ cannot be obtained. Hence, solving a crystal structure from a diffraction pattern requires identification of a set of $\Phi(hkl)$ values that produce the correct image of the electron density *via* the Fourier summation. Finding values for $\Phi(hkl)$ is a massively complex problem. Even for a centrosymmetric structure, where $\Phi(hkl)$ can only take the values 0 or π (Section 7.3), there are 2^N possible combinations for N structure factors. If N is in the thousands, there are far too many combinations to consider an exhaustive search. However, the problem can be overcome. Broadly speaking, two approaches can be identified as most relevant for small-molecule pharmaceutical crystallography.

11.3.1 Direct Methods

A general approach to solving the phase problem is *direct methods*, which works by considering how the structure factors contribute to the Fourier summation. The physical basis is that we know what a reasonable electron density should look like: it should be greater than or equal to zero at all positions and it should be gathered into discrete concentrated regions corresponding to atoms. These expectations can be used to deduce probable relationships between the phases of specific structure factors. For example, consider $F(001)$ and $F(002)$ (Figure 11.6). These contribute cosine terms to the Fourier summation that run parallel to the (001) planes and have one and two cycles, respectively, within the unit cell. For the purposes of this illustration, assume that the structure is centrosymmetric, so $\Phi(hkl)$ must equal 0 or π, and that $F(001)$ has its maximum at the origin. If both $F(001)$ and $F(002)$ have large amplitude they must have a significant influence on the final value for $\rho(xyz)$. Since we know that $\rho(xyz)$ should not be negative, it is not likely that $F(001)$ and $F(002)$ should combine to produce large negative regions. If $F(001)$ has its maximum at the origin, it has a large negative value at $z = \frac{1}{2}$. To counter this, $F(002)$ would be expected to have its maximum at $z = \frac{1}{2}$. This means that $\Phi(002) \approx 0$, where "$\approx$" means "probably equal to". If $F(001)$ were instead to have its minimum at the origin, $F(002)$ would be expected to counter that with its maximum at the origin, so again $\Phi(002) \approx 0$. The conclusion is that we should always expect $\Phi(002) \approx 0$. This is an example of a *pair relationship*: whenever there are two strong diffracted beams with indices h,k,l and $2h,2k,2l$ for a

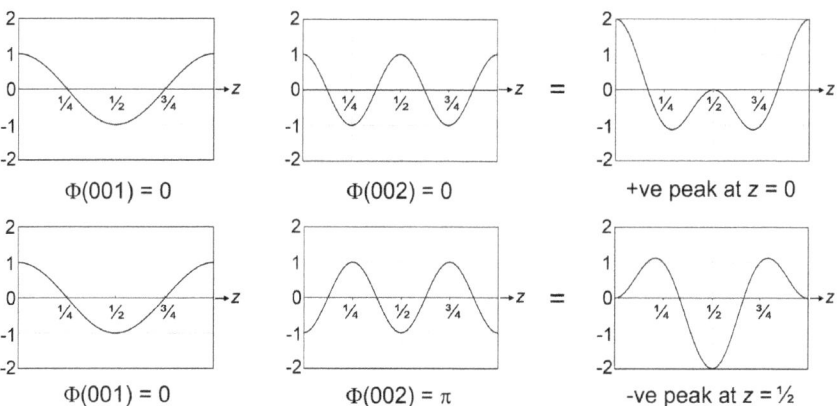

Figure 11.6 Direct methods: a pair relationship illustrated for $F(001)$ and $F(002)$. The plots show the values of the cosine terms at the specified z coordinates. If $F(001)$ has its maximum at the origin, $F(002)$ must have its maximum at the origin to avoid significant negative regions in the resulting sum. If $F(001)$ were to have its minimum at the origin (not shown), $F(002)$ would still have to have its maximum at the origin to avoid producing a large negative peak at $z = 0$.

centrosymmetric structure, $\Phi(2h,2k,2l) \approx 0$. The probability that the expected phase is correct increases as the amplitude increases for both structure factors.

Similar *triplet relationships* can be deduced for sets of three related structure factors and *quartet relationships* for sets of four. The concepts are equally applicable to non-centrosymmetric structures, where the expected phase relationships are expressed more generally as sums. For a measured single-crystal data set, an extensive set of relationships can be constructed between the various structure factors, which enables phases to be assigned in a "trickle-down" manner from only a few starting phases. Setting up the relationships and assigning the phases is ideally suited to a computer algorithm so the entire process can be automated. The probable phase relationships are more likely to be correct for stronger reflections, so a limited number of the strongest reflections are usually chosen for initial phase assignment. The structure factor amplitudes are "normalised" (dividing through by the sum of the atomic scattering factors) so that higher-angle reflections are also sufficiently strong to participate in the phase relationships. The sequential nature of the assignments makes it highly sensitive to an incorrect choice at an early stage, so the process is usually run many times with different starting phase sets. Each resulting phase set can be applied to make the Fourier summation then the best image of $\rho(xyz)$ must be identified. There are numerous criteria on which to assess the resulting $\rho(xyz)$ but in principle the best result is the one that looks most like a correct crystal structure, *i.e.* concentrated regions of electron density with chemically-sensible distances between these regions, no negative regions, *etc.*

11.3.2 Dual-space Methods

"Dual-space" methods refer broadly to a group of structure-solution methods that iterate between information in real space (the electron density) and reciprocal space (the structure factors). Like direct methods, they make use of the expectation that $\rho(xyz)$ should not be negative and should be gathered into atoms. An illustrative variant is the "charge flipping" algorithm (Figure 11.7). $\rho(xyz)$ is calculated using the measured $|F(hkl)|$ values and a current set of phases, then the result is modified so that any negative value of $\rho(xyz)$ is multiplied by -1. The modified all-positive $\rho(xyz)$ is used to calculate new phases and the process is iterated until the procedure converges on

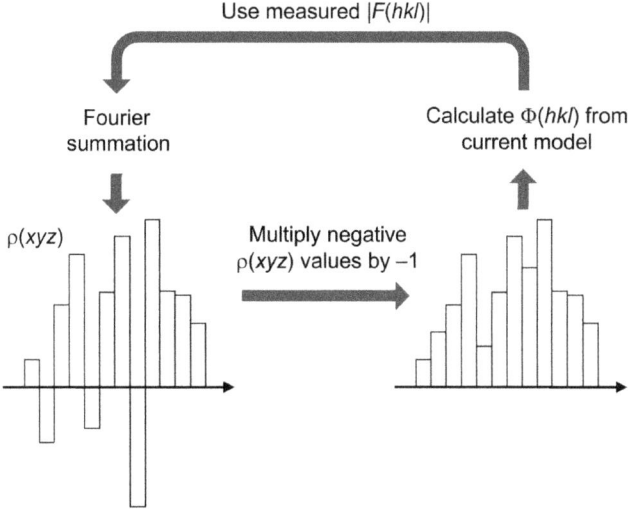

Figure 11.7 Schematic representation of a dual-space recycling algorithm. In the very first step, the phases could be essentially chosen at random or a more sophisticated algorithm can be applied. A 1-D grid of $\rho(xyz)$ is shown for simplicity; a real structure has values on a 3-D grid.

an all-positive result for $\rho(xyz)$. The current $\rho(xyz)$ is used only to calculate the phases, which are then combined with the *measured* $|F(hkl)|$ values to provide the modified $\rho(xyz)$; this crucial concept will be discussed in more detail in Section 11.4.

At the beginning of these iterative algorithms, the "current set of phases" can be chosen at random or some more sophisticated procedure can be applied to produce a starting point. Like direct methods, the iterative dual-space algorithms are highly likely to reach "false minima", where they converge on an incorrect solution. The chances of reaching a correct solution can be increased by making various modifications to $\rho(xyz)$ along the way. One is based on the concept of "atomicity", whereby $\rho(xyz)$ is scanned for maxima, then a set of spherical functions is defined at the positions of those maxima. $\rho(xyz)$ is multiplied by this "atomic mask" to force it to take on the expected atomic form (Figure 11.8) before re-calculating the phases. This turns out to be highly effective for properly "atomic" electron densities, but it might be less effective if there is considerable atomic displacement or disorder. Like direct methods, it is necessary for dual-space algorithms to make many attempts to identify the phase set that produces the best structure.

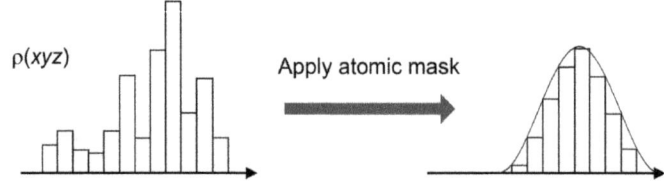

Figure 11.8   Imposing an "atomic mask" on the electron density. This step can be applied to modify the structure model as part of a dual-space recycling procedure.

11.4   Structure Solution in Practice

In practice, the average user can (and probably does) view structure-solution as a "black box". The brief descriptions given here hopefully provide some insight into how the algorithms work, but the details are not so important in a practical context. What is important is to realise what is required for success. Clearly, methods based on expectations of how $\rho(xyz)$ should look require sufficient data to produce an adequate image of $\rho(xyz)$. This means that data must be collected to adequate resolution. A typical guideline for direct methods or dual-space recycling is that data should be observed to at least 1 Å. Lower-resolution images of the electron density can cause problems because "atomicity" might not be so clearly defined and series termination errors might produce negative regions of electron density even for the correct phase set, especially in the presence of heavy atoms. Problems can also arise if the unit cell is too large, since there will be a very large number of structure factors contributing to the Fourier summation. For direct methods, this decreases the probability that any particular phase relationship will hold. If the probabilities become too low, the process is little better than a random phase assignment. This particularly affects macromolecular crystallography. For typical small-molecule pharmaceutical structures with data to 1 Å resolution or better, both dual-space recycling and direct methods are usually highly effective. Which one to choose will probably depend on practical preferences (*e.g.* how the algorithms are incorporated in the local data collection and processing packages). If both dual-space recycling and direct methods algorithms fail for a given data set, it is almost always indicative of some problem with the crystal. The source of the problem should be sought in the primary diffraction images. For most "standard" structure determinations carried out on an adequate single crystal, modern programs for structure solution should produce a more-or-less complete image of the structure with little or no user intervention.

11.4.1 Choosing a Space Group for Structure Solution

We saw in Chapter 7 that the space group of a structure can be identified from the diffraction pattern by considering the point group of the intensity-weighted reciprocal lattice and any systematic absence conditions. For direct methods, it is usually required to specify the space group during structure solution so that it is possible to set up relationships between symmetry-related phases. For dual-space recycling methods, it is common to run the initial phasing process on the entire unit-cell contents, without assuming any symmetry relationships. The resulting set of structure factors can then be examined to identify the most probable space group(s). When using such methods, the user should keep an eye on which space groups are actually being considered since the search will probably be restricted to groups that are compatible with the apparent Laue group (Table 7.1). For the triclinic, monoclinic and orthorhombic crystal systems, which are most commonly observed for pharmaceutical structures, there is only one choice of Laue group, so this is less likely to be a problem. However, structures in higher-symmetry crystal systems may need to be investigated in several Laue groups. If structure solution should fail, it should be considered whether all potential Laue groups have been considered. For example, a data set that fails to yield a solution in Laue group mmm might actually be monoclinic (Laue group $2/m$) with a lattice that just happens to have $\beta \approx 90°$. The "last resort" is usually to run the structure solution algorithms in space group $P1$ or specifying Laue group $\bar{1}$. This may provide an image containing several recognisable molecules that can often be updated to a more appropriate description in a higher-symmetry space group.

11.5 Developing the Structure Model

The first efforts to solve a structure will often provide most, but perhaps not all atomic positions. The next step is to consider how a partial model can be developed into a complete structure. This impacts every stage from structure solution to the completed refinement so it is vital to understand the concept. It is summarised in Figure 11.9. The overall aim is to produce a complete image of $\rho(xyz)$

Measured $|F(hkl)|_{obs}$

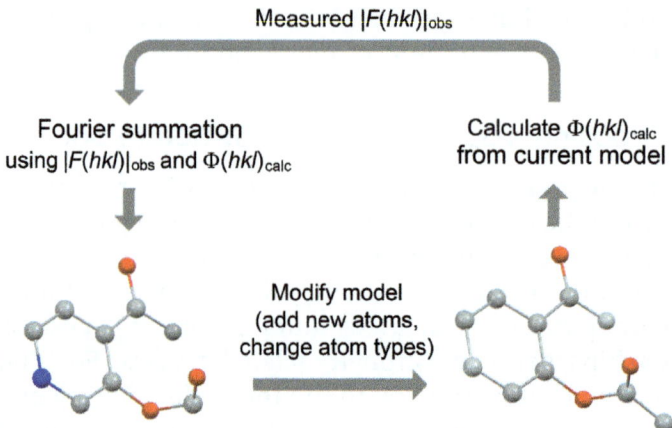

Fourier summation
using $|F(hkl)|_{obs}$ and $\Phi(hkl)_{calc}$

Calculate $\Phi(hkl)_{calc}$
from current model

Modify model
(add new atoms,
change atom types)

Figure 11.9   Schematic illustration of the procedure to develop a structure model.
The key step is to calculate a Fourier map using the measured
$|F(hkl)|_{obs}$ values and the $\Phi(hkl)_{calc}$ values calculated from the current
structure model.

by making a Fourier summation using the correct values of $|F(hkl)|$
and $\Phi(hkl)$. The correct $|F(hkl)|$ values are known because they are
obtained from the measured intensities; they are labelled $|F(hkl)|_{obs}$
("obs" = observed). Values for $\Phi(hkl)$ cannot be measured. For any
partial structure model, $|F(hkl)|_{calc}$ and $\Phi(hkl)_{calc}$ can be calculated
using the structure factor equation. If a Fourier summation is made
using $|F(hkl)|_{calc}$ and $\Phi(hkl)_{calc}$, the result will be an exact repro-
duction of the partial structure model used to calculate those values
(that is what the inverse Fourier transform does), so no progress
would be made by going around this circuitous route. Crucially,
however, a Fourier summation using $|F(hkl)|_{obs}$ and $\Phi(hkl)_{calc}$ should
produce an image of $\rho(xyz)$ that is closer to the correct structure be-
cause the correct $|F(hkl)|_{obs}$ values have been introduced. Hopefully,
it will be possible to identify new atoms within the resulting electron
density that can be added to the partial structure model, then the
process can be iterated. As the structure develops, the $\Phi(hkl)_{calc}$ values
calculated from the current structure model will gradually approach
the correct $\Phi(hkl)$ values. At some point, the structure model will be
complete, which means that the $\Phi(hkl)_{calc}$ values will be correct.
A Fourier summation using $|F(hkl)|_{obs}$ and $\Phi(hkl)_{calc}$ will then simply
reproduce the correct structure, so the iterative process converges on
the complete solution.

   In practice, the procedure can be adjusted in various ways to
produce maps that emphasise different features. For example, the
difference density, $\Delta\rho(xyz)$, is produced from a Fourier summation

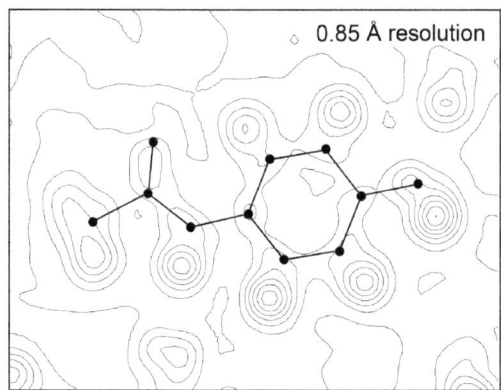

Figure 11.10 Difference density, $\Delta\rho(xyz)$, within the plane of a paracetamol molecule with H atoms not yet included in the model. The positions of all H atoms in the molecule are clearly visible. Contours are shown in the range $0–0.8$ e $Å^3$ in steps of 0.1 e $Å^{-3}$.

using $\{|F(hkl)|_{obs} - |F(hkl)|_{calc}\}$ with $\Phi(hkl)_{calc}$ (Figure 11.10). Peaks with positive $\Delta\rho(xyz)$ represent electron density not yet accounted for by the structure model, while "holes" with negative $\Delta\rho(xyz)$ appear where the model contains too much electron density. When a structure model is complete, there should be no significant peaks or holes remaining in $\Delta\rho(xyz)$. For a perfect scenario, $\Delta\rho(xyz)$ would be zero throughout the unit cell. In practice, series termination and other measurement errors mean that $\Delta\rho(xyz)$ oscillates around zero. A correct and complete structure should have average $\Delta\rho(xyz)$ close to zero and absolute values of $\Delta\rho(xyz)$ less than *ca.* 1 e $Å^{-3}$. Series termination errors may lead to more substantial residual electron density close to heavy atoms, but any significant peaks in $\Delta\rho(xyz)$ in the vicinity of lighter atoms may be indicative of an incorrect or incomplete structure. $\Delta\rho(xyz)$ plays a crucial role in structure refinement, to be discussed in Chapter 12.

11.6 Summary of Key Points

- In mathematical language, the structure factor equation is a Fourier transform. Any Fourier transform has an associated inverse, which in this case describes how the diffracted beams can be recombined to produce the electron density within the unit cell.
- The electron density at any point in the unit cell is obtained by making a Fourier summation. In principle, an infinite number

of structure factors is required to reproduce the electron density exactly, so any value obtained for $\rho(xyz)$ must be an approximation.

- The Fourier summation can be visualised as addition of 3-dimensional periodic cosine functions. Each cosine has a planar wave front parallel to the planes specified by (hkl) (perpendicular to the reciprocal lattice vector hkl) and there are $h + k + l$ repeat cycles of the cosine per unit cell.

- The cosine terms with smaller values of hkl contribute broad features to the image of $\rho(xyz)$. Finer details are added by terms with larger hkl, which correspond to higher-angle diffracted beams. The resolution of the image of $\rho(xyz)$ is usually quoted as the smallest d-spacing (highest diffraction angle) that contributes to the image.

- Atomic positions in the crystal structure are identified as the centroids of concentrated regions in the electron density. Lower resolution images of $\rho(xyz)$ can provide a coarse image of a structure, but higher resolution images are required for more accurate atomic positions and meaningful discussion of bond lengths and angles.

- The Fourier summation requires both the amplitudes, $|F(hkl)|$, and phases, $\Phi(hkl)$, of the structure factors. Only $|F(hkl)|$ can be obtained from the measured diffracted intensities, so structure solution amounts to recovering the lost phase information.

- Direct methods and dual-space methods are generally applied to solve small-molecule pharmaceutical structures. Both are based on the physical expectation that the electron density should be greater than or equal to zero at all positions and should be gathered into discrete concentrated regions corresponding to atoms.

- A partial structure model can be developed by making a Fourier summation using the observed $|F(hkl)|$ values and $\Phi(hkl)$ values calculated from the current structure model. This should produce an image of $\rho(xyz)$ that allows missing atoms to be identified. The process can be iterated until it converges on a complete structure.

- The difference density, $\Delta\rho(xyz)$, shows peaks representing electron density that is not yet accounted for by the structure model and holes where the model contains too much electron density. For a complete structure model, $\Delta\rho(xyz)$ should not show any significant peaks or holes.

11.7   Case Study: Sulfathiazole

Sulfathiazole (Figure 11.11) is one of the most studied polymorphic pharmaceutical compounds. To date, crystal structures are known for five polymorphs of the pure API, all of which crystallize in space group $P2_1/c$. The literature is extensive and sometimes confusing, with different authors applying different labelling schemes to the same set of structures. This Case Study considers polymorph IV to illustrate some aspects of structure solution, and particularly Fourier summations. Refinement of the structure is described in the Case Study of Chapter 12. The data for this discussion are taken from ref. 2, which includes careful intensity measurements made for sulfathiazole to 0.46 Å resolution. This goes far beyond the standard expectations and provides an opportunity to compare $\rho(xyz)$ at very high resolution to the commonly applied limit of 0.85 Å.

The measured unit cell at 100 K is primitive with dimensions $a = 10.789$ Å, $b = 8.484$ Å, $c = 11.398$ Å, $\alpha = 90°$, $\beta = 91.64°$, $\gamma = 90°$. Analysis of systematic absences indicates that the space group is $P2_1/n$, which is an alternative setting of space group $P2_1/c$ (see the Case Study in Chapter 4). It has 2_1 screw axes running parallel to the b axis, and perpendicular n-glides. The structure can be solved routinely by running a standard direct methods or dual-space algorithm, so there is little to illustrate in that respect. The structure solution produces a set of phases, $\Phi(hkl)$, to go with the measured $|F(hkl)|$ values obtained as the square root of the measured $I(hkl)$ values. Each $\Phi(hkl)$ must be either 0 or π because the structure is centrosymmetric.

The main point of interest is the resulting Fourier summation, which produces an image of $\rho(xyz)$ using the measured $|F(hkl)|$ values and the established set of phases. The result in the plane of the aminobenzene ring is shown in Figure 11.12. The figure shows $\rho(xyz)$ calculated using data up to the exceptionally high resolution of 0.46 Å, then using data to the more standard resolution of 0.85 Å, and data truncated to the artificially low resolution of 1.20 Å. The images are "contour maps", where points having the same value of $\rho(xyz)$ are

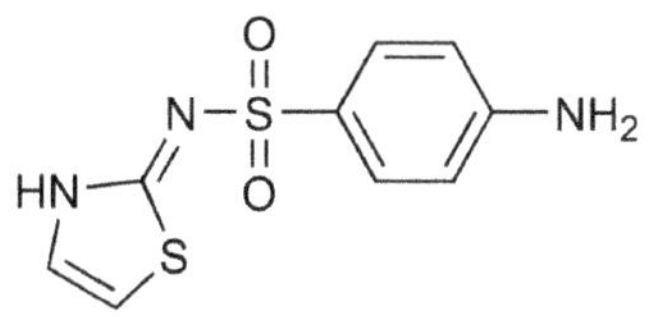

Figure 11.11   Molecular structure of sulfathiazole, $C_9H_9N_3O_2S_2$.

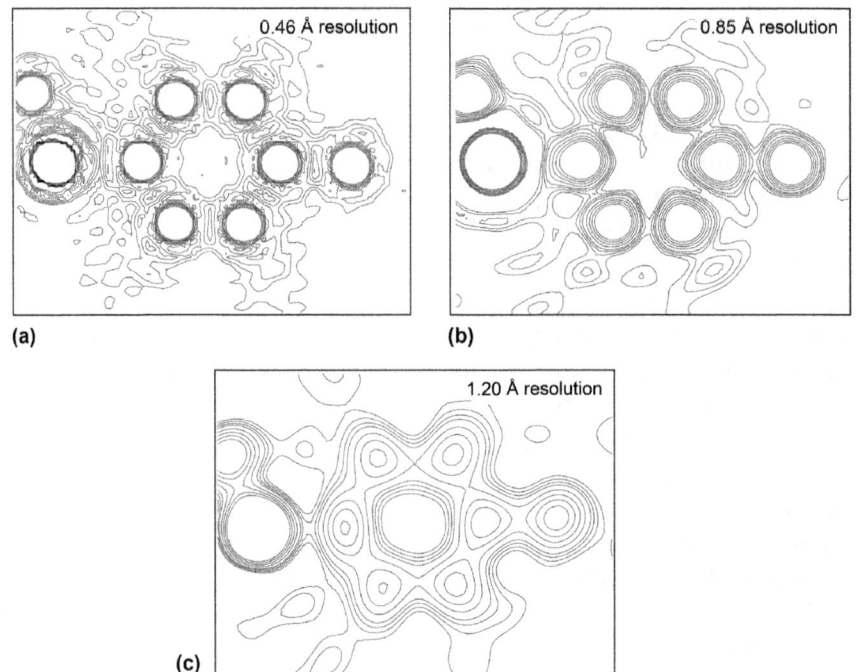

Figure 11.12 Fourier map, $\rho(xyz)$, calculated for sulfathiazole in the plane of the aminobenzene ring at different resolutions. The contours are shown at 0.5, 1, 1.5, 2, 3, 4, 5, 6 e Å^{-3}.

joined. At 0.46 Å resolution, the image contains clearly separate peaks that can be identified as atoms. For each atom, the contours are very close together, indicating steep features emerging prominently from the background. Clear features are also seen in the positions expected for the H atoms on the ring. The H atoms on the NH_2 group are not visible in the diagrams because the geometry around N is pyramidal, so the H atoms project out of the defined plane. This can be viewed as an ideal example of $\rho(xyz)$, and it is obvious how the positions of the atoms should be defined in this projection. By contrast, $\rho(xyz)$ at 1.20 Å resolution has larger spacing between the contours, indicating more gradual slopes. The positions of C atoms of the ring can still be defined, but the peaks merge together into a continuous ring with less well-defined maxima. The shapes of the C atoms bonded to H are distorted towards the positions where the H atoms would be expected. Since the C atoms are defined at the maxima of these features, the derived positions will be shifted away from the positions that would be obtained from the 0.46 Å image. This is an illustration of how the resolution of the data can affect the final atomic model and derived

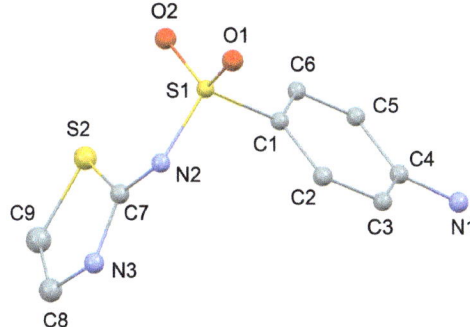

Figure 11.13 Molecular structure of sulfathiazole obtained from a dual-space algorithm with automatic interpretation of the atom types. The size of each atom reflects its isotropic displacement parameter, as discussed in Chapter 12.

bond lengths and angles. The image of $\rho(xyz)$ to the standard resolution of 0.85 Å is of course somewhere between the two extremes: the atoms are separated, and there is only a small hint of distortion towards the H atom positions.

Software for structure solution usually will not display $\rho(xyz)$ directly but will produce an interpreted atomic model. For sulfathiazole, using data to 0.85 Å, a dual-space algorithm with automatic interpretation of the atom types gives an essentially complete picture of the molecule (Figure 11.13). This structure is taken forward for refinement in Chapter 12.

References

1. Z. Bocskei, K. Simon, R. Rao, A. Caron, C. A. Rodger and M. Bauer, *Acta Crystallogr., Sect. C: Struct. Chem.*, 1998, **C54**, 808, [CSD: NOZWII].
2. I. Sovago, M. J. Gutmann, J. Grant Hill, H. M. Senn, L. H. Thomas, C. C. Wilson and L. J. Farugia, *Cryst. Growth Des.*, 2014, **14**, 1227, [CSD refcode family: SUTHAZ].

12 Refining X-ray Crystal Structures

Summary

The purpose of structure refinement is to provide the best fit between the structure model and the measured diffracted intensities. A set of measured diffraction data produces a list of structure factor equations, one for each measured diffracted beam. These are simultaneous equations involving a single set of structural parameters from which the best set of parameter values must be found using least-squares methods. During refinement, additional parameters are introduced to model atom displacement, which includes both thermal and static displacement. It may be necessary to introduce restraints to ensure that the structure remains physically and chemically reasonable. For non-centrosymmetric structures with significant anomalous scattering, the absolute structure can be determined by refining the Flack parameter. The refined crystal structure is assessed quantitatively by the agreement between the measured intensities and those calculated from the structure model and also by considering features remaining in the difference electron density.

12.1 Introduction

Methods for structure solution may provide a more-or-less complete picture of the atoms within the unit cell. The purpose of structure refinement is to develop the structure model to achieve the best possible fit to the measured diffracted intensities. The refinement fits the atom types and positions within the unit cell to the measured intensities. Additional parameters must be introduced to describe

Pharmaceutical Crystallography: A Guide to Structure and Analysis
By Andrew Bond
© Andrew Bond 2019
Published by the Royal Society of Chemistry, www.rsc.org

thermal and static displacement of atoms in real crystal structures. For non-centrosymmetric structures, the absolute structure must also be considered carefully. In practice, structure refinement is closely tied to available software packages and it is necessary to gain experience working with those packages. This chapter aims to provide the background knowledge to support that practical exercise.

12.2   Structure Refinement in Theory

For a given structure model, any structure factor can be calculated using the structure factor equation. The positions and types of all atoms in the unit cell contribute to every structure factor by way of the atomic scattering factors and fractional coordinates. A set of measured diffraction data produces a list of structure factor equations, one for each measured diffracted beam. These are simultaneous equations involving a single set of parameters from which the best set of parameter values must be found. For N atoms in the unit cell, there are N atom types to be chosen and $3N$ fractional coordinates (some of which might be fixed if the atoms lie on special equivalent positions; see Section 4.2.2). By defining the symmetry relationships of the space group, it is only necessary to describe parameters for the atoms within the asymmetric unit. For a typical API structure, the number of atoms in the asymmetric unit is probably around 20–50, so the number of parameters to be refined is typically in the hundreds. The number of measured diffracted beams is usually in the thousands. Hence there are many more structure factor equations than parameters to be determined. In mathematical terms, the problem is *overdetermined*. Typically, the data-to-parameter ratio for a single-crystal analysis is around 10, and it is this feature that allows a crystal-structure refinement to provide robust and reproducible results.

12.2.1   Least-squares Fitting

A general method to fit an overdetermined system of equations is the least-squares method. A simple example is the fitting of a straight line between a set of data points on 2-D graph (Figure 12.1). The result in this case gives the best values for two parameters, namely the gradient (m) and intercept (c) of the line defined by the equation $y = mx + c$. The difference between the observed value at each point and the value calculated using the defined equation with the fitted values of m and c is called the *residual*. The best solution is considered to be the one

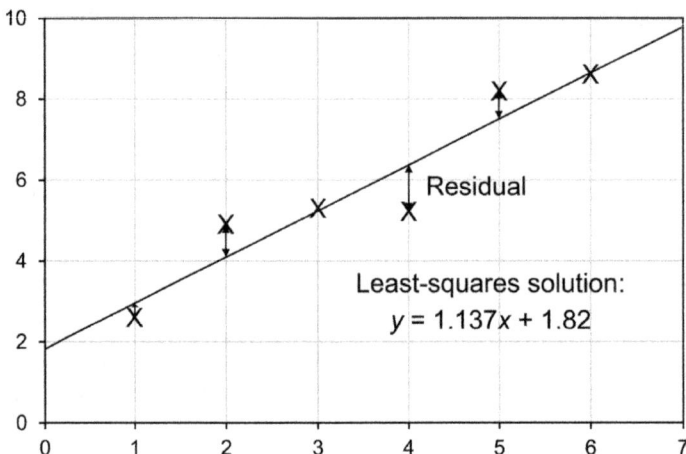

Figure 12.1 Linear least-squares fitting through six data points. The line shows the linear least-squares solution, which minimises the sum of the squared residuals between the data points and the line.

that minimises the sum of the squares of the residuals over all data points.

Least-squares fitting of this type is implemented in most spreadsheet packages. Behind the scenes, the calculations are carried out using matrix algebra. Problems based on linear equations can be solved in this way without requiring any prior knowledge of the parameters to be determined. For crystal-structure refinement, however, the structure factor equations are not linear. The concepts of least-squares fitting can still be applied but the process is more complicated. It is not possible to work directly with the structure factor equations, but it is possible to approximate them in a way that is linear with respect to *changes* in the parameters. In this approximated form, the least-squares process can indicate how an existing structure model should best be changed. Applying those changes does not immediately provide the best fit to the *actual* structure factor equations, but the model should move closer to the best fit. The procedure can be repeated until it converges on a situation where the suggested changes are negligibly small. Hence, crystal-structure refinement is an iterative process that must be continued to convergence. Crucially, the non-linear process requires reasonable starting estimates for the parameters. Hence, the structure-solution methods described in Chapter 11 must provide an adequate starting structure before refinement can begin.

The appropriate least-squares residual is in principle the difference between a measured structure factor and the structure factor calculated from the structure model. However, since the phases of the structure factors cannot be measured, the residual that can be used in practice is the difference between the observed and calculated intensity of each diffracted beam, which corresponds to $\{|F(hkl)_{\text{obs}}|^2 - |F(hkl)_{\text{calc}}|^2\}$. The sum of the squares of the residuals is then:

$$\sum \{|F(hkl)_{\text{obs}}|^2 - |F(hkl)_{\text{calc}}|^2\}^2$$

When making this calculation, it is necessary to define a scale factor to match the calculated intensities to the observed intensities (one scale factor for the whole structure), which is another parameter to be determined during structure refinement. Where $|F(hkl)_{\text{obs}}|^2$ and $|F(hkl)_{\text{calc}}|^2$ are compared in the following sections, it is implicit that this scale factor has been suitably applied.

12.2.2 Weighted Least-squares Fitting

The method described above will provide the best fit to the data on the assumption that all observations are of equal importance. In the context of an experimental measurement, this would mean that all measured intensities are subject to the same degree of error so there is no reason to trust one measurement more than any other. In a genuine scenario, it is likely that the measured intensities will have different degrees of uncertainty, as estimated during the integration process (Section 9.2.4). The best overall fit should be expected to prioritise the most certain measured intensities and tolerate worse fits to the intensities with greater uncertainty. This can be achieved by "weighting" the importance of each observation.

Weighted least-squares fitting is carried out by multiplying each individual residual by a suitable weighting factor, $w(hkl)$. The sum of the squares of the residuals becomes:

$$\sum w(hkl)\{|F(hkl)_{\text{obs}}|^2 - |F(hkl)_{\text{calc}}|^2\}^2$$

In principle, a suitable weight is simply the reciprocal of the error (where the error in statistical language means the *variance* of the observation; see Chapter 14). In practice, more complicated empirical expressions are usually used that also take account of the relative magnitude of each structure factor. The idea is that systematic measurement errors (for example, absorption or different illuminated volume due to a poorly centred crystal) are likely to be more significant for stronger diffracted beams, so these structure factors should be downweighted. The mathematical details of the weighting scheme are not so important. But it is necessary to be aware that a weighting scheme will be applied during refinement and that some associated empirical parameters must be optimised before the refinement is completed.

12.2.3 R-factors and Goodness-of-fit

To quantify how well the structure model fits the data, the sum of the squared residuals, with their weights, is divided by the sum of the squares of the observed intensities, then square rooted to bring the scale back to the residuals (rather than their squares). The result is called the *R-factor*:

$$wR2 = \left[\frac{\sum w(hkl)\{ |F(hkl)_{\text{obs}}|^2 - |F(hkl)_{\text{calc}}|^2 \}^2}{\sum w(hkl)\{ |F(hkl)_{\text{obs}}|^2 \}^2} \right]^{\frac{1}{2}}$$

A perfect fit between the structure model and the measured data would mean that all $|F(hkl)_{\text{obs}}|^2 = |F(hkl)_{\text{calc}}|^2$, which would give $wR2 = 0$. As the differences between $|F(hkl)_{\text{obs}}|^2$ and $|F(hkl)_{\text{calc}}|^2$ increase, $wR2$ increases. In principle, there is no maximum for $wR2$, but in practice, refinement of the scale factor relating $|F(hkl)_{\text{obs}}|^2$ and $|F(hkl)_{\text{calc}}|^2$ will generally restrict $wR2$ to a value less than 1.0. A typical value for a correct crystal structure refined against good data is in the region of 0.10–0.20. The symbol "$wR2$" is used to denote that the R-factor is weighted and that it is based on $|F(hkl)|^2$. An alternative measure, given the symbol "$R1$", is based on $|F(hkl)|$:

$$R1 = \frac{\sum \{ | |F(hkl)_{\text{obs}}| - |F(hkl)_{\text{calc}}| | \}}{\sum \{ |F(hkl)_{\text{obs}}| \}}$$

$R1$ was predominantly used in earlier refinements, so it is well established in the literature and retained today for comparative purposes. A typical value for a correct crystal structure with good data is in the region 0.03–0.10. Due to difficulties assessing the errors in $|F(hkl)|$ for weak reflections, $R1$ is usually calculated only for reflections with $|F|$ greater than some multiple of the associated error in $|F|$. Hence, $R1$ is calculated using only stronger data while $wR2$ is calculated from all data. A typical result might be written: $R1 = 0.045$ (1234 data with $I > 3\sigma(I)$), $wR2 = 0.146$ (all 1678 data). In this expression, $\sigma(I)$ refers to the standard deviation of the measured intensity, as described in Chapter 14.

Although R-factors provide an indication of the quality of a crystal-structure determination, they should not be considered in isolation and should not be viewed too rigidly. With a modern X-ray diffractometer and an adequate single crystal it is common to obtain $R1 < 0.05$. However, $R1 > 0.05$ does not immediately mean that the structure determination is inadequate. The results should be viewed against numerous criteria, discussed in the remainder of this chapter. Under no circumstances should a threshold value of $R1$ or $wR2$ be considered to determine whether a structure is "correct" or (worse) "publishable".

An additional measure to assess the least-squares refinement is the *goodness-of-fit*:

$$\text{GooF} = \left[\frac{\sum w(hkl)\{ |F(hkl)_{\text{obs}}|^2 - |F(hkl)_{\text{calc}}|^2 \}^2}{(n-p)} \right]^{\frac{1}{2}}$$

where the numerator is the weighted sum of the squares of the residuals, n is the number of data (observations) and p is the number of refined parameters. In the denominator, $(n-p)$ indicates the number of *degrees of freedom for error*, which means the number of observations that are available to assess the precision of the established parameters. In principle, p parameters could be assessed from p simultaneous equations, but there would be no further equations available to assess the quality of those parameters. As n increases relative to p, it becomes possible to assess how well the determined parameters fit many observations and therefore to assess the uncertainties for the parameters. For an appropriate weighting scheme, the GooF should be around 1.0.

12.2.4 Considering the Electron Density

The Fourier synthesis described in Chapter 11 produces an image of the electron density, $\rho(xyz)$, within the unit cell. The refined structure model, comprising atoms of specified types at specified positions, aims to reproduce that electron density as closely as possible. During refinement, the difference density, $\Delta\rho(xyz)$ (Section 11.5), can indicate how the structure model should be modified to improve the fit to $\rho(xyz)$. The most obvious example is an atom missing from the model, which will produce a clear peak in $\Delta\rho(xyz)$ at the position where the new atom should be added. Depending on the refinement software, it may be rare to look directly at $\Delta\rho(xyz)$. It is more common to be presented with an interpretation showing the positions of the largest maxima (Figure 12.2). Maxima in $\Delta\rho(xyz)$ are generally referred to as "*Q peaks*" because Q is a convenient label that does not correspond to

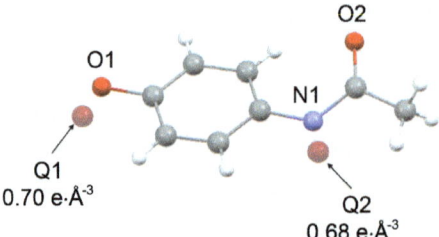

Figure 12.2   Peaks in the difference electron density illustrated for paracetamol.
H atoms on O1 and N1 have been omitted from the model and show
up clearly as "Q peaks" in $\Delta\rho(xyz)$.

any atomic symbol. The associated peak height is given in electrons
per cubic angstrom (e$\mathring{A}^{-3}$).

The final $\Delta\rho(xyz)$ is at least as important as the R-factors and GooF
to assess the quality of a structure refinement. Residual features in
$\Delta\rho(xyz)$ should generally be as small as possible, but they must be
considered in context. Routine refinement of a data set to 0.85 Å
resolution (or better) will generally produce maximum residual peaks
substantially less than 1 e$\mathring{A}^{-3}$. For a particularly good structure, the
maxima are often located between bonded atoms, indicating electron
density in the chemical bonds. Lower-resolution structures could have
more significant residual features in $\Delta\rho(xyz)$ due to series termination
errors, particularly close to heavy atoms. It may be reasonable for a
well-refined structure to have significant residual electron density
close to a heavy atom, but a peak in $\Delta\rho(xyz)$ located at a chemically
reasonable position probably indicates that an atom should be added
to the model. Hence, the heights *and positions* of maxima and minima
in $\Delta\rho(xyz)$ are important, and the expected features will depend on the
resolution of the data set. It does not make sense simply to define
some global threshold for an acceptable residual in $\Delta\rho(xyz)$.

12.3   Structure Refinement in Practice

In practice, structure refinement is an iterative process that alternates
between cycles of least-squares refinement and checking/modifying
the structure model. After initial structure solution, the R-factors will
probably be much larger than the guidelines indicated in Section
12.2.3, but they should decrease, hopefully smoothly, as the refinement
progresses. In the early stages, the main task is to ensure that all non-H
atoms are included with the correct atom types defined. H atoms have a
special status (described in Section 12.3.3) and can be added later.
When checking or modifying the model, the difference electron density

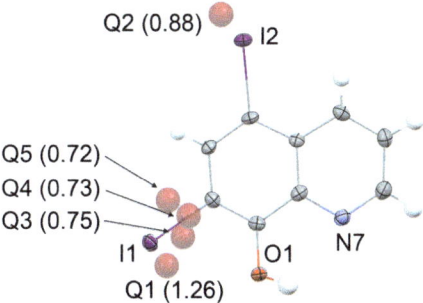

Figure 12.3 The five highest peaks (e Å^{-3}) in $\Delta\rho(xyz)$ for diiodohydroxyquinone are close to the heavy iodine atoms (using data from ref. 1). It does not make physical or chemical sense to add atoms in these positions.

must be considered, but always in the context of physical and chemical sense. For example, $\Delta\rho(xyz)$ may indicate a peak of several electrons but it only makes sense to include a new atom if the peak lies at a chemically reasonable position (Figure 12.3). A good knowledge of bond distances/angles and the conformations of molecules is therefore helpful. Significant peaks in $\Delta\rho(xyz)$ that are too close to other atoms may be indicative of disorder, considered in Chapter 13.

12.3.1 Atomic Displacement Parameters

In the structure factor equation, the atomic scattering factor f quantifies the combined scattering from all electron density associated with each atom. The function for f is produced from a calculated electron density for a static isolated atom. In real crystal structures there must be some thermal motion, which will act to distribute an atom's electron density over a larger volume compared to a static atom. We saw in Section 6.4.4 that the value of the atomic scattering factor drops off more rapidly for atoms with larger volume, so thermal motion acts to decrease f in this way. To accommodate this in the structure factor equation, an additional multiplier is included. Assuming that the atomic displacements are *isotropic* (*i.e.* the displacement can be viewed on average to define a sphere), a suitable functional form is:

$$\exp\left\{-8\pi^2\, U_{\mathrm{iso}}\left((\sin\theta)/\lambda\right)^2\right\}$$

where U_{iso} is the mean value of the squared displacements (in Å) from the specified atomic position. The function includes $(\sin\theta)/\lambda$, as plotted for the atomic scattering factor in Figure 6.13, and the negative sign in the exponent means that the value of the function decreases as U_{iso} increases. The maximum possible value is 1.0 if $U_{\mathrm{iso}} = 0$ (*i.e.* the

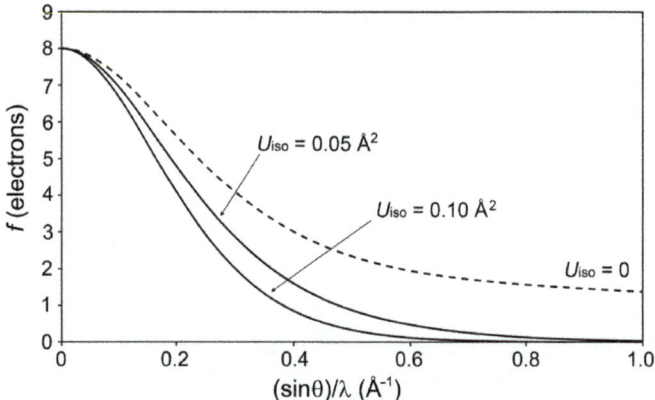

Figure 12.4 Drop-off of atomic scattering factors at different U_{iso} values for an O atom. The drop-off becomes more rapid as U_{iso} increases. The dashed line shows the case with zero displacement.

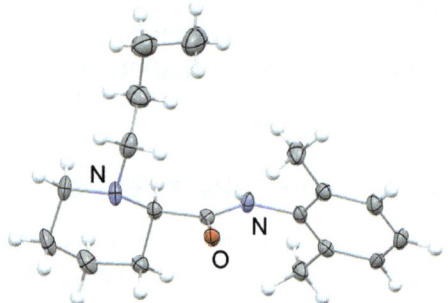

Figure 12.5 Molecular structure of bupivacaine showing displacement ellipsoids at 50% probability for non-H atoms. The ellipsoids represent anisotropic ADPs, comprising six parameters (U_{11}, U_{22}, U_{33}, U_{23}, U_{13}, U_{12}). H atoms are shown as spheres of arbitrary radius.

function has no effect if there is no displacement). The effect on the atomic scattering factor for an O atom is illustrated in Figure 12.4. The consequence of thermal motion is seen to be a more rapid decrease of the atomic scattering factor as a function of diffraction angle.

In most refinements, the description is extended for non-H atoms to an *anisotropic* model, where the displacement of each atom can be viewed on average to define an ellipsoid (Figure 12.5). Six parameters are required to define an ellipsoid, which can be viewed physically to represent the three lengths of the principal displacement axes and their orientation. The six parameters are called the *atomic displacement parameters* (ADPs) and given the symbol U_{ij}, where the subscripts i and j refer to the position within a symmetrical 3×3 matrix. In the representation usually used for small-molecule crystal structures, U_{ij} values have dimensions Å^2. Anisotropic refinement brings the total number of parameters to be refined to 9 for each atom (3 coordinates and 6 U_{ij} values). Hence, for N atoms in the unit cell, defined with anisotropic ADPs, the maximum number of parameters to be refined is $9N + 1$ (the additional parameter being the scale factor). This may be reduced if some atoms occupy special equivalent positions but it is a good estimate to keep in mind.

The physical size of a displacement ellipsoid is not immediately straightforward to define because it actually represents a *probability density function*. It is most common to visualise displacement ellipsoids "at the 50% probability level", which means that there is 50% probability that the atom is located within the defined volume. Crystallographic diagrams showing displacement ellipsoids should always report the applied probability level. The shape of an ellipsoid is sometimes described by considering the relative lengths of its three principal axes. A sphere has three principal axes of the same length (Figure 12.6(a)). If two of the principal axes have comparable length but one is much shorter or longer, the ellipsoid is referred to as *oblate* or *prolate*, respectively (Figure 12.6(b) and (c)). A quantitative measure to compare anisotropic ADPs is the *equivalent isotropic displacement parameter*, U_{eq}, which is a spherical equivalent of the ellipsoid, comparable to U_{iso}.

There is another source of atom displacement that has the same effect as thermal motion. Exploiting the symmetry of the crystal to produce one set of atomic coordinates within the asymmetric unit of the specified space group assumes that the atomic positions conform *exactly* to the symmetry through the entire crystal. The atomic positions must be very close to symmetrical through the crystal, else

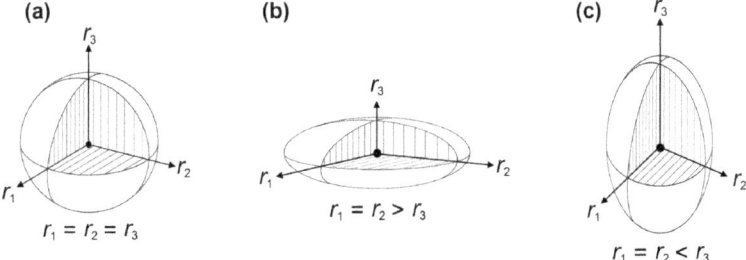

Figure 12.6   Displacement ellipsoids: (a) a sphere has three principal axes of the same length; (b) an oblate ellipsoid has two principal axes of comparable length and one shorter; (c) a prolate ellipsoid has two principal axes of comparable length and one longer.

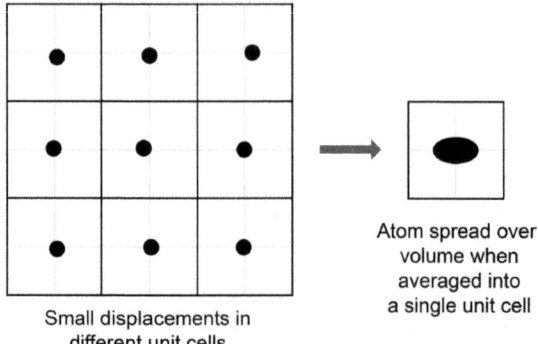

Small displacements in
different unit cells

Atom spread over
volume when
averaged into
a single unit cell

Figure 12.7 Static displacements from exact symmetry-equivalent positions produce a distribution similar to thermal motion.

the entire discussion on symmetry would not apply, but there will
generally be subtle displacements from the exact positions. When
averaging into a single asymmetric unit, it will appear as if each atom
is spread over a volume, exactly as described for thermal motion
(Figure 12.7). Hence, ADPs account for both thermal motion (a
dynamic phenomenon) and subtle variations in the atomic positions
that come from averaging the whole crystal into one asymmetric unit
(a *static* phenomenon). For this reason, the descriptive term *atomic
displacement parameter* is preferred over *thermal displacement
parameter*. From one measurement of a single-crystal X-ray structure,
static and dynamic atomic displacement cannot be distinguished. It
may be possible if several measurements are made at different temperatures, because dynamic effects should change with temperature
while static effects probably will not. However, ADPs are generally the
least reproducible of crystallographic parameters (Section 14.3.2) so
caution must be applied when interpreting them.

In the course of a structure refinement, ADPs convey important
information. Generally, the expectation is that all atoms should have
approximately the same degree of static displacement and thermal
motion, so all displacement ellipsoids should look approximately the
same. Together with physical and chemical knowledge, this provides
a basis to confirm atom types. Consider a scenario where an O atom
has been mistakenly included as an S atom. In the structure factor
equation, the value of f for the incorrect S atom will be too large, so
the refinement must seek to reduce it to produce the best fit to the
observed intensities. The least-squares procedure cannot change the
atom type, because these are set when defining the model, but it can
optimise the ADPs. According to Figure 12.4, larger ADPs serve to

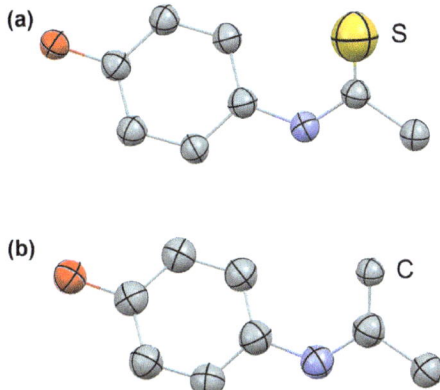

Figure 12.8 Displacement ellipsoid plot showing an incorrect atom in paracetamol. The atom should be O. (a) Refined as S gives an ellipsoid that is too large. (b) Refined as C gives an ellipsoid that is too small. Displacement ellipsoids are drawn at the 50% probability level.

reduce the scattering from an atom. Hence, the refined displacement ellipsoid of the incorrect S atom will appear larger than those of other correctly-identified atoms (Figure 12.8(a)). In general, a displacement ellipsoid that is too large suggests that the site has been given too much electron density and should be changed to an atom type with fewer electrons. Conversely, a displacement ellipsoid that is too small suggests that the site requires more electron density and should be changed to an atom type with more electrons. For example, the effect of labelling the same O atom as a C atom is shown in Figure 12.8(b). The effect in this case is quite subtle because the two elements differ only by two electrons, but it becomes easy to spot with practice. Of course, chemical knowledge (bond valence, bond distances/angles) provides an independent indication of the validity of atom types.

12.3.2 Restraints and Constraints

The purpose of refinement is not just to produce a structure that fits the X-ray data, but to produce a *physically and chemically reasonable structure* that fits the X-ray data. This means that the chemical connectivity and all interatomic distances and angles should be consistent with established expectations. For many single-crystal X-ray analyses, this is achieved "automatically" because $\rho(xyz)$ clearly defines the correct atomic positions and types. For some structures, however, the image obtained for $\rho(xyz)$ may not allow the refinement to converge automatically on a physically reasonable structure, so it becomes necessary to intervene. This might seem dangerous: if we impose our expectations on a structure, how do we know that these expectations are correct? The fact is that after more than 100 years of structure determination by X-ray diffraction, combined today with knowledge from quantum-chemical calculations, we know rather well what is chemically and physically reasonable. If a new X-ray structure suggests some unusual molecular geometry, particularly for the types of API molecules generally relevant in the pharmaceutical sciences, we should be suspicious of the X-ray result.

If the diffraction data cannot independently produce sensible values for all parameters to be refined, there are two options: (1) specify exact values for some parameters and do not attempt to refine them or (2) introduce new information that will help to limit the refined parameters to acceptable values. Option (1) is referred to as a *constraint* and option (2) as a *restraint*. Constraints are exact requirements such as "the x coordinate for this atom must be 0.5". Restraints are suggestions such as "the distance between atom X and atom Y should be close to 1.54 Å". A restraint is given a tolerance that specifies an acceptable degree of error, which can range from a "tight" restraint to a "loose" restraint. Sometimes, restraints are confusingly referred to as "soft constraints". This is misleading because constraints and restraints are applied quite differently. Constraints remove parameters from the refinement while restraints add new observational equations (in addition to the list of structure factor equations). The mathematical details are not so important, but the conceptual difference between the two approaches should be appreciated.

A common example of a constraint arises when an atom occupies a special equivalent position (SEP) within the space group. In that case, one or more of the atomic coordinates must equal the coordinate(s) of the corresponding symmetry element. For example, an atom on an inversion centre at the origin in space group $P\bar{1}$ must have $xyz = 0,0,0$, so these coordinates cannot be refined. Likewise, anisotropic displacement ellipsoids of atoms on SEPs must conform to the symmetry of the site, so it may be that some of the U_{ij} values should be constrained to equal others. Matters related to symmetry will be taken care of automatically by refinement software, but the user should be aware that these constraints are imposed. Aside from atoms on SEPs, there are few other circumstances where constraints are appropriate. One is for H atoms bonded to C atoms, since we can usually be quite certain of the chemical geometry around C. Another may be to control the hexagonal geometry of a benzene ring in a disordered situation (Chapter 13), again because we can be certain that a benzene ring has a regular hexagonal geometry. There are few (if any) other situations

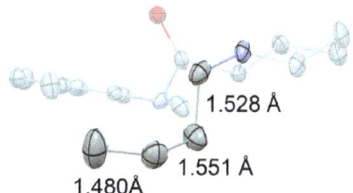

1.528 Å

1.551 Å

1.480Å

Figure 12.9 Displacement ellipsoids along the *n*-butyl chain in bupivacaine. H atoms are omitted. The displacement ellipsoids get gradually larger along the chain and the terminal C atom shows an elongated (prolate) ellipsoid. The terminal C–C bond is shorter than expectations, so a restraint might be applied.

where we can be so certain, so other constraints on atomic coordinates should generally be avoided. Constraints on displacement parameters are even less likely to be appropriate because it is difficult to have any consistent expectations of how ADPs should behave.

An example of a situation where restraints may be beneficial could be an *n*-alkyl chain, where the C atoms are likely to have relatively large displacement ellipsoids. Often, the displacement ellipsoids increase in size along such a chain, and the terminal C atom can have a particularly distorted ellipsoid due to "cumulative" lateral displacement along the chain (Figure 12.9). Each C atom produces a stretched-out feature in $\rho(xyz)$ that must be approximated by one atomic site. The resulting C–C distances are likely to deviate from well-established values (C–C = 1.54 Å and C$\cdots$C = 2.52 Å), typically appearing shorter towards the end of the chain. Clearly, this is just a consequence of the way that the atomic positions are deduced from $\rho(xyz)$ so it is reasonable to add restraints to encourage the atoms to converge on positions that better reflect the true molecular geometry. After introducing restraints, the resulting atomic positions probably will not give the *optimal* fit to $\rho(xyz)$, so the *R*-factors and residual electron density may increase. However, the atomic positions now produce a physically reasonable model that fits $\rho(xyz)$ well enough. Crucially, we understand why the fit to $\rho(xyz)$ in this region may be slightly less than optimal and accept it as a consequence of a decision to prioritise chemical and physical sense.

12.3.3 H Atoms

Locating H atoms can often be vitally important for pharmaceutical solids, for example to distinguish between salts and co-crystals. Since H atoms have only one electron, they scatter X-rays weakly. Viewing the structure factor equation as a vector summation on the Argand diagram (as in Figure 6.12), the length of a vector associated with an H atom is small, so it will make little difference to the resultant structure factor, regardless of which direction it points. This means that the coordinates of an H atom could be changed significantly with only minimal effect on the fit to the diffracted intensities. Similarly, features in $\rho(xyz)$ or $\Delta\rho(xyz)$ will be relatively small, perhaps even smaller than the average background level for a low-resolution data set. In that case, it may not be possible to obtain any information on H atom positions. There is an even more fundamental issue: atomic positions are derived from the positions of maxima in $\rho(xyz)$. The assumption that the atomic nucleus lies at the centroid of the electron density is fair for atoms with many electrons, but it does not apply for H atoms, for which the single electron must be involved in a covalent bond (Figure 12.10). The true position of the H-atom nucleus is therefore not the same as the maximum in $\rho(xyz)$. There is nothing to be done to "correct" for this—indeed, we do not want to correct for it because X-ray diffraction fits a structure model to the image of the *electron density*, so the optimal position to include an H atom is at the position of the peak in $\rho(xyz)$. The consequence, however, is that bond lengths involving H atoms are always systematically underestimated in X-ray crystal structures. This is considered further in Chapter 14.

None of this means that H atoms can be ignored. A typical pharmaceutical crystal structure will include many H atoms, which collectively make a significant contribution to each structure factor. To omit them would cause a substantial systematic error in the fit to the

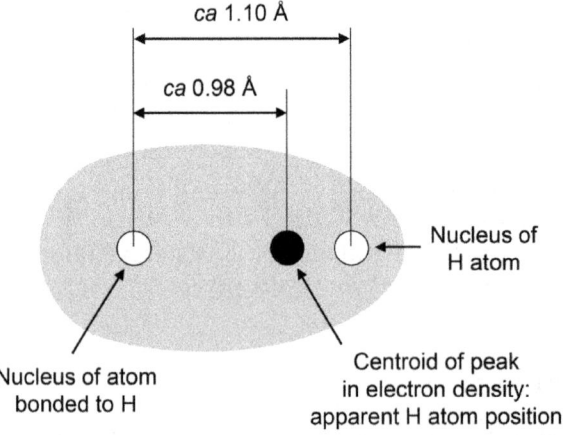

ca 1.10 Å

ca 0.98 Å

Nucleus of
H atom

Nucleus of atom
bonded to H

Centroid of peak
in electron density:
apparent H atom position

Figure 12.10 Schematic illustration of the nuclear and apparent X-ray position of an H atom. Because the electron of H is involved in the bond, the centroid of the electron density is closer to the bonded atom than is the H atom nucleus. The distance values are typical for H bonded to C.

diffracted intensities, which would be detrimental to every aspect of the refinement. However, it influences the way in which H atoms are handled during refinement. For H atoms bonded to C atoms, the geometry around C is generally so well known that there is nothing to be learned from refining H-atom positions. It is more beneficial to eliminate the parameters. The universal strategy is to introduce H atoms at calculated positions then allow them to "ride" on their parent C atom during subsequent refinement cycles. The coordinates of the H atoms are not refined and the C–H bond lengths and angles are constrained to standard values. The positions are usually re-calculated after every few refinement cycles. For a normal refinement against X-ray data, H atoms always have isotropic ADPs. For constrained H atoms, the value of U_{iso} is fixed at a multiple of U_{eq} for the C atom to which it is bonded. A typical multiple is 1.2 for CH and CH_2 groups, increasing to 1.5 for terminal CH_3 groups. For terminal CH_3 groups, it is common to allow the group to rotate around its $C-CH_3$ bond, so that the added H atoms rotate to occupy any maxima in $\Delta\rho(xyz)$.

For H atoms bonded to atoms such as N or O, the geometry can be harder to predict, especially if those groups are involved in hydrogen bonding within the structure. Hence, it is not so easy to place H atoms in calculated positions. With modern X-ray instruments and low-temperature data collections to 0.85 Å resolution, it is often possible to identify H atoms in $\Delta\rho(xyz)$ and to refine their coordinates freely with an isotropic ADP. If such an approach converges on a reasonable chemical geometry, this is the best approach for H atoms in hydrogen bonds. Often, however, the maxima in $\Delta\rho(xyz)$ are quite poorly defined in the vicinity of hydrogen bonds, so the resulting positions produce unreasonable X–H bond lengths. In that case, it is common to restrain X–H distances to reasonable values. In some cases, particularly for solvent molecules, there may be no reasonable method to locate H atoms, so they are just omitted from the model. Omission should probably be a last resort, however, because it can cause confusion about chemical identity.

Dealing with H atoms is an area where the user can have significant influence, so it is particularly important to pay close attention. Methods for automatic addition usually consider the bond lengths around C when deciding how many H atoms to add, and mistakes can frequently occur when the bonds lie around borderline values. Ideally, and certainly in cases of potential ambiguity, $\Delta\rho(xyz)$ should be examined before and after auto-generation of H atoms to ensure that the added atoms are consistent with all available indications in $\rho(xyz)$. For molecules with known chemical composition, it is advisable to have an independent count of the empirical formula and to check carefully that the correct number of H atoms are added. It is definitely not advisable to deduce the chemical formula solely on the basis of auto-generated H atoms. For H atoms involved in H-bonding, it is important to report whether the H atoms were located in $\rho(xyz)$ or placed automatically and to describe clearly how they were refined. Do the resulting positions actually reflect $\rho(xyz)$, or have the H atoms been "forced" into their positions on the basis of some assumptions? This might have significant consequences if the crystal structure is later retrieved from a crystallographic database. Finally, conclusions drawn from H atom positions (*e.g.* salt or co-crystal?) must take very careful account of how those positions were produced and what can actually be deduced from the experimental image of $\rho(xyz)$. This is considered further in Chapter 14.

12.4 Absolute Structure

The theory behind absolute structure determination was described in Section 7.3. For non-centrosymmetric structures, it is necessary to determine which of two inversion-related structures provides the best fit to the measured intensities. One way to achieve this is simply to try separate refinements for the two alternatives to see which provides the best results. More efficient, however, is to introduce a refinable parameter to express the absolute structure. The key information lies in the difference between $I(hkl)$ and $I(\bar{h}\bar{k}\bar{l})$, which can be written as follows:

$$I(hkl)_{\text{obs}} = (1-x)\, I(hkl)_{\text{calc}} + x\, I(\bar{h}\bar{k}\bar{l})_{\text{calc}}$$

The equation states that the observed intensity of each diffracted beam can be represented as a linear combination of the intensities calculated from the structure model for the reflection hkl and its Friedel pair $\bar{h}\bar{k}\bar{l}$. If the model has the correct absolute structure, $I(hkl)_{\text{obs}} = I(hkl)_{\text{calc}}$, which means that $x = 0$. If the model is inverted, $I(hkl)_{\text{obs}} = I(\bar{h}\bar{k}\bar{l})_{\text{calc}}$, which means that $x = 1$. Determining the absolute structure of a non-centrosymmetric crystal can be achieved by refining the value of x, which is known as the *Flack parameter*. When the Flack parameter is determined from experimental data, the errors associated with the intensity measurements propagate through to an uncertainty on the refined value of x. Details on errors and the reliability of an absolute structure determination are considered in Chapter 14.

Some extensions of the basic Flack parameter refinement are based on combinations of Friedel pairs,[2] specifically differences (D) or quotients (Q):

$$D(hkl) = I(hkl) - I(\bar{h}\bar{k}\bar{l})$$

$$Q(hkl) = \{I(hkl) - I(\bar{h}\bar{k}\bar{l})\}/\{I(hkl) + I(\bar{h}\bar{k}\bar{l})\}$$

Using the definition of the Flack parameter, it can be established that a plot of $D(hkl)_{\text{obs}}$ *vs.* $D(hkl)_{\text{calc}}$ or $Q(hkl)_{\text{obs}}$ *vs.* $Q(hkl)_{\text{calc}}$ should be linear with a slope of $(1 - 2x)$. The value of x obtained by a least-squares fit to the "Q plot" generally has a much smaller associated uncertainty compared to conventional assessment of the Flack parameter. This has a significant impact in the pharmaceutical sciences because it improves the prospects of absolute structure determination for light-atom structures (containing only C, H, N, O) using a standard Cu X-ray source. The method is implemented post-refinement, and does not impose any additional requirements during the refinement stage. To obtain a meaningful Q plot, a *complete* set of Friedel pairs should be measured. An example is given in the Case Study at the end of Chapter 14.

12.5   Summary of Key Points

- A set of measured diffraction data produces a list of structure factor equations, one for each measured diffracted beam. These are simultaneous equations involving a single set of parameters that describe the structure. The purpose of refinement is to obtain the best possible set of parameter values.
- Mathematically, refinement is an overdetermined problem, which can be handled using least-squares methods. The structure factor equations cannot be handled directly using linear least-squares methods, but they can be approximated in a form that is linear with respect to changes in the parameters. An iterative process is applied until convergence.
- The intensities are weighted to prioritise those that have been measured with the most certainty. An empirical weighting scheme is usually applied, with parameters that must also be refined.
- The fit of the structure to the measured intensities is quantified by the R-factors and goodness-of-fit (GooF). It is also important to consider the heights and positions of maxima and minima in the difference electron density, $\Delta\rho(xyz)$.
- Atomic displacement parameters (ADPs) are introduced to model thermal motion and subtle static displacement of atoms

from exact positions. Anisotropic ADPs are usually applied for non-H atoms, where the displacement is viewed as an ellipsoid. Isotropic ADPs view the displacement as a sphere. The ADPs serve to modulate atomic scattering factors so that they drop off more rapidly.

- The refinement must produce a *physically and chemically reasonable* structure. To achieve this, it may be necessary to introduce restraints, which are additional suggestions to guide the refinement. In some cases, usually related to symmetry requirements, parameters are constrained, which means that they are not refined.

- H atoms require special attention because they scatter X-rays only very weakly. H atoms on C atoms are usually placed in idealised positions and tied to their parent C atom. H atoms on other atoms can often be located in $\rho(xyz)$, but they are often refined with restraints on their bond distances.

- Bond lengths involving H atoms refined against X-ray data are always underestimated because maxima in $\rho(xyz)$ corresponds to electrons in the covalent X–H bond, rather than the position of the H nucleus.

12.6   Case Study: Sulfathiazole (Continued)

Continuing the sulfathiazole case study of Chapter 11, all non-H atoms were identified and the structure can now be refined. For this discussion, we use data to 0.85 Å resolution. The current state of the structure is as shown in Figure 11.13. All atoms have displacement spheres roughly the same size, indicating that the atom types are correctly defined. At this stage, $R1 = 0.088$ and $wR2 = 0.393$, and the largest peak in $\Delta\rho(xyz)$ is 2.44 e Å^{-3}, situated 0.53 Å from S2 in the thiazole ring (Figure 12.11).

Introducing anisotropic ADPs makes a significant difference. The values of $R1$ and $wR2$ decrease to 0.053 and 0.262, respectively, as the model now fits more effectively the electron density in the region of each atom. The most distorted displacement ellipsoid is seen for S2, which is elongated perpendicular to the ring plane (Figure 12.12). This ellipsoid effectively "mops up" the peaks in $\Delta\rho(xyz)$ that previously existed close to S2. The largest peak in $\Delta\rho(xyz)$ is reduced to 1.02 e Å^{-3}, now situated 0.94 Å from atom C6 in the benzene ring. At this stage, the empirical parameters of the weighting scheme can be optimised to give $R1 = 0.047$, $wR2 = 0.146$ and GoOF = 1.03. Applying

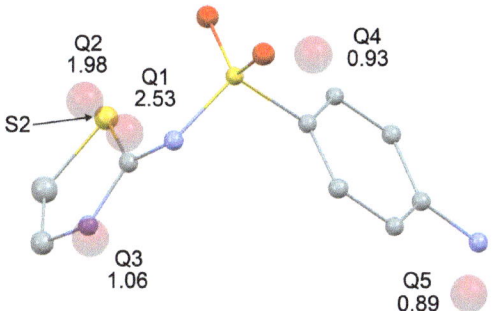

Figure 12.11 Molecular structure of sulfathiazole after initial structure solution. Displacement spheres are at 50% probability. The diagram is the same as Figure 11.13, except that it also shows the five largest peaks in the residual electron density (e Å^{-3}). The largest peak (Q1) is 0.53 Å from S2.

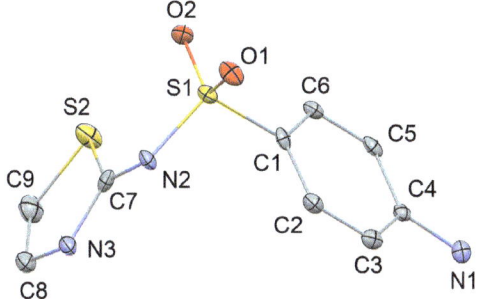

Figure 12.12 Displacement ellipsoids (drawn at 50% probability) for sulfathiazole after the first cycles of anisotropic refinement.

the weights has an obvious effect on $wR2$, since this includes $w(hkl)$ in its definition (Section 12.2.3). However, it also reduces $R1$, which does not involve $w(hkl)$, because the whole refinement is improved by prioritising the most reliable measured intensities in the least-squares process. The number of parameters being refined is 145, corresponding to 3 coordinates and 6 U_{ij} values for each of 16 non-H atoms plus one scale parameter relating $|F(hkl)_{\text{calc}}|^2$ to $|F(hkl)_{\text{obs}}|^2$. The number of measured intensities is 1845, giving a data-to-parameter ratio of 12.7 : 1.

The next step is to look for H atoms by considering $\Delta\rho(xyz)$, produced from a Fourier summation using $\{|F(hkl)|_{\text{obs}} - |F(hkl)|_{\text{calc}}\}$ with $\Phi(hkl)_{\text{calc}}$, where "calc" refers to the structure factor amplitudes and phases calculated from the current model. For the sulfathiazole structure in its current state, the ten largest maxima in $\Delta\rho(xyz)$ are indicated in Figure 12.13. Peaks Q1–Q9 clearly correspond to the

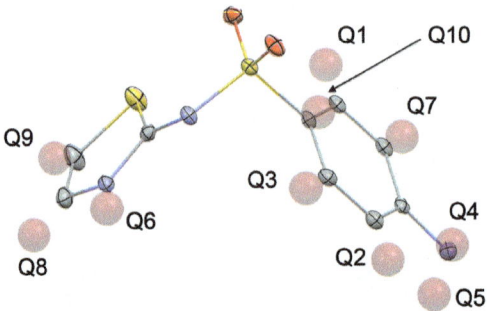

Figure 12.13   Ten largest maxima in the difference electron density of sulfathiazole after refining all non-H atoms with anisotropic ADPs. Peaks Q1–Q9 (range 1.01–0.69 e Å^{-3}) correspond to the anticipated positions for H atoms. Peak Q10 (0.33 e Å^{-3}) lies at the middle of the C1–C6 bond.

anticipated positions for H atoms, while the next highest peak (Q10) is significantly smaller and corresponds to electron density in the middle of one of the C–C bonds. This shows what can be expected for carefully measured data to sufficient resolution for a good crystal at low temperature (in this case 100 K). The H atoms bonded to C atoms in the benzene and thiazole rings can be placed in calculated positions and allowed to ride on their parent C atoms (Section 12.3.3). Adding these six H atoms to the model and optimising the weighting scheme gives $R1 = 0.034$ and $wR2 = 0.101$, without introducing any further refined parameters. The three largest maxima remaining in $\Delta\rho(xyz)$ now have heights 0.93–0.86 e Å^{-3}, compared to 0.30 e Å^{-3} for the fourth highest peak. Clearly, as the model moves closer to completion and the $\Phi(hkl)_{calc}$ values improve, the image of $\rho(xyz)$ becomes increasingly well defined. The three peaks, which correspond to the H atoms on N1 and N3, can be included in the positions identified from $\Delta\rho(xyz)$ and refined freely with their own U_{iso} values. This adds a further 12 parameters to the refinement, bringing the total to 157. The result (Figure 12.14) has $R1 = 0.025$, $wR2 = 0.072$ and GooF = 1.15.

The resulting H atom positions give N–H bond lengths of 0.86–0.89 Å, which are typical for X-ray data. The refined U_{iso} values are comparable to those of the other H atoms, which is an important sign that the data support free refinement of H atoms in this way. Recall from Section 12.3.3 that the N–H bond lengths will be shorter than the true bond lengths, because the H atoms are optimised against the positions of maxima in the electron density rather than nuclear positions. Interestingly (and unusually), the literature paper from which this example is taken also reports neutron diffraction measurements, which reveal true nuclear positions because neutrons

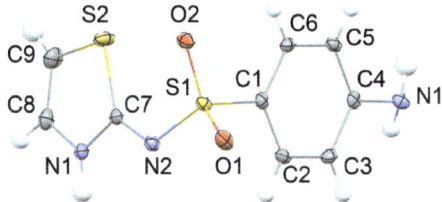

Figure 12.14 Final molecular structure for sulfathiazole (0.85 Å resolution), including all H atoms. The H atoms on C are riding, while those on N are refined freely with U_{iso} values. Displacement ellipsoids/spheres are shown at 50% probability.

are scattered from atomic nuclei. The neutron-derived positions match those of the X-ray analysis, with corresponding N–H bond lengths extended to 1.02–1.05 Å.

The refinement is now completed. The chemical model is complete: there are no further significant features in $\Delta\rho(xyz)$ and the atom sites sum to the expected empirical formula of $C_9H_9N_3O_2S_2$. The refinement software reports that the weighting scheme cannot be optimised any further. Any further cycles of least-squares refinement indicate that suggested changes for the parameters are smaller than the anticipated errors, which means that the refinement has converged.

References

1. M. S. Butt, Z. Akhter, M. Bolte and H. M. Siddiqi, *Acta Crystallogr., Sect. E: Crystallogr. Commun.*, 2007, **E63**, o15, [CSD: NEVMOR].
2. S. Parsons, H. D. Flack and T. Wagner, *Acta Crystallogr., Sect. B: Struct. Sci., Cryst. Eng. Mater.*, 2013, **B69**, 249.

13 Disorder and Twinning

Summary

Often, crystals deviate to some extent from the idealised description. Disorder breaks the symmetry of the space group at a local level so the structure must be described as an average of local alternatives. In severe cases of disorder, such as solvent molecules occupying several positions and orientations, it may be difficult to produce any sensible atomic model. An alternative approach in that case is solvent-masking, which optimises electron density values on a grid within a defined solvent-accessible volume. Twinning refers to the circumstance where a measured crystal comprises misaligned domains of the same structure. The measured diffraction pattern is an overlay of the diffraction patterns for each domain. Two cases can be identified which impose different practical problems on measurement and processing of X-ray data. Merohedral twinning refers to the circumstance where the diffraction patterns overlap exactly, while non-merohedral twinning refers to non-overlapping diffraction patterns. Inversion twinning is a special case of merohedral twinning which impacts on determination of absolute structure.

13.1 Introduction

The discussion of refinement in Chapter 12 focusses on well-ordered crystals, following the idealised description in Chapters 2–4. Many pharmaceutical crystals conform to this picture so refinements regularly proceed in the straightforward manner described. Often, however, crystals are found to deviate to some extent from the idealised description. This is frequently seen for flexible regions of molecules,

Pharmaceutical Crystallography: A Guide to Structure and Analysis
By Andrew Bond
© Andrew Bond 2019
Published by the Royal Society of Chemistry, www.rsc.org

which may have numerous conformations with similar energy, or for solvent molecules which may interact only weakly with the "host" API. Any deviation of a crystal structure from the ideal symmetrical situation is called *disorder*. Disorder breaks the symmetry in the region of individual molecules so the structure must be described as an average of local alternatives. At a macroscopic level it is common for crystals to grow with more than one domain, which may mean that a measured diffraction pattern comprises an overlay of patterns in different orientations. This is called *twinning*. Often the domains adopt a specific relative orientation which can be deduced by careful analysis of the diffraction pattern. This Chapter gives an overview of the concepts and provides case studies to illustrate the two principal themes.

13.2 Disorder

During refinement, the developing model is an interpretation of the underlying electron density. In an ideal case, $\rho(xyz)$ will comprise well-defined, approximately spherical peaks that are clearly identifiable as atoms. Thermal displacement and subtle static variation of the atomic positions will generally lead to some distortion from spherical peaks, which can be modelled using anisotropic ADPs. Sometimes, $\rho(xyz)$ might show more distorted features or a collection of peaks that cannot immediately be interpreted as a sensible chemical structure. If the separation between apparent atomic sites is less than a reasonable distance for a chemical bond or non-bonded contact between atoms, the two sites cannot possibly be occupied simultaneously. Hence, the model must represent an average, where an atom might occupy one position or the other (Figure 13.1). Such a

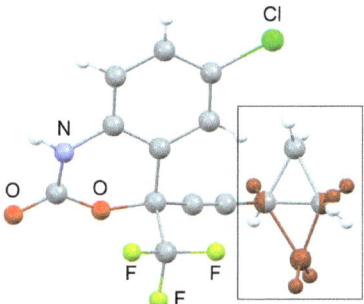

Figure 13.1 Disorder in the crystal structure of the anti-HIV drug efavirenz (using data from ref. 1). The cyclopropyl ring has alternative orientations (shaded and unshaded), both having site occupancy 0.5. The half-shaded atoms are common to both orientations.

structure is said to be *disordered*. Disorder refers to any deviation from the ideal situation described in Chapters 2–4.

To model disordered structures, *site occupancy factors* (SOFs) are introduced. These are scalar multipliers in the structure factor equation that specify the extent to which each atomic site is occupied:

$$F(hkl) = \sum_{n=1}^{N} (\text{SOF})_n f_n \exp\{2\pi i(hx_n + ky_n + lz_n)\}$$

The value of the SOF for each atom indicates an average over the entire crystal for the duration of the measurement. In a fully ordered structure, all SOFs are fixed at 1.0. In disordered structures, the SOFs of some atoms will be less than 1.0. If a given atom occupies either one atomic site or the other, the SOFs of the two alternative sites should sum to 1.0. In some cases, atoms may be refined with a SOF less than 1.0 but with no corresponding alternative site. The physical interpretation is that the atomic site is occupied sometimes but unoccupied otherwise. This might be applied to solvent molecules in crystalline solvates and hydrates, for example. In the structure factor equation, a SOF less than 1.0 has an effect similar to a large ADP, serving to reduce the scattering contribution from a given atomic site. Hence, it may be difficult to distinguish partial site occupancy from significant displacement.

When modelling disorder, physical and chemical sense should always be paramount. For a complex region in $\rho(xyz)$, it may be tempting to add a large number of atomic sites, each with fractional site occupancy and perhaps with large or distorted ADPs. Proceeding in this way will generally cause *R*-factors to decrease because it introduces more parameters that can be adjusted to improve the fit to the data. However, the result will have little physical meaning. A good disorder model consists of a limited number of specific sets of atomic positions, for example two sets labelled "A" and "B" (Figure 13.2). At any position in the crystal, either set A is occupied or set B is occupied. All atoms in set A have the same SOF (often refined as a single variable, v) and all atoms in set B have the same SOF (constrained to equal $1 - v$). Crucially, both sets A and B have sensible chemical structures so the structure is reasonable whichever set is occupied. In some cases, it may be appropriate to include more than two sets, with the SOFs of all components summing to 1.0, but increasing the number of disorder components generally risks "over-interpretation" of $\rho(xyz)$. In essence, this means losing sight of the aim to describe a periodic, ordered crystal structure in the manner of Chapters 2–4. If it

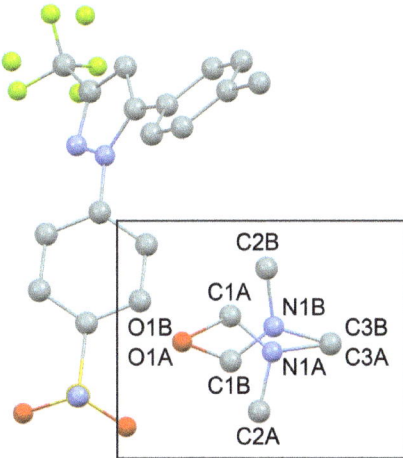

Figure 13.2 Disordered dimethylformamide (DMF) molecule in the structure of a celecoxib–dimethylformamide (1 : 1) solvate. Either atoms with suffix A (SOF 0.80) or suffix B (SOF 0.20) are occupied at a given DMF site. H atoms are omitted.

appears necessary that most of the atoms in a structure should be disordered, it should be considered whether the space group or unit cell has been chosen appropriately or whether the crystal actually conforms to the expectations of long-range order.

Almost certainly, it will be necessary to control the geometry of disordered regions using restraints (Section 12.3.2). When there is more than one disorder component, it is reasonable to expect that all chemically-equivalent components should have comparable bond lengths and angles, although they may have different torsion angles defining alternative molecular conformations. It is usually important also to check *intermolecular* contacts. In circumstances where one disorder component has SOFs significantly larger than those of another component and its geometry appears to be well defined, the distances and angles in the minor component might be restrained to equal those in the major component. In this way, the geometry of the minor component is controlled by the major component without imposing outside expectations of what the geometry should be. In some cases (for example, a C–CF$_3$ group disordered by rotation about the C–C bond; Figure 13.3), there might be numerous equivalent bonds that should have the same length, which could all be restrained to a single refined parameter. In circumstances where the geometry simply cannot be controlled, or where there might be only one troublesome atom, specific values can be given for selected restraints. Where possible, these should be derived from an analysis of comparable

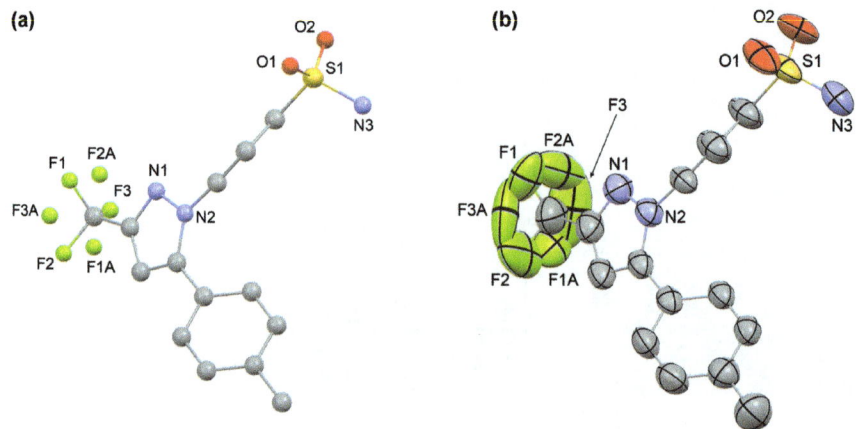

Figure 13.3 A disordered CF_3 group in celecoxib, taken from the (1:1) dimethyl-formamide solvate shown in Figure 13.2. The ball-and-stick model in (a) shows two disorder components for the CF_3 group, having SOFs 0.57 (F1–F3) and 0.43 (F1A–F3A). The displacement ellipsoids shown in (b) suggest comprehensive rotational disorder of the CF_3 group. H atoms are omitted.

structures in a crystallographic database (Section 14.4.3). It is obviously dangerous to assume restraint values without any clear basis for the decision. As noted in Section 12.3.2, it is likely that the addition of restraints to a disordered region will increase *R*-factors. This is generally an acceptable consequence of imposing physical and chemical sense, but it should not be too severe. When adding restraints, a sudden dramatic rise in the *R*-factors probably indicates that the restraints are not appropriate. Refinement software usually lists a summary of applied restraints and the extent to which the refined model conforms to the expectations. If the refined structure continues to deviate significantly from suggested restraints, it should be considered carefully whether those restraints have been correctly applied (have the atoms involved in the restraints been specified correctly?) and whether they are actually appropriate.

13.3   Solvent Masking

It is common, especially for solvates or hydrates, to find that some parts of the structure are well ordered, but other parts are disordered. This could be expected for a solvate if the interactions between the solvent molecules and API molecules are non-specific and weak. Since every atom in the unit cell contributes to every structure factor, failing to model adequately one part of the structure is detrimental to the entire refinement. In other words, difficulties describing disordered solvent molecules will also compromise the description of the well-ordered API. Given that the structure factor equation is a summation, it is always possible to separate atoms conceptually into groups. We can imagine a group containing the well-behaved API molecules and a separate group for the disordered solvent molecules. The API component can then be described in the usual way, but the question is how to model the disordered solvent part.

We saw in Section 13.1 that disorder manifests itself as an electron-density distribution that cannot be modelled by a single set of discrete atoms. For heavily disordered solvent molecules, it may be necessary to add several sets of atomic positions to account for all peaks in the electron density, each with fractional site occupancy factors and in close proximity to other atoms (Figure 13.4). Such circumstances are likely to require numerous restraints to ensure that the defined molecules retain a reasonable geometry. If it becomes necessary to build a model so complicated, it means that the underlying electron density is not well described by the "atomic approximation". In such circumstances, it may be preferable to abandon the idea of building an atomic model. It is clear that the solvent is disordered (else we wouldn't be having the discussion) and we are not going to learn anything by continuing to add partially-occupied, tightly-controlled (or worse, uncontrolled) molecules to the model. The aim is to provide the best fit to the diffraction data, and it is possible to achieve that in another way.

The alternative is to "mask out" the solvent area. This process is commonly referred to as *SQUEEZE*, after the software package in which it was first implemented.[2] The basic principle is that the

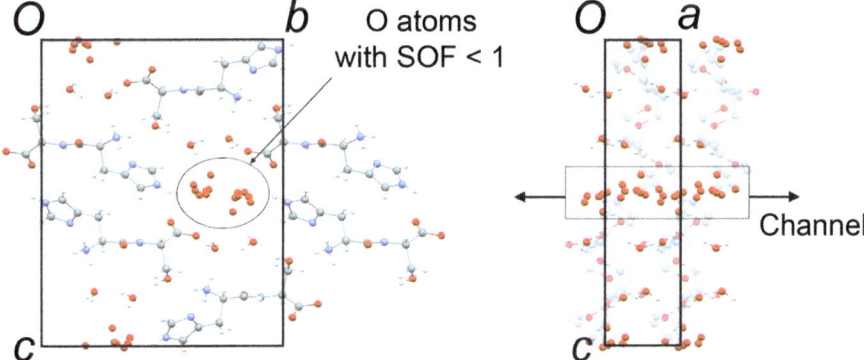

Figure 13.4 Continuous disorder in a hydrate structure of L-histidine-L-serine dipeptide.[3] The structure contains large water-containing channels running along the *a* axis. Two unique water molecules are well ordered, forming hydrogen bonds directly to the host dipeptide. The remaining water molecules are modelled as numerous O atom sites, each with fractional SOFs.

structure factors are the Fourier transform of the electron density, which is a continuous function within the unit cell. In Chapter 6, we moved rapidly to an atomic picture, where atoms lie at specific positions and atomic scattering factors describe the total scattering contribution of all electron density attributed to each atom. For a "non-atomic" electron density, we can take a step back and consider the complete electron density, evaluated at points on a defined grid in *xyz*. The structure factor equation in this region can be written:

$$F(hkl)_{\mathrm{solv}} = V_{\mathrm{g}} \sum \rho(xyz)_{\mathrm{g}} \exp\{2\pi i(hx + ky + lz)\}$$

where the summation is over all grid points, $\rho(xyz)_{\mathrm{g}}$ is the electron density value at each grid point and V_{g} is the volume of each region on the grid. This is a logical progression of adding more and more partially-occupied atoms within a complicated disorder model. Multiplication by V_{g} is applied because $\rho(xyz)_{\mathrm{g}}$ is the electron density rather than the required value in electrons. The calculation is more intensive than the standard structure factor equation because the summation is taken over many grid points rather than a limited number of atoms. However, it is usually only applied once towards the end of the refinement, then the contribution from the solvent region is added to the structure factors calculated in the usual way from the atomic part of the structure.

Practically, solvent masking requires an initial definition of the volume to be masked. The *solvent-accessible volume* is calculated from defined atomic radii, considering where it would be reasonable to accommodate additional atoms (Figure 13.5). Proper assessment of this volume requires the ordered part of the structure to be completed, including all H atoms. The grid is then defined within that volume and the values of $\rho(xyz)_{\mathrm{g}}$ at each grid point are deduced from $\Delta\rho(xyz)$. The procedure must be iterated because the calculation of $\Delta\rho(xyz)$ will initially be made using structure factor phases calculated from a model with an empty solvent region. As electron density is introduced into the solvent-accessible volume, the calculated phases of the structure factors change, so $\Delta\rho(xyz)$ will change. When the process converges, the number of electrons in the solvent volume is obtained as the sum over all grid points. If the chemical identity of the solvent molecule is known, an indicative number of solvent molecules per unit cell can be deduced.

Solvent masking can be extremely effective. It can reduce *R*-factors dramatically and improve the description of the ordered atomic parts of the structure. However, it must not be applied indiscriminately to manipulate *R*-factors or just to avoid work involved with modelling

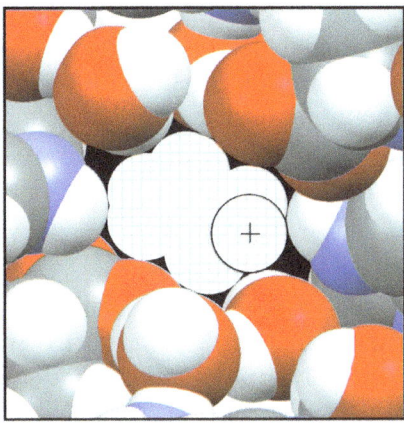

Figure 13.5 Definition of the solvent-accessible region in one channel of the L-histidine-L-serine dipeptide structure shown in Figure 13.4. The atoms in the ordered part of the structure are shown with their van der Waals radii. The solvent region includes all grid points where it would be possible to accommodate an atom of van der Waals radius 1.2 Å, as indicated by the open circle.

disorder. On the experimental side, the measured intensity data should be as complete as possible to adequate resolution (as ever, 0.85 Å is a reasonable guide). Any poorly-measured strong reflections, for example low-angle reflections obscured by the beamstop on the diffractometer, can seriously affect the electron count and must be identified and omitted before running the algorithm. On the chemical side, it must be made clear what has been done to the structure and why. After applying a solvent mask, the resulting structure will contain substantial voids. Anyone retrieving the structure from the literature or from a database must be in no doubt that these voids are artificial. Simulating a PXRD pattern from the resulting structure will produce inaccurate intensities, which could be highly misleading if the reason is not made clear. Care should also be taken to avoid removing parts of the structure that may have significant bearing on the chemical interpretation. For example, it is probably not appropriate to eliminate counter-ions, whether disordered or not, since the resulting structure will not be charge-balanced. The best advice when considering solvent masking is to try the procedure *in addition to attempts at disorder modelling*. This can demonstrate that an unusually high R-factor can be attributed to difficulties modelling a disordered region rather than some other problem with the data. A decision might be made to retain an atomic model of a heavily disordered counter-ion so as to maintain charge balance, but concurrent reporting of the solvent-masked structure can provide assurance that the disorder is the sole source of refinement difficulties. Key conclusions regarding the geometry of the ordered part of the structure, particularly in relation to uncertainties (Chapter 14), might consider both the conventional and solvent-masked refinements to draw the best conclusions.

13.4   Twinning

Twinning refers to the circumstance where a measured crystal comprises domains of the same structure misaligned in some way (Figure 13.6). Section 6.3.5 described *mosaicity*, which is subtle misalignment of crystalline domains causing broadening of the diffracted beams. This is always present for real crystals and requires no special attention during integration and refinement. Twinning is a far more substantial effect where domains have entirely different orientations, generally related by a local symmetry operation that is not part of the space group. In that case, the diffraction pattern comprises an overlay of diffraction patterns with some symmetry relationship between them. Clearly, this will have a substantial influence on the measured intensities, and possibly also on the measured geometry. If the relationship between the overlaid diffraction patterns can be established, it is possible to handle the situation during data processing and structure refinement and there is every possibility to produce reliable crystallographic results from twinned crystals. In principle, circumstances could be imagined with numerous twin components having several different symmetry relationships (especially for crystal

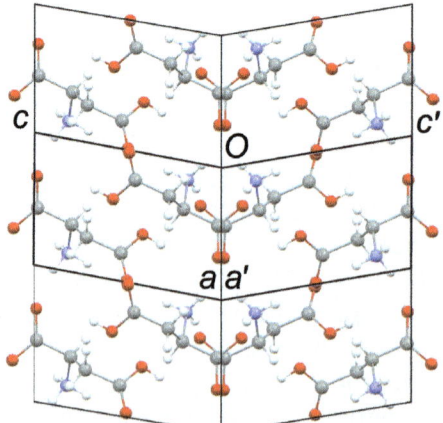

Figure 13.6   Schematic representation of twinning for ʟ-aspartic acid, as discussed in Case Study 2. The unit cells describing each twin domain are related by a 2-fold rotation around the *a* axis.

systems with higher symmetry), but in practice it is most common to encounter two components only. The discussion here is limited to the two-component case.

13.4.1 Merohedral *vs.* Non-merohedral Twinning

For the diffraction pattern of a twinned crystal with two components, two limiting circumstances can be imagined: (1) the diffraction patterns of the two components overlap completely; (2) the diffraction patterns of the two components do not overlap completely. We will refer to the first case as *merohedral* twinning and the second as *non-merohedral*. The exact definitions of these terms are possibly more complicated and several other categories of twinning might also be defined. We will use *merohedral* and *non-merohedral* as descriptive terms in the manner above and avoid any more complicated defin-itions. The main reason to distinguish the two cases is that they impose different practical requirements. The following sections emphasise how twinning impacts the various stages of a single-crystal analysis.

13.4.2 Merohedral Twinning

In order for the diffraction patterns of two twin components to overlap completely, the reciprocal lattice must have appropriate geometry. Merohedral twinning can arise when the metric symmetry of the reciprocal lattice is higher than the actual point symmetry defined by the intensities. For example, consider a monoclinic crystal with Laue group $2/m$ but with $\beta \approx 90°$ (Figure 13.7). The metric sym-metry of the reciprocal lattice is effectively orthorhombic, which means that there are metric 2-fold rotation axes parallel to each of the reciprocal lattice axes. The 2-fold rotation axis parallel to $\mathbf{b}^*$ also ap-plies to the intensities, but the 2-fold axes parallel to $\mathbf{a}^*$ and $\mathbf{c}^*$ do not. Hence, if the crystal were to be twinned so that it had domains related by 2-fold rotation around $\mathbf{a}^*$, the two components of the diffraction pattern would be overlaid, and the intensities would be a mixture of $I(hkl)$ in one diffraction pattern and $I(h\bar{k}\bar{l})$ in the other. Since $I(hkl)$ and $I(h\bar{k}\bar{l})$ are not equivalent according to the Laue group, the overlaid intensities must be unravelled if the structure is to be solved.

When making the initial assessment of the lattice parameters in a single-crystal analysis, there will be no indication of exact merohedral twinning. In the case described above, the lattice parameters would be determined (correctly) with all angles *ca.* 90°. After collecting the intensity data (hopefully at least a hemisphere), the integration would

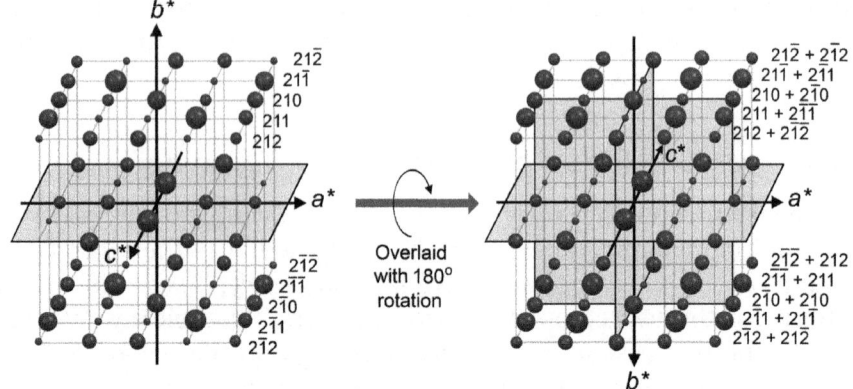

Figure 13.7   Overlay of an intensity-weighted reciprocal lattice with point group
2/*m* with a copy generated by 2-fold rotation around *a**. A few lattice
points are labelled, showing inequivalent reflections *I*(*hkl*) and *I*(*hkl̄*)
overlaid as a result of the twinning. If the two orientations are overlaid in
a ratio close to 50 : 50, the Laue group will appear to be *mmm*.

be carried out in the usual way, but the resulting intensities could
appear to be consistent with Laue group *mmm*, as illustrated in
Figure 13.7. However, difficulties will arise when attempting to solve
the structure in orthorhombic space groups because the correct
structure is actually monoclinic. If structure solution is attempted in
all orthorhombic groups, some recognisable molecule may be obtained in some space groups but it is likely to be overlaid with other
molecules because too many symmetry operations are applied.

To deal with merohedral twinning, the first step is to obtain some
reasonable image of the structure. In the case above, it would be
necessary to try structure solution in the monoclinic system. It may be
difficult to identify which of the three axes is unique (*i.e.* aligned with
the genuine 2-fold symmetry) so the unit cell and *hkl* indices of the
intensities may need to be transformed to try each axis as the *b* axis. If
none of these attempts produce a reasonable structure, solution could
be tried in the triclinic groups *P*1 and *P*1̄. The point is that the true
symmetry of the structure and its diffraction pattern must be lower
than it seems and we need to find a reasonable picture of the unit-cell
contents by trying lower-symmetry possibilities. It may be possible
subsequently to identify a higher-symmetry description, but first we
need a starting point. If a reasonable structure solution can be obtained, merohedral twinning can be handled during refinement.
Conceptually, we need to specify which intensities are overlapped and
define a scale factor to quantify the ratio of the two twin components.

This requires the *twin law*, which is a 3×3 matrix describing the relationship between the overlapped *hkl* values. The twin law is actually a matrix representation of the symmetry operator relating the two reciprocal lattices. In the case of rotation about **a*** in Figure 13.7, the relationship would be:

$$\begin{bmatrix} h' \\ k' \\ l' \end{bmatrix} = \begin{bmatrix} 1 & 0 & 0 \\ 0 & -1 & 0 \\ 0 & 0 & -1 \end{bmatrix} \begin{bmatrix} h \\ k \\ l \end{bmatrix}$$

According to the rules of matrix multiplication, this says $h' = h$, $k' = -k$ and $l' = -l$. Each observed intensity is then compared to the calculated overlapped intensity:

$$I(hkl)_{calc(total)} = k\, I(hkl)_{calc} + (1-k)\, I(h'k'l')_{calc}$$

where k is the scale factor to be refined (one scale factor for the whole structure, not an individual scale factor for each reflection). Handling merohedral twinning in this way is practically straightforward, as long as the twin law can be identified. Various automated algorithms exist to suggest possible twin laws. These might be tried for refinements that look hopeful but get "stuck" with unexpectedly high *R*-factors. There are also a number of common twin laws (for example the case described above), which become recognisable with experience.

13.4.3 Non-merohedral Twinning

Non-merohedral twinning requires a different approach because the diffraction pattern comprises two separate diffraction patterns. The lattice parameters of each diffraction pattern are the same (because they describe domains of the same crystal structure), but their orientations are different and they give separate spots on the diffraction images. Two unit cells must be defined which have the same dimensions but different orientations. Essentially, we have to describe the diffraction images twice. In practice, it is almost inevitable that some beams in the two diffraction patterns will overlap, so the diffraction images and the derived representation in reciprocal space can be complex and potentially misleading.

After collecting and harvesting the first images of the diffraction pattern, the reciprocal lattice viewer should give an indication of non-merohedral twinning because the derived reciprocal lattice points will not look like a genuine lattice (Figure 13.8). In the best scenario, it might be possible by eye to identify a lattice in more than one orientation so

248 Chapter 13

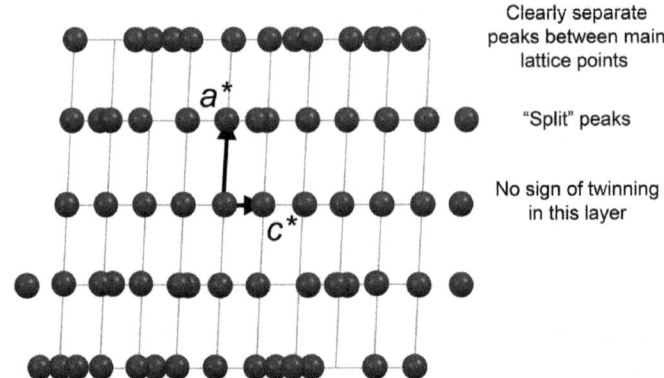

Clearly separate
peaks between main
lattice points

"Split" peaks

No sign of twinning
in this layer

Figure 13.8 Example of a non-merohedral twin within the reciprocal lattice viewer. The correct reciprocal lattice axes are indicated. The first plane up from the origin shows peaks that appear to be closely split, while the next plane up reveals clearly separate peaks that are not indexed.

the existence of twinning is at least recognised. In some cases, the construction will just look like a mess. Running standard indexing algorithms on the data set will produce *some* result, but it is likely to fit only a fraction of the data and to give poor results under least-squares refinement. Often, the fact that there are more reciprocal lattice points due to the overlay means that standard indexing algorithms will suggest unreasonably large unit cells. In this context, it is helpful to have some idea of the expectations for the unit-cell volume and an appreciation of the likely number of molecules in the unit cell for the most probable space groups (*e.g.* 2 for space group $P2_1$, 4 for $P2_1/c$, *etc.*).

Special algorithms are required to index non-merohedral twins. Usually, they make an exhaustive search of the vectors between reciprocal lattice points then consider unit cells based on the shortest and most frequently observed vectors. Unit cells that fit a significant fraction of the reciprocal lattice points within some defined tolerance can be considered reasonable candidates. For each candidate, the algorithm can then consider reciprocal lattice points that are *not* fitted. The aim is to find a second orientation of the same unit cell to fit the remainder of the points. The best scenario is to identify two orientations of a unit cell with a sensible symmetry relationship between them, which together fit all of the measured reciprocal lattice points. In this context, a "sensible" relationship is almost always a 2-fold rotation around one of the real or reciprocal lattice axes. An apparently random rotation is much less likely to be correct. As ever, the result should be assessed by looking at the diffraction images, which must now be compared to two sets of predicted spots (Figure 13.9).

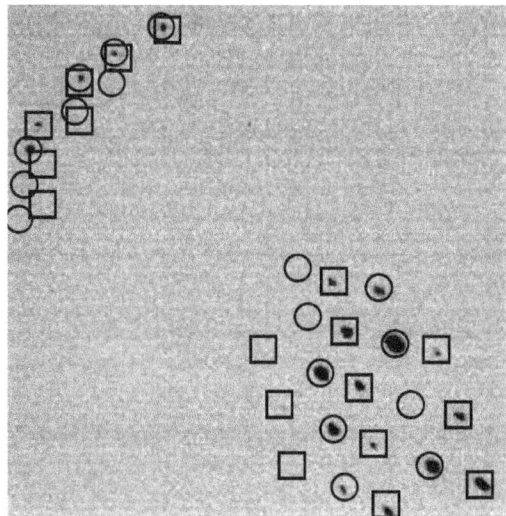

Figure 13.9 Enlarged section from a diffraction image of a non-merohedral twin. The predicted spot positions for two orientations of the unit cell are shown as circles and squares. Some spots are clearly separate for the two domains, while others overlap.

Assuming it is possible to deduce a sensible unit cell in different orientations, it is sufficient to collect complete data for one twin component and then to handle whatever fraction of the other component is collected with it. Preferably, the "primary" unit cell should be the strongest contributor to the diffraction pattern. Even better is to measure the whole of reciprocal space, and this approach to data collection is particularly recommended in twinned cases. However, this will mean spending maximum time on a data collection that ultimately may not provide any satisfactory result, so it is important to make a sensible early assessment of the chances of success. After data collection, the integration and multi-scan correction proceed largely as for any other data collection, except that two components are defined. During the process, there are two categories of reflection: (1) reflections belonging solely to one component and (2) overlapped reflections containing contributions from more than one component. In the final reflection list, isolated reflections from each component are effectively separate observations of the same $I(hkl)$, scaled by the fraction of the twin component. During refinement, the calculated $I(hkl)$ value is compared to the scaled sum of the two measured values. For reflections that overlap, the situation is identical to that described for merohedral twinning. The calculations will of course be taken care of appropriately by the refinement program, but it may be necessary to apply different practical procedures or file formats.

The additional complications of non-merohedral twinning will impact every stage of data processing so it is often necessary to explore different approaches to find the optimal result. If the intensities from one twin component are much stronger than the other, the intensities of the major component are likely to be measured more reliably than those of the minor component. In that case, the best approach may be to ignore the individual intensities of the minor component and proceed only with the major component and any overlapped reflections. If a second component is clearly very weak, it might be better just to ignore the twinning altogether and to carry out a normal integration on the major component. Obviously, intensities will be compromised if they have some contribution from the minor component, but this may turn out to have less severe consequences than the cumulative effect of integrating and refining a two-component data set.

13.4.4   Merohedral *vs.* Non-merohedral Twinning Revisited

In practice, the distinction between merohedral and non-merohedral twinning is seldom so clear-cut. Returning to the example of a monoclinic crystal with $\beta \approx 90°$, it could be that $\beta = 91°$, which means that the reciprocal lattice points will not overlay exactly, but they will be very close. Moreover, the spacing between the points from the two reciprocal lattices will increase further from the origin. In real space, this means that low-angle reflections probably will appear as single (broad) reflections, but higher-angle reflections might clearly be split (Figure 13.10). In this case, the best practical way to proceed can be uncertain. If the splitting between reflections is small, it could just be ignored, and the integration could be allowed to count all pixels in the split spots as a single spot. In other words, the situation could be handled as a merohedral twin. Alternatively, the problem could be approached as a non-merohedral twin with the unit cell defined in two different orientations. In that case, very closely spaced reflections that are meant to be separate might cause instabilities during the integration. In these borderline situations, it will often be necessary to try several approaches to determine which works best. Hopefully, the preceding discussion provides the background understanding to support that practical exercise.

13.4.5   Inversion Twinning

Inversion twinning is a special case, where the structure is non-centrosymmetric but twin domains are related by an inversion centre.

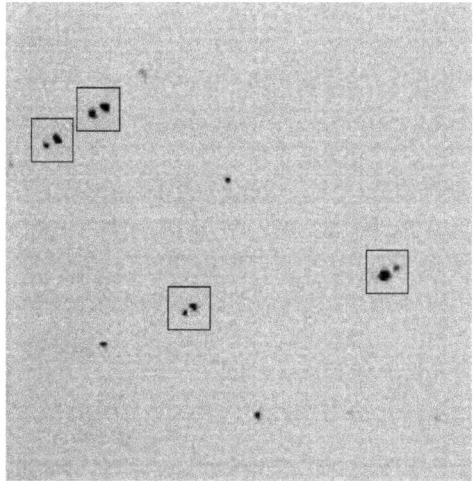

Figure 13.10   Enlarged section from a diffraction image of a non-merohedral twin. Some reflections appear normal, while those highlighted by squares appear to be split.

Clearly, this is the same as any other merohedral twinning. It must be merohedral because a diffraction pattern always has metric inversion symmetry, but the inversion symmetry does not apply to the intensities. Inversion twinning affects the determination of absolute structure because $I(hkl)$ and $I(\bar{h}\bar{k}\bar{l})$ are mixed together. We have already seen how to handle this (Section 12.4): refinement of the Flack parameter (x) is equivalent to refinement of the scale factor between inversion-related twin components:

$$I(hkl)_{\text{obs}} = (1 - x)\, I(hkl)_{\text{calc}} + x\, I(\bar{h}\bar{k}\bar{l})_{\text{calc}}$$

Refining the Flack parameter is therefore a refinement of inversion twinning. If there is no twinning, the Flack parameter should refine to 0 (indicating that the absolute structure of the model is correct) or 1 (indicating that the absolute structure of the model should be inverted), while any intermediate value indicates the presence of domains with different absolute structure. Physical conclusions about the value of the Flack parameter must consider the associated experimental uncertainty, to be discussed in Chapter 14. For now, the point is to appreciate the relationship between absolute structure determination and the methods described to handle merohedral twinning.

13.5 Summary of Key Points

- Disordered structures represent an average of local alternatives. A good disorder model commonly consists of a limited number of specific sets of atomic positions, each of which define a chemically and physically reasonable local structure.
- Disordered atoms have site occupancy factors (SOFs) less than 1. In the structure factor equation, this serves to reduce the scattering contribution from a given atomic site by scaling the atomic scattering factor.
- The geometry of disordered regions must usually be controlled with restraints. If external restraints are imposed, these should as far as possible be derived from analysis of comparable structures in a crystallographic database.
- Disordered regions that are too complex to be modelled using discrete atoms may be handled by "solvent-masking", where electron density values are included at points on a defined grid. Improving the fit to the disordered region improves the outcome of the entire refinement, including the well-ordered parts, but potentially compromises chemical understanding.
- Twinning refers to the circumstance where a measured crystal comprises domains of the same structure related to each other by a local symmetry operation that is not part of the space group. The diffraction pattern comprises an overlay of diffraction patterns from all domains.
- Merohedral twinning, by which the diffraction patterns of the domains are overlaid exactly, can arise when the metric symmetry of the reciprocal lattice is higher than the actual point symmetry of the intensities. It may not be noticed during data collection.
- Non-merohedral twinning requires two separate unit cells to be defined during the data collection and processing, with the same dimensions but different orientations. The diffraction images in this case will be complex and potentially misleading.
- Inversion twinning refers to domains of a non-centrosymmetric structure related by the inversion operation. This is of relevance for absolute structure determination.

13.6 Case Study 1: Griseofulvin–Nitroethane (1 : 1)

This case study concerns the antifungal medication griseofulvin (Figure 13.11) and a study of some nitromethane and nitroethane

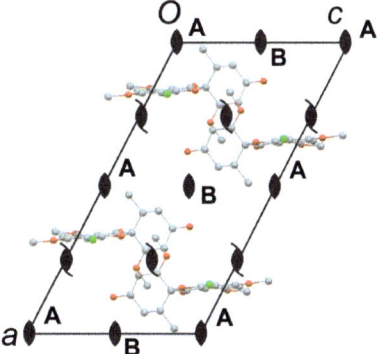

Figure 13.11 Molecular structure of griseofulvin, $C_{17}H_{17}ClO_6$.

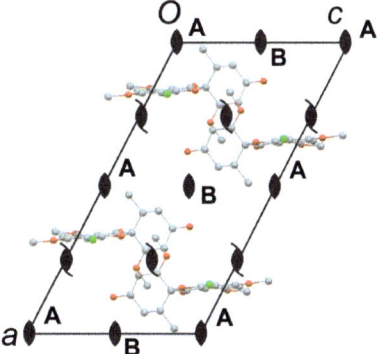

Figure 13.12 Unit cell and symmetry elements (space group *C2*) for griseofulvin–nitroethane (1 : 1), showing ordered griseofulvin molecules and 2-fold symmetric sites where the nitroethane molecules are accommodated (labelled **A** and **B**). H atoms are omitted.

solvates.[4] The 1 : 1 solvate with nitroethane crystallises in the monoclinic space group *C2* (Section 4.4.2) with unit-cell parameters $a = 21.931(4)$, $b = 8.6541(17)$, $c = 11.797(2)$ Å, $\beta = 117.57(3)°$. The unit cell contains four griseofulvin molecules, related by the symmetry operations of the space group (Figure 13.12). The interest here is in the nitroethane molecules, which occupy positions close to 2-fold rotation axes between griseofulvin molecules. Since a molecule of nitroethane does not have 2-fold rotational point symmetry, it must be disordered. This scenario is opposite to the discussion of formoterol fumarate dihydrate in Chapter 3, where individual fumarate anions exhibit point symmetry elements that do not extend continuously through the entire crystal. Here, the 2-fold rotation symmetry does extend through the entire crystal, but it does not apply to the individual nitroethane molecules. There are two distinct nitroethane molecules within the unit cell, which are aligned in different ways relative to the 2-fold axes (Figure 13.13). We may think, perhaps, that if the 2-fold rotation axes do not apply to the nitroethane molecules,

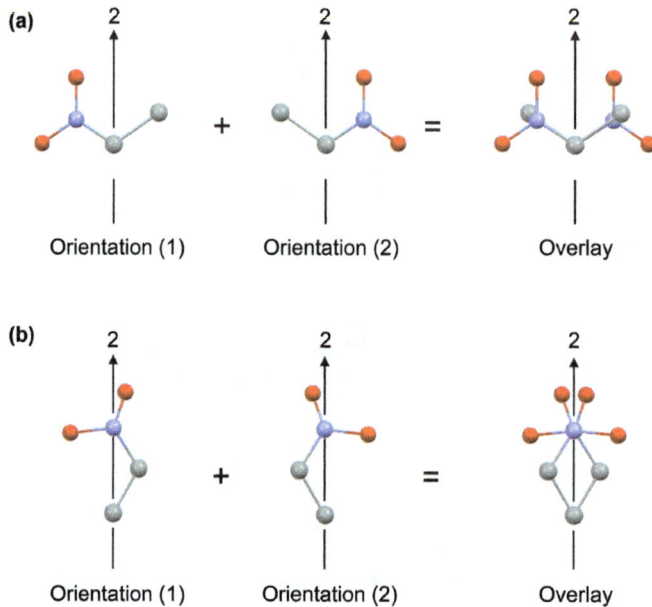

(a) Orientation (1)   Orientation (2)   Overlay

(b) Orientation (1)   Orientation (2)   Overlay

Figure 13.13   Sites **A** and **B** for the nitroethane molecules. H atoms are omitted. (a) In site **A**, the orientation of the molecule relative to the 2-fold rotation axis produces an overlaid image that could be mistaken for dinitromethane. (b) In site **B**, the overlaid image is difficult to recognise as a nitroethane molecule. Note that the N atom and one C atom lie on the 2-fold axis.

they should not be included in the symmetry of the space group. However, it is clear that the griseofulvin molecules conform to space group $C2$, so the nitroethane molecules occupy sites with genuine 2-fold rotational symmetry. In such a site, each nitroethane molecule could adopt one of two positions, related by the 2-fold axis, with entirely equivalent surroundings (and therefore identical interaction energy). Assuming that the choice of position in one site does not affect the choice of position in another site, we can expect an even distribution between the two possibilities. When this situation is averaged into a single unit cell, the electron density at the nitroethane sites will appear to be a 50:50 overlay of the two orientations, giving an overall electron density that conforms to the 2-fold symmetry. The challenge during structure refinement is to unravel the "overlaid" electron density to reveal the individual molecules.

   In the original publication of the structure,[4] the authors include some atomic sites to account for the electron density close to the 2-fold axes. For the molecule in position **A**, the sites grow into a molecule that could be mistaken for dinitromethane, $O_2N{-}CH_2{-}NO_2$ (Figure 13.13(a)). We know that it is not dinitromethane because that was not present in the sample, but there is clearly a chance for mis-interpretation if this structure were to be viewed in isolation (this is not the fault of the authors). The molecule in orientation **B** is harder to model and the published atomic sites are not recognisable as nitroethane. In fact, a recognisable molecular structure corresponding to situation **B** can be produced from this data set by careful modelling with some restraints applied to control the geometry (Figure 13.13(b)). It is even possible to refine anisotropic ADPs, with only one O atom showing any significant distortion. This updated refinement produces $R1 = 0.039$ and $wR2 = 0.095$.

Since the original structure refinement did not produce any recognisable molecule for orientation **B**, and a potentially misleading picture for orientation **A**, the authors decided to apply the *SQUEEZE* procedure. After removing the atoms of the nitroethane molecules, *SQUEEZE* identifies two voids in the unit cell, each with volume 223 $Å^3$. The electron count in the unit cell is 157, which matches well to the expected four $CH_3CH_2NO_2$ molecules (40 electrons per molecule). Including the contributions from *SQUEEZE* to the structure factors, the refinement converges to give $R1 = 0.033$ and $wR2 = 0.077$. The fact that the R-factors are marginally improved compared to the atomic model shows that even the improved disorder molecule does not ideally fit the electron density in the disordered regions. In this case, however, the improvement with *SQUEEZE* is marginal so the benefits of having a genuine atomic model probably outweigh the benefits of an improved overall fit to the electron density. For example, it is interesting to compare PXRD patterns simulated from the two cases (Figure 13.14), which look rather different. The peak positions are the same in the two cases (because they depend only on the unit-cell geometry) but the intensities are very different (because they depend on the atom types and positions within the unit cell). Hence, it will be difficult to compare the simulated pattern to any experimental patterns if *SQUEEZE* is used. In cases where disorder cannot be so readily interpreted, the benefits of *SQUEEZE* on the overall refinement will be greater, but authors should still provide the intensity data prior to application of *SQUEEZE* to allow others to investigate potential atomic descriptions.

13.7 Case Study 2: L-Aspartic Acid

Aspartic acid is a chiral non-essential amino acid used in biosynthesis of proteins. Crystal structures have been reported for the pure

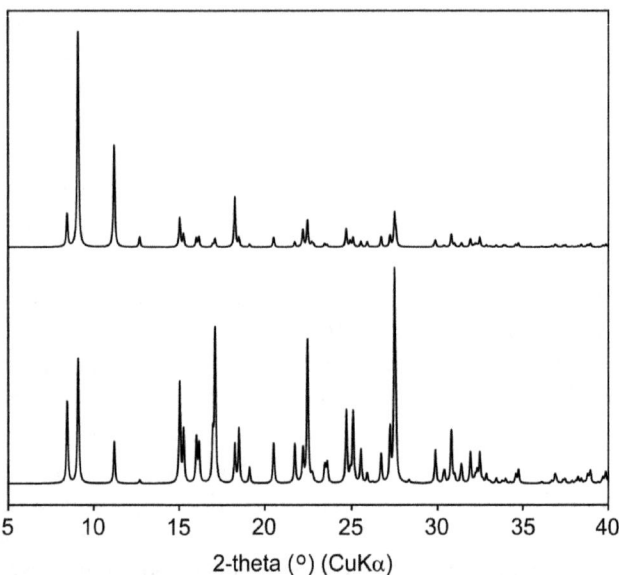

Figure 13.14 Simulated PXRD patterns produced from griseofulvin–nitroethane (1:1): (bottom) with an atomic representation of the disordered nitroethane molecules; (top) omitting the nitroethane molecules and applying *SQUEEZE*.

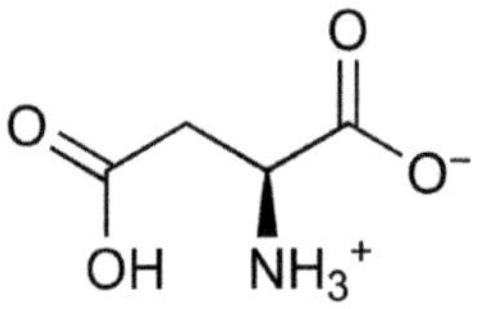

Figure 13.15 Molecular structure of L-aspartic acid, $C_4H_7NO_4$, shown in its zwitterionic form.

enantiomer and racemic mixture.[5,6] This case study considers the L-enantiomer (Figure 13.15) and is based on diffraction data collected to 0.84 Å resolution at 298 K using Cu radiation. Initial measurements yielded the positions of 614 reflections, which could be indexed fitting 96% of reflections to give the following potential Bravais lattices:

Bravais Lattice	FOM	a (Å)	b (Å)	c (Å)	α (°)	β (°)	γ (°)	Vol (Å^3)
Orthorhombic C	0.72	5.14	29.98	6.97	89.88	90.01	90.06	1076
Monoclinic C	0.83	29.98	5.14	6.97	90.01	90.12	89.94	1076
Monoclinic P	0.70	5.14	6.97	15.20	90.12	99.68	90.01	538
Triclinic P	1.00	5.14	6.97	15.20	90.12	99.68	90.01	538

The metric situation is reminiscent of aspirin form I (Case Studies 8 and 9), where the lattice is approximately consistent with ortho-rhombic C but we should probably be cautious because the mono-clinic and triclinic crystal systems are overwhelmingly more popular for molecular crystals. Collecting and integrating the data using the triclinic P cell (*i.e.* making no assumptions about the symmetry) al-lows the Laue group to be properly assessed using the intensities. This yields $R_{int} = 0.025$ for monoclinic P and $R_{int} > 0.26$ for all C-centred options. Hence, the listed monoclinic P unit cell and Laue group $2/m$ seem appropriate.

Proceeding with monoclinic P, the systematic absences ($0k0$ absent for odd values of k) indicate space group $P2_1$ and dual-space methods identify a structure in $P2_1$ with two molecules in the asymmetric unit (Figure 13.16). However, initial cycles of least-squares refinement converge with $R1 = 0.384$ and $wR2 = 0.823$. Clearly, something is wrong.

The clue on how to proceed is indicated in Figure 13.16. The un-usual feature is that the two molecules within the asymmetric unit appear to be related by *translation*. Hence it appears that the chosen lattice does not properly describe the translational symmetry. Meas-uring the apparent translation within a structure viewer, the a and b axes of the unit cell look correct but the c axis looks to have length *ca.* 7.6 Å and form a different angle to the ab plane. This exercise takes us back to the beginning of Chapter 2: we have produced an image of the crystal structure so we can look at its translational symmetry to identify the underlying lattice. Now we must look again at the dif-fraction pattern to see whether the proposed lattice can be fitted to it. We can imagine taking the measured reciprocal lattice and looking for vectors within it that correspond to the expected set of unit-cell lengths. In this case, we suggest unit-cell lengths $a \approx 5.14$, $b \approx 6.97$ and $c \approx 7.61$ Å (leaving the angles unspecified) and allow appropriate

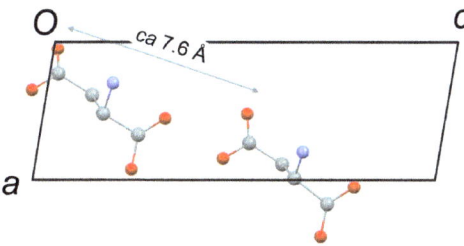

Figure 13.16 Asymmetric unit for the initial structure solution of L-aspartic acid in space group $P2_1$ shown within the defined unit cell. The two mol-ecules are clearly related by translation.

software to search the reciprocal lattice. Crucially, the process should not be expected to fit *all* reflections because we suspect that there may be another crystal domain contributing to the diffraction pattern. The search identifies a lattice with primitive unit cell $a = 5.14$, $b = 6.97$, $c = 7.60$ Å, $\alpha = 90.1$, $\beta = 99.9$, $\gamma = 90.0°$ that fits 67% of reflections within a standard tolerance. Searching again on the reflections that have not been fitted finds a second orientation of the same lattice, related to the first by a 180° rotation around the a axis. The two orientations of the unit cell together account for 610 out of the initial 614 measured reflections: 222 reflections belong only to domain 1, 198 belong only to domain 2, and 190 are overlapped. The situation is illustrated in Figure 13.17.

The reason for the difficulties is now clear. The two orientations of the reciprocal lattice overlap in such a way that layers in the b^*c^* planes alternate between exact overlap and offset so that c^* appears to be halved. This leads to the apparent doubled translation (*ca.* 15.2 Å) in the initial suggested lattice. With the complete picture, it can be seen how the reflections are distributed between the two domains. Every other layer of reflections (those with even h) overlap fully. In the layers that do not overlap fully (those with odd h), each domain fits half of the reflections. Hence, $\frac{1}{3}$ of the reflections are unique to domain 1, $\frac{1}{3}$ are unique to domain 2, and $\frac{1}{3}$ overlap. This is roughly the ratio obtained from the indexing software $(222:198:190)$. Once the diffraction pattern is indexed in this way, it can be integrated as a non-merohedral twin with two components and then the structure can be solved and refined easily in space group $P2_1$ with one molecule

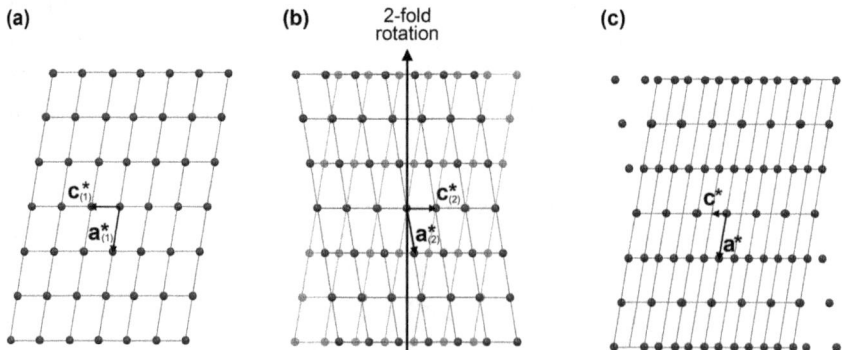

Figure 13.17   Projection of the reciprocal lattice measured for L-aspartic acid along b^*, showing the a^*c^* plane: (a) shows the lattice for one domain; (b) shows also the lattice for the second domain, rotated with respect to the first one by 180° around the a axis; (c) shows the observed overlay with the incorrect lattice vectors that were initially obtained.

in the asymmetric unit. The refinement converges to $R1 = 0.044$, $wR2 = 0.137$, with a refined scale factor indicating twin domains in a $75:25\%$ ratio for this specific crystal.

References

1. S. Cuffini, R. A. Howie, E. R. T. Tiekink, J. L. Wardell and S. M. S. V. Wardell, *Acta Crystallogr., Sect. E: Crystallogr. Commun.*, 2009, **E65**, o3170, [CSD: AJEYAQ].
2. A. L. Spek, *Acta Crystallogr., Sect. C: Struct. Chem.*, 2015, **C71**, 9.
3. C. H. Görbitz, *Acta Cryst.*, 2010, **C66**, o531, [CSD: LAMDEK].
4. S. Aitipamula, P. S. Chow and R. B. H. Tan, *Acta Crystallogr., Sect. B: Struct. Sci., Cryst. Eng. Mater.*, 2013, **B70**, 54, [CSD: PINNEH].
5. J. L. Derissen, H. J. Endeman and A. F. Peerdeman, *Acta Crystallogr., Sect. B: Struct. Sci., Cryst. Eng. Mater.*, 1968, **B24**, 1349, [CSD: LASPRT].
6. A. Sequeira, H. Rajagopal and H. Ramanadham, *Acta Crystallogr., Sect. C: Struct. Chem.*, 1989, **C45**, 906, [CSD: DLASPA02].

14   Crystallographic Results

Summary

Structure refinement produces a set of parameters describing the crystal structure, comprising a unit cell, space group and list of atom types, coordinates and displacement parameters. Each refined parameter has an associated uncertainty, obtained from the least-squares refinement, which carries through to uncertainties in derived parameters such as interatomic distances and angles. These must be considered carefully when assessing crystallographic results. The nature of crystallographic data is well suited to a structured representation and the primary format for reporting crystallographic data is the electronic Crystallographic Information File (CIF). The CIF contains the entire record of the analysis after integration of the primary diffraction images, and it can be checked for consistency using automated validation systems. Published crystal structures are collected and distributed in extensive databases containing hundreds of thousands of structures. The reasons to determine a crystal structure and the ways in which structural information might be interpreted can vary widely. This chapter gives some examples relevant to pharmaceutical crystals, including a case study concerning absolute-structure determination.

14.1   Introduction

Structure refinement produces a set of parameters describing the crystal structure. As for any measurement, the established parameters must have associated uncertainties that come from experimental

Pharmaceutical Crystallography: A Guide to Structure and Analysis
By Andrew Bond
© Andrew Bond 2019
Published by the Royal Society of Chemistry, www.rsc.org

error, which can be estimated within the least-squares refinement. The interest in a crystal structure is usually not in primary parameters such as the atomic coordinates but rather in derived results such as interatomic distances and angles. Uncertainties in the unit-cell parameters and atomic coordinates carry over into uncertainties in these derived results, which must be considered carefully when comparing crystal structures. The reasons to determine a crystal structure and the ways in which structural information might be interpreted can vary widely. This chapter gives a few examples relevant to pharmaceutical crystals, including an illustration of absolute-structure determination. Whatever the reason for determining a crystal structure, it is important to consider the extent to which the established primary and derived parameters conform to physical and chemical expectations and the validity of any conclusions in relation to the uncertainties that come from experimental error. The aim of this chapter is to introduce an appropriate degree of scepticism to enable critical assessment and comparison of crystallographic results. The first sections are a little abstract, but they are vital to build up a physical picture of what it means to compare crystallographic values.

14.2 Uncertainties

The best estimate for any experimental parameter is the mean value obtained from many measurements, quoted with an associated uncertainty related to the variation of the measurements around the mean. Assuming that experimental errors are random, repeated measurements of the same parameter should tend towards a *normal distribution* (Figure 14.1), which indicates the probability of obtaining a given value when a measurement is made. The shape of the normal distribution shows that any measurement of a parameter is most likely to yield its mean value and that the probability of obtaining some other value decreases as it gets further from the mean. The distribution is symmetrical around the mean with width quantified by the *standard deviation*, σ. If the total area under the curve is normalised to 100%, the area between defined limits gives the percentage probability of measuring a value between those limits. 68.3% of the area is contained within $\pm\sigma$ of the mean, 95.5% is contained within $\pm2\sigma$ of the mean, and 99.7% is contained within $\pm3\sigma$ of the mean. Hence, any measurement of a parameter is almost guaranteed to produce a value within $\pm3\sigma$ of its mean.

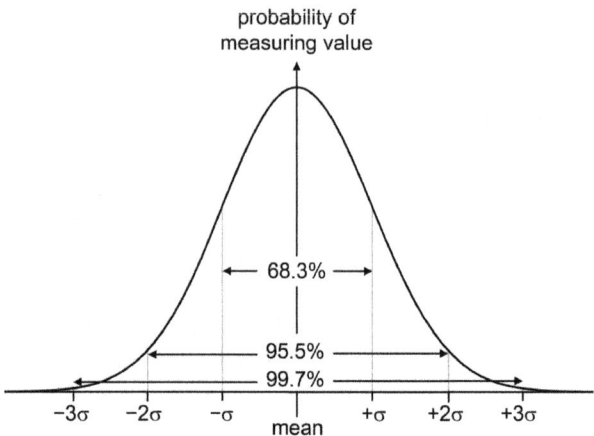

Figure 14.1   Normal distribution around the mean value of a parameter with standard deviation σ. The percentage values refer to the area under the curve between the $\pm\sigma$ limits.

For a single-crystal X-ray analysis, the parameters are not measured *directly*. For example, it is not possible to make repeated experimental measurements of a specific atomic coordinate. However, as noted in Chapter 12, every parameter contributes to every structure factor. Since there are many more structure factor equations than parameters, there *are* many indications of each parameter but they are mixed together in the various equations. It is possible to unravel the equations in the least-squares process to extract a standard deviation for each refined parameter. The mathematical details can be left to the refinement software, which will output the results as estimated σ values. Since errors on X-ray measurements come from many sources, not just random statistical error, the more general term *standard uncertainty* (s.u.) is preferred over *estimated standard deviation* (e.s.d.). The *precision* of each parameter is inversely related to its standard uncertainty and the overall "precision of the crystal structure" refers qualitatively to the magnitudes of the uncertainties. When reporting crystallographic results, the convention is to include the standard uncertainty in parentheses referring to the least-significant digit. For example, 1.6020(8) indicates a measured value of 1.6020 with an s.u. of 0.0008, while 0.40610(12) indicates a value of 0.40610 with an s.u. of 0.00012. The established convention is to limit the s.u. to the range 2–19. Thus, a value 2.10674 with s.u. 0.00382 should be reported as 2.107(4).

14.2.1 Accuracy *vs.* Precision

The standard uncertainty refers to statistical variation of a measured value around its mean due to random errors. This does not give any indication of whether the mean is actually anywhere near the true value because the mean could be displaced by *systematic errors* (Figure 14.2). The agreement between the measured mean and the true value is the *accuracy* of the parameter. If a value is measured precisely, it should be reproducible within the specified uncertainty under the same measurement conditions. If the value is measured precisely and accurately, it should be reproducible within the specified uncertainty under any measurement conditions. It is possible for a parameter to be measured precisely but entirely inaccurately. For example, unit-cell parameters are optimised at the end of the integration by least-squares refinement against thousands of measured reflections. Since there are very few parameters, the data-to-parameter ratio is huge and the resulting s.u. values are often very small. Reported s.u. values are typically of the order of 10^{-4} Å (*i.e.* 1×10^{-14} m) for unit-cell lengths and 10^{-3} degrees for unit-cell angles. If these values are measured accurately, multiple reports of the same crystal structure should produce identical values within the reported precision. A cursory glance of unit-cell parameters in a crystallographic database will demonstrate that this is not always the case. The level of agreement is very good (else we would have no faith in

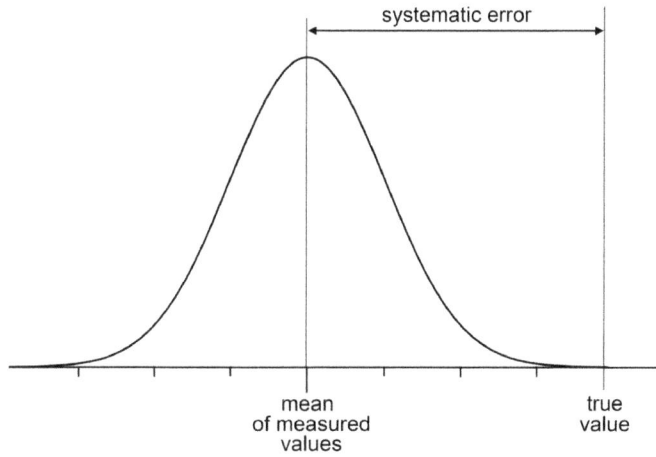

Figure 14.2 Accuracy *vs* precision. The measured values show a normal distribution whose mean is offset from the true value as a result of systematic error. The normal distribution may be narrow (*i.e.* the measurement is precise), but the result is inaccurate.

crystallographic results), but it is not quite as good as suggested by the precision that is often reported.

14.2.2 Comparing Values with Uncertainties

Each standard uncertainty produced from the least-squares refinement can be viewed to define a normal distribution around the true value of the parameter. The actual parameter values obtained from the refinement are *measured values*, which fall somewhere within the corresponding normal distributions. To compare parameters, we are interested in the difference between their true values, which we will denote Δ, but we can only compare measured values. We must understand how the uncertainty related to each measured value contributes to the uncertainty in the assessment of Δ. When we measure the two parameters to be compared, the most probable outcome is the mean value for both, so the most probable value for Δ is the difference between the two means. However, there are many combinations of measured values that produce the same Δ, so the total probability of obtaining the mean difference is not just the probability of measuring both means. It is actually the sum of the probabilities of measuring any two individual values separated by the mean value of Δ. The probability of obtaining any other value for Δ will depend similarly on the probabilities of obtaining two measured values with that difference. Again, this will be a sum over all possible combinations giving that value of Δ. Without showing it explicitly, this line of thinking builds a normal distribution for Δ around its mean value (Figure 14.3). The remaining question is how the standard uncertainty of the difference, σ_Δ, is related to the individual standard uncertainties, σ_1 and σ_2. It turns out that the squares of the standard deviations (corresponding to the *statistical variance*) are additive, so $\sigma_\Delta^2 = \sigma_1^2 + \sigma_2^2$. Hence, σ_Δ is larger than either of the individual uncertainties (as we might expect), but it is smaller than their sum.

14.2.3 Practical Consequences for Comparing Crystallographic Parameters

The most basic question when comparing two measured parameters is whether they are the same or different. If the true values of the parameters are the same, their true difference is zero and the measured difference should fall within a normal distribution centred on zero with standard deviation σ_Δ. Two measured parameters can be

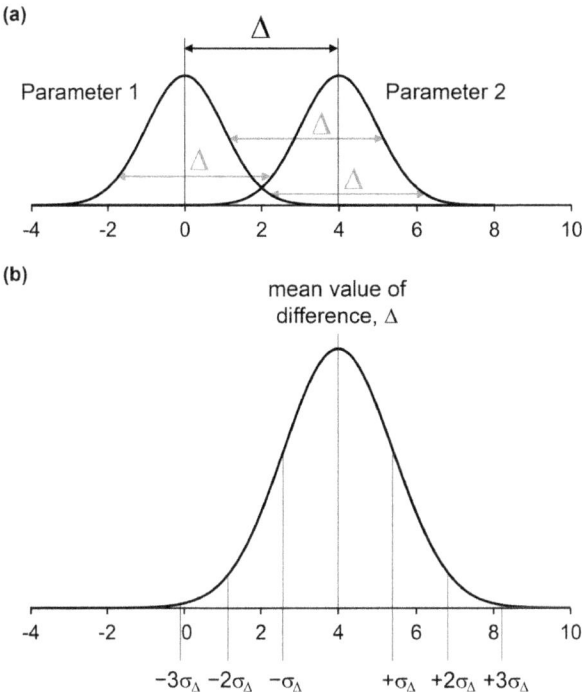

Figure 14.3 Building a normal distribution for the difference (Δ) between two parameters from the normal distributions of the individual parameters: (a) the difference between the two means is indicated, with examples of other combinations giving the same Δ; (b) the normal distribution for Δ is centred on the mean difference and has σ_Δ larger than σ for either parameter.

said to be different if the measured difference is greater than $\pm 3\sigma_\Delta$ away from zero. Two examples are as follows:

Value(1)	Value(2)	Δ	σ_Δ	$3\sigma_\Delta$	Conclusion
1.367(4)	1.382(5)	0.015	$\sqrt{(0.004^2 + 0.005^2)}$ $= 0.0064$	0.019	$\Delta < 3\sigma_\Delta$ Values cannot be said to be different
1.367(3)	1.382(2)	0.015	$\sqrt{(0.003^2 + 0.002^2)}$ $= 0.0036$	0.011	$\Delta > 3\sigma_\Delta$ Value (2) is greater than value (1)

In the second case, we cannot say *exactly* how much greater value (2) is than value (1) because we don't know either true value, but we can say for sure that the true value of (2) is greater than the true value of (1) (at the confidence level implied by 3σ). This shows the crucial role of uncertainties when comparing crystallographic values: only a slight increase in the uncertainties in the first case eliminates the possibility to conclude that value (2) is different from value (1). The uncertainties in the primary refined parameters and unit-cell parameters carry through to uncertainties in derived results such as interatomic distances and angles. Larger uncertainties for any of the primary parameters produce larger uncertainties in the derived results.

Some parameters of a crystal structure will not have associated uncertainties. For example, unit-cell parameters constrained by symmetry during the least-squares refinement. Similarly, constrained parameters in the structure refinement will not have uncertainties. The most common example is H atoms riding on their parent C atoms (Section 12.3.3), for which the C–H distance and constrained U_{iso} value for H will be exact. The refinement software should deal with this appropriately but the user should realise why some output results do not have uncertainties. When comparing to a value without an uncertainty it must simply be assumed that the stated value is the true value without any associated normal distribution ($\sigma = 0$). Unfortunately, it is common in the literature to see reported values that should have uncertainties but do not. This often arises for non-bonded distances or angles that may not be output automatically by the refinement software so are measured later from the completed structure. To obtain uncertainties, these should be specified as a result to be output during the refinement since this is the only way that the least-squares equations can properly calculate the values.

14.3 Interpreting Crystallographic Results

The reasons to determine a crystal structure, and the ways in which structural information might be interpreted, can vary widely. However, some basic comments apply to all crystallographic results. The main considerations should always be the extent to which the primary and derived parameters conform to physical and chemical expectations, and the validity of any conclusions in relation to the uncertainties. The following sections give some examples.

14.3.1 Resolution

The resolution of the measured intensity data has a significant impact on all refined parameters and their uncertainties. As the resolution decreases (*i.e.* the maximum measured diffraction angle decreases), the image of $\rho(xyz)$ produced by the Fourier summation is degraded and the data-to-parameter ratio decreases in the least-squares refinement. The Case Study in Chapter 11 showed $\rho(xyz)$ for sulfathiazole, calculated using data to different resolution levels. It is interesting to observe the consequences for the refined atomic coordinates as the resolution is reduced. The coordinates of the N atom of the NH_2 group are as follows:

	x	y	z
0.46 Å	0.98736(3)	0.19077(4)	0.57026(3)
0.85 Å	0.98746(12)	0.19057(16)	0.57057(12)
1.00 Å	0.98735(18)	0.1905(3)	0.5710(2)

The uncertainties increase significantly, from *ca.* 3×10^{-5} at the best resolution to *ca.* 2×10^{-4} at the worst resolution. Hence, the overall precision of the crystal structure is crucially dependent on the resolution of the data. The effects are carried over into the derived bond lengths. For example, the uncertainties on the bond lengths of the aminobenzene ring increase from *ca.* 4×10^{-4} Å to 3×10^{-3} Å (Figure 14.4). Although the values of matching bond lengths in Figure 14.4 appear to change, the differences are mostly not significant in relation to the associated uncertainties. In one case, however, this is not true: comparing the 0.46 Å and 1.00 Å data sets, 1.3873(5) Å is significantly longer than the matching value of 1.376(3) Å ($\Delta = 0.0113$, $3\sigma_\Delta = 0.009$). Hence, changing the resolution of the data set in this case leads to a genuinely different result.

This discussion should be put in the context of the aims for determining the crystal structure. Collecting intensity data to 0.46 Å resolution is an extreme example, which most crystallographers will never need to contemplate. At 1.00 Å resolution, the image of the sulfathiazole molecule and its arrangement in the crystal structure is clear and unambiguous, so that could be sufficient if those are the points of interest. However, data to 0.85 Å resolution are probably required to obtain reasonable uncertainties on the bond distances and angles, especially if there is any intention to compare these to other crystal structures.

14.3.2 Atomic Displacement Parameters

Atomic displacement parameters represented as ellipsoids (or spheres for isotropic ADPs) provide an informative picture of a refined crystal structure, and they should always be examined when assessing a

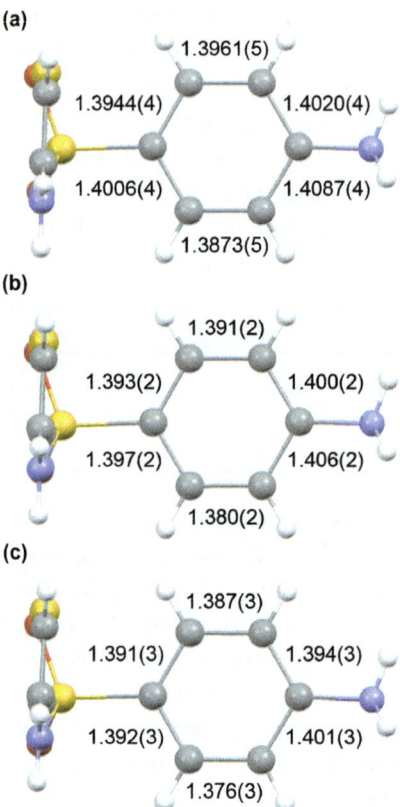

(a)

1.3961(5)

1.3944(4)   1.4020(4)

1.4006(4)   1.4087(4)

1.3873(5)

(b)

1.391(2)

1.393(2)   1.400(2)

1.397(2)   1.406(2)

1.380(2)

(c)

1.387(3)

1.391(3)   1.394(3)

1.392(3)   1.401(3)

1.376(3)

Figure 14.4 Geometry of the aminobenzene ring in sulfathiazole for data sets measured to 0.46, 0.85 and 1.00 Å resolution. Literature reference as in Chapter 11.

crystallographic result. It has already been mentioned that displacement ellipsoids should be expected to have broadly similar sizes for all atoms in a molecule. Disordered regions or solvent molecules within the structure can be expected to be more erratic, but ellipsoids that are clearly different from their neighbours should generally raise suspicion (as in Figure 12.8). One circumstance that was not mentioned in Chapter 12 is the possibility for a displacement ellipsoid to become "negative". Ordinarily, the U_{ij} values define an ellipsoid with three positive principal axis lengths. However, the refined U_{ij} values can produce a matrix that does not conform to this condition (*i.e.* one or more axis is calculated to be negative). Such a matrix is referred to as *non-positive definite* (n.p.d.) and the conclusion is that the ADPs have lost any physical meaning. It may be just that the specific atom type is wrong, but more likely it indicates wider problems with the measured intensities, such as significant anisotropic absorption or twinning that has not been accounted for. This is a clear sign that something is wrong so a crystal structure with non-positive definite ADPs should not be trusted.

Since there are twice as many U_{ij} values as coordinates for each anisotropic atom, anisotropic ADPs often comprise a significant majority of the parameters to be refined. When there are few physical expectations of how the resulting displacement ellipsoids should look, there is a risk that these parameters might "soak up" errors in the data without any recognisable warning signs. If this was to occur for the atomic coordinates, it would be obvious from the resulting molecular geometry. It is much harder to recognise unreasonable ADPs. For this reason, the ADPs are sometimes referred to as the potential "dustbin for errors". Aside from the idea that the ellipsoids should be broadly comparable in size and not deviate too far from spherical, what other physical expectations might be applied? One concept is that atoms bonded to each other should have similar displacement along the direction of the bond. The physical basis is that chemical bonds are by far the stiffest interactions between atoms in a crystal structure. If a bond is approximated as rigid, the bonded atoms must have equivalent displacements along the bond axis. This leads to the *Hirshfeld rigid bond test*, which is typically implemented in automated structure validation systems (Section 14.4.2). The test involves calculating the displacement of each bonded atom along the direction of the line between them and checking that this is equivalent within the associated uncertainty. The purpose of such tests is not necessarily to draw physical conclusions from the ADPs, but rather to hunt for warning signs that the ADPs might be compensating for deficiencies in the measured intensities or structure model.

14.3.3   H Atoms

There is always reason to be suspicious of H atoms in X-ray crystal structures. This does not mean that H atoms cannot be reliably identified, just that it is necessary to be aware of the issues surrounding H atoms and the variety of refinement approaches that might be taken. H atoms bonded to C atoms are generally likely to be placed in sensible positions by refinement software, although it is possible that this could go wrong if the user was inattentive. Mistakes of that type will usually be obvious because they produce unexpected chemical formulae, unusual chemical connectivity and residual features in $\Delta\rho(xyz)$. The U_{iso} values of riding H atoms will be tied to those of their parent atoms

(Section 12.3.3), so they have little meaning on their own. The C–H bond lengths, and any other measurements involving constrained H atoms, will not have standard uncertainties because the H atom co-ordinates have not been assessed as part of the least-squares refinement. The values must have *some* degree of uncertainty, which instinctively would be expected to be greater than the rest of the structure, but this is not obtained from the refinement.

H atoms involved in hydrogen bonds are more difficult to treat consistently and to assess later. The best scenario is that the H atoms were refined without any restraints and with their own isotropic ADPs that converge to values comparable to other atoms in the structure. Refined U_{iso} values for H atoms can be expected to be a little larger, but they should not be excessive. If this approach produces reasonable X–H bond distances (a reasonable range probably being 0.8–1.0 Å) and a sensible H-bonding pattern, then it can broadly be trusted. If restraints have been applied to the X–H bond distances, it may be an indication that the data did not support free refinement of the H atoms or it may just be an indication of the preferences of the person carrying out the refinement. If H atoms involved in hydrogen-bonding are *constrained*, the results must be taken with a degree of scepticism. Hopefully, the reasons for choosing where to add H atoms will be explained (in the CIF or associated journal article), but it may well be just a modelling exercise without any real regard for the data. Obviously, such an approach cannot form the basis for an important chemical conclusion such as whether a multi-component crystal should be viewed as a salt or co-crystal. Even for refined H atoms, the uncertainties on the co-ordinates and derived interatomic distances are likely to be larger than those involving non-H atoms. For example, in sulfathiazole, using data to 0.85 Å, the H atoms of the NH_2 group are refined freely and the geometrical features of the hydrogen bonds are as follows:

	N–H (Å)	H··· N/O (Å)	N··· N/O (Å)	N–H··· N/O (°)	Symmetry operator applied to acceptor N/O
N11–H11A ···N10	0.88(2)	2.36(2)	3.1791 (18)	155.9 (19)	$\frac{1}{2}+x, \frac{1}{2}-y, -\frac{1}{2}+z$
N11–H11B ···O12	0.89(2)	2.14(2)	2.9888 (18)	159.3 (17)	$\frac{3}{2}-x, \frac{1}{2}+y, \frac{3}{2}-z$

The uncertainties on the distances involving H are 0.02 Å, compared to 0.0018 Å for the distance between the non-H atoms. Hence, if seeking to compare H-bonds, the separation between the non-H atoms is far more reliable than any value derived from refined H atom positions.

The message is that conclusions involving H atoms in X-ray crystal structures require *thought*. If the details of hydrogen bonding or H-atom transfer are vital points of interest, it is advisable to make a specific crystallographic analysis where every step is designed to optimise the possibility to make a valid conclusion. For example, make a low-temperature data collection on the best possible crystal to the highest possible resolution. For published studies where the intensity data are available, the last stages of the refinement can be looked at again. Deleting H atoms from the model then calculating $\Delta\rho(xyz)$ can indicate the extent to which H atoms can actually be identified. As described in Section 12.3.3, bond lengths involving H atoms will always be systematically short in an X-ray crystal structure. Ordinarily this is just accepted, but it is important to correct for it if the structure is to be used as the basis for some energy calculation where nuclear positions are required. The correction is achieved by moving the H atoms along their bond vector to standard distances derived from neutron diffraction (since neutrons are scattered from atomic nuclei rather than electrons). Quantum-chemical calculations can also be extremely useful to augment X-ray results because they provide an independent assessment of whether the established H-atom positions are physically and chemically reasonable. Especially for low-resolution data sets, or structures determined from PXRD data, extra confidence can be gained by demonstrating that the structure remains stable if its geometry is optimised to a position of minimum energy.

14.3.4 Absolute Structure

When the Flack parameter (x) is determined from experimental data, the errors associated with the intensity measurements propagate through to an uncertainty on the refined value of x. A crucial question is how precisely x must be determined to make a valid conclusion about the absolute structure. A useful concept is the "inversion-distinguishing power", which can be considered to be the reciprocal of the standard uncertainty on x. The smaller the value of the uncertainty of x, the higher the inversion-distinguishing power of the data set. Essentially, this depends on the magnitudes of the differences between $I(hkl)$ and $I(\bar{h}\bar{k}\bar{l})$, which are enhanced by overall stronger scattering and a greater anomalous contribution. Hence, the best chances of success should be obtained for strongly-scattering crystals with a substantial anomalous contribution. For any measured data set, we must establish that the data have sufficient inversion-distinguishing power before interpreting the value of x. Some guidelines are developed as follows.

For a non-centrosymmetric structure, the physical range of x is 0–1, where 0 corresponds to the correct absolute structure, 1 corresponds to the inverted structure, and any intermediate value indicates inversion twinning (domains of opposite handedness within the same crystal; see Section 13.4.5). We can imagine normal distributions centred on 0 and 1 and expect the measured value to lie within $\pm 3\sigma$ of either 0 or 1. This defines a "statistical range" of $-3\sigma \le x \le 1 + 3\sigma$ in which the value of x must lie. Any refined value of x outside this range indicates a serious problem. Then we can define what we would consider to be strong inversion-distinguishing power. For a crystal that is known to be enantiopure, for example because it was grown from a solution known to contain only one enantiomer of a molecule, the true value of x can only be 0 or 1. A data set with strong inversion-determining power should be able to distinguish between these two cases unambiguously, which means that the refined value of x should lie within $\pm 3\sigma$ of either 0 or 1, and *not* lie within another normal distribution centred at $x = 0.5$. If the three distributions just touch each other (Figure 14.5), the range 0–1 spans 12σ, so σ can take a maximum value of $1/12$ (≈ 0.08) before the distributions would start to overlap. A general guideline for satisfactory determination of the

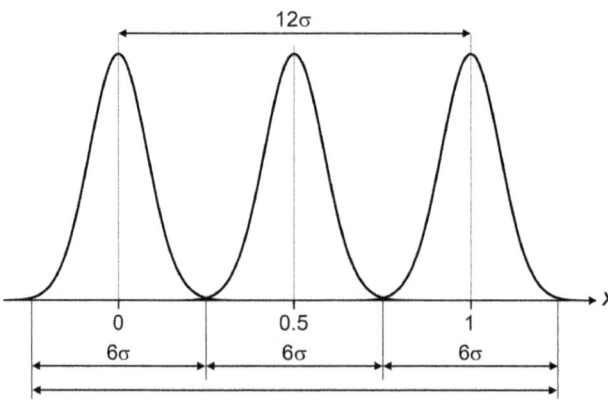

Statistical range of Flack parameter

Figure 14.5 Normal distributions with standard deviation σ centred on values of the Flack parameter (x) corresponding to the correct absolute structure $(x = 0)$, incorrect absolute structure $(x = 1)$ and inversion twinning $(x = \frac{1}{2})$. If the distributions just touch, the range 0–1 spans 12σ.

absolute structure of a crystal *that is known to be enantiopure* is therefore that the refined value of x should be within $\pm 3\sigma$ of 0, and $\sigma \leq 0.08$. If the structure cannot be known to be enantiopure, more stringent criteria must be imposed and the maximum allowed value for σ is $1/24 \approx 0.04$. If the Flack parameter is determined with a standard uncertainty around the values specified but the refined value is significantly different from 0 or 1, inversion twinning is implied.

It turns out that the standard uncertainty on the Flack parameter is commonly *overestimated* when it is obtained from the least-squares refinement. The reason for this is probably that the weighting scheme applied to produce the best fit of the structure to the intensities is not necessarily optimal to extract the Flack parameter. The diffraction data are actually better able to assess the absolute structure than might be indicated from refinement of x as a least-squares parameter. Two alternative methods are available to yield more appropriate estimates of the precision after the structure has been refined. One is based on quotients, $Q(hkl) = \{I(hkl) - I(\bar{h}\bar{k}\bar{l})\}/\{I(hkl) + I(\bar{h}\bar{k}\bar{l})\}$, as described in Section 12.4. A plot of $Q(hkl)_{obs}$ *vs.* $Q(hkl)_{calc}$ has an expected slope of $(1 - 2x)$, so x can be extracted with its uncertainty by a simple linear least-squares fit to the data points. A typical example is shown in Figure 14.6. To obtain a reliable fit, it is important to collect a complete set of Friedel pairs and to filter out any obvious "outliers" which could skew the result. Outliers can be identified automatically using various criteria, but it is advisable always to check the actual Q plot.

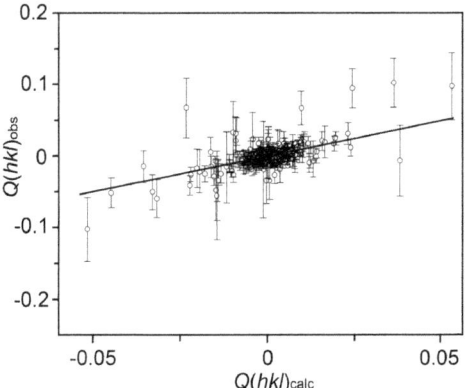

Figure 14.6 Example of a Q plot. The line shows the least-squares fit to all data points and has slope $(1 - 2x)$, where x is the Flack parameter. The error bars are derived from the uncertainties on the measured intensity values. Obvious outliers have been removed.

An alternative method uses a Bayesian statistical approach, which assigns a probability that the absolute structure is correct or wrong given the measured data and the assumption that the crystal is enantiopure. The test can be extended to consider also the probability of the crystal being a racemic twin if it cannot be known to be enantiopure. The methods construct an entire probability distribution function (*i.e.* defining the probability that x takes each value between 0 and 1) from which the most probable value of x can be established with its uncertainty. Usually, comparable results are obtained using either quotients or Bayesian methods, and both yield a value for x that is more precise than refinement of the Flack parameter in the least-squares process. The post-refinement methods thereby enable more confident determination of absolute structure for typical small-molecule APIs.

14.3.5 Missed Symmetry

Missed symmetry refers to the circumstance where the refined crystal structure could be represented more efficiently using a space group with additional symmetry operations. Probably the most common example in the literature is missed inversion symmetry, which leads to bogus indications of non-centrosymmetric structures and sometimes mistaken conclusions over absolute structure. The obvious consequence of missed symmetry is that there are too many molecules in the asymmetric unit. Why does that matter? The structure description may not be as efficient as it could be, but applying the symmetry operations of the space group to the molecules in the asymmetric unit will generate the correct unit-cell contents. In most cases, there will still be many more measured intensities than parameters and the refinement will appear to proceed entirely normally. However, problems can occur in the least-squares process due to correlation between parameters, and the results can be erratic. For example, crystals of sulfodimethoxine (Figure 14.7) adopt space group $P\bar{1}$ with one molecule in the asymmetric unit. One report of the crystal structure in the Cambridge Structural Database uses space group $P1$ with two independent molecules. The consequences of the missed inversion symmetry are revealed by significant distortions of the molecular geometry, as illustrated for the pyrimidine ring in Figure 14.7. It would be highly unusual for two chemically identical molecules to display such variation in their bond lengths, and this result is purely an artefact of the wrong choice of space group.

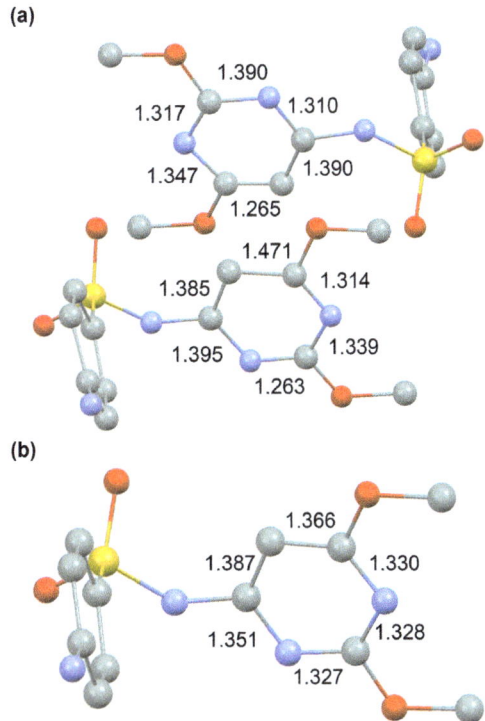

Figure 14.7 Crystal structure of sulfodimethoxine reported in (a) space group *P*1 and (b) the correct space group *P*1̄. Note the distortion of the pyrimidine rings in (a). Using data from ref. 1a.

Due to the development and widespread use of automated validation systems (Section 14.4.2), reported occurrences of missed symmetry should now be quite rare. Finding relationships between sets of co-ordinates is an ideal task for automation and available algorithms are highly effective. Identifying missed symmetry can never be an exact art because there is always a question of the tolerance applied to match atoms. For the sulfodimethoxine example in Figure 14.7, the geometry of the two molecules is quite distorted so applying an inversion operation to one molecule will not match it to the other molecule *exactly*. Some distance tolerance is usually applied so that the structure can be said to conform to a given space group within (say) 0.5 Å tolerance, but perhaps not within 0.1 Å tolerance. During an actual analysis, it should be a habit to check for possible higher symmetry and to try the refinement in any potential higher-symmetry group. Does the higher-symmetry group produce an acceptable result? In some cases, this might be hard to decide, especially where the alternatives might be an ordered description in a lower-symmetry space group or a disordered description in a higher-symmetry group. The best advice in ambiguous cases is probably just to try the various possibilities and report the conclusions for each.

14.3.6 Bonding and Molecular Geometry

Pictures of molecules in crystal structures invariably show atoms joined by lines indicating chemical bonds. Most crystallographic software will assign a *covalent radius* to each atom type, and two atoms are considered to be connected if they are separated by less than the sum of their covalent radii plus some defined tolerance (Figure 14.8). The radii are usually well optimised, but there will inevitably be ambiguities close to the borderlines. Hence, the connectivity shown within a crystallographic viewer should be treated with caution, especially for less-common atom types. For disordered structures, where atomic sites may be closer than any reasonable bonding distance, the representation within a crystallographic viewer might be a mess.

Bond lengths are frequently used to infer bond types (single, double, *etc.*). For example, a typical C–C single bond length is around 1.54 Å, a C=C double bond is around 1.35 Å, and a C≡C triple bond is around 1.20 Å. Ambiguities around the borderlines are typically responsible for incorrect placement of H atoms in automated procedures. Conclusions regarding connectivity and bond types must be affected by the uncertainties on the interatomic distances and by the resolution of the data set. The sulfathiazole example in Figure 14.4 shows how a typical C–C bond length varies for data sets at different resolution, and how the uncertainties can change by an order of

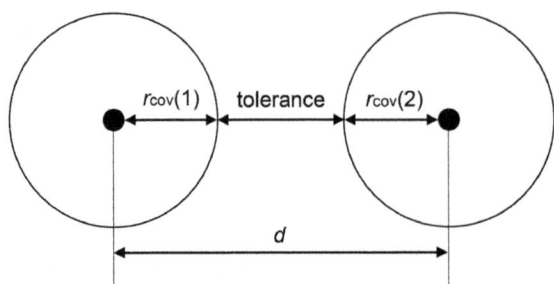

Figure 14.8   Schematic illustration of connectivity assignment for two atoms with covalent radius $r_{cov}(1)$ and $r_{cov}(2)$. The atoms are connected if $d < (r_{cov}(1) + r_{cov}(2) + \text{tolerance})$.

magnitude. Bond angles typically have uncertainties of the order of hundredths of a degree, and torsion angles typically have uncertainties of the order of tenths of a degree.

14.3.7 Same Structure or Different Structure?

On several occasions, we have seen how the same crystal structure might be reported with different unit cells or different origins that result in entirely different sets of fractional coordinates. The Case Study in Chapter 4 showed an example where the monoclinic form of paracetamol was explicitly transformed from one unit-cell setting to another. This chapter introduces a further element of doubt by showing how parameters must be compared within the limits of their uncertainties. With all of these potential reasons for crystal structures to be different, what criteria can be used to establish that they are actually the same? Numerous quantitative measures exist, such as root-mean-square deviations of corresponding atomic positions, but all of them are complicated to some degree by the need to match atoms in the structures being compared. The best advice is probably just to *look carefully* at the structures. The possibility to compare structures side-by-side or overlaid in a 3-D visualiser is extremely valuable (Figure 14.9). The structures should be expanded ("packed") to fill the

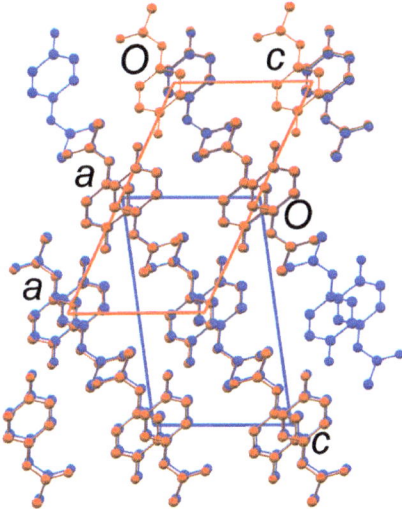

Figure 14.9 Overlay of the reported structures of the monoclinic form of para-cetamol described in the Case Study of Chapter 4 (CSD: HXACAN01 and HXACAN07). The structures are seen to be the same, but the unit cells are different. The projection is along the *b* axis, which must be identical if two monoclinic structures are to be the same.

unit cell or several unit cells in order to see the entire crystal structure. Although structures typically look clearest when projected along the unit-cell axes, it can sometimes be helpful to omit the unit cells when looking at the structures so that they do not immediately influence the comparison. If the structures can be overlaid initially without considering the unit cells, turning on a view of the unit cells can then indicate any required transformations. Putting aside the details of the space group, potential missed symmetry, uncertainties on parameters, *etc.*, the key question is whether the molecules have comparable conformations and lie in comparable positions and orientations.

An independent way to compare structures is to simulate diffraction patterns. The PXRD pattern is an accessible fingerprint of the crystal structure that is not influenced by any particular choice of unit cell or space-group setting. Of course, the question then becomes how similar two PXRD patterns must be to indicate the same crystal structure; this was touched upon in Section 10.5.1, and more sophisticated quantitative measures also exist.

14.4   Reporting Crystallographic Results

Although the primary result of a single-crystal X-ray analysis is the set of measured diffraction images, these are rarely reported directly. The interest is naturally in the refined crystal structure and in the conclusions that can be drawn from it. Generally, it must be assumed that the primary diffraction images have been interpreted appropriately, *i.e.* the practical analysis described in Chapters 8 and 9 should have been carried out correctly. The refined crystal structure is then reported and distributed. The following sections highlight some aspects of this process.

14.4.1   The Crystallographic Information File (CIF)

The highly-ordered nature of crystallographic results is well suited to electronic representation. The standard format for reporting small-molecule crystallographic results has for some years been the *Crystallographic Information File* (CIF). This is a dictionary-based format with a defined set of data names representing crystallographic parameters and descriptive fields covering aspects of the measurement and refinement. Each data name has an associated data value or collection of data values in a defined loop. For example, the unit-cell parameters are represented by the following data names and values:

```
_cell_length_a          10.7891(2)
_cell_length_b          8.4836(2)
_cell_length_c          11.3978(2)
_cell_angle_alpha       90
_cell_angle_beta        91.643(2)
_cell_angle_gamma       90
_cell_volume            1042.82(3)
```

The atom types and coordinates are presented in a loop such as:

```
loop_
  _atom_site_label
  _atom_site_type_symbol
  _atom_site_fract_x
  _atom_site_fract_y
  _atom_site_fract_z
  _atom_site_U_iso_or_equiv
S1  S  0.48192(2)  0.11488(2)  0.76289(2)  0.01086(2)
S2  S  0.34374(2)  0.34275(2)  0.56714(2)  0.02029(2)
```

The CIF is designed to be machine-readable and is produced automatically by refinement software. Hence, there should be no need to have any particular knowledge of the file format. In practice, it is frequently necessary to update values, especially descriptive fields which are intended to record special features of the measurement or refinement. This can be done within a text editor, or (more reliably) by specific software for CIF editing. As for any electronic file, it is vital to maintain the strict syntax and of course the refinement results should not be manually modified in any way. In addition to the refined parameters and derived results, CIFs now generally contain the list of measured intensities plus checksum elements that detect unexpected modification of the file.

In principle, the CIF contains the entire record of the crystallographic analysis after integration of the primary diffraction images. There should be no need to provide additional technical information. Publication of a crystal-structure determination in the chemical or pharmaceutical literature *must* be accompanied by a CIF (even if a journal does not ask for it). It can also be useful to provide summary information in an article, such as a table or footnote containing the unit-cell parameters and some key indicators of the refinement, but there is little value in printed lists of atomic coordinates or displacement parameters in journal articles or associated supporting information documents—these are leftovers from the pre-electronic era. Tables of selected bond lengths and angles may be relevant, but only if there is some reason to refer to them. The structured CIF format makes it easy to produce any such table automatically, so it is more efficient simply to publish the CIF and provide software tools to summarise the information if it is needed. If molecular diagrams are to be included in any publication, it is always helpful to show displacement ellipsoids (with a stated probability level) since they convey far more information than "ball-and-stick" models.

14.4.2   Validation Systems

It has become standard practice for small-molecule crystal structures to be validated using automated systems such as the online *checkCIF* service provided by the International Union of Crystallography.[2] The aim is to identify problems or inconsistencies in crystallographic results and to advise when action might need to be taken. Validation systems are extremely good at identifying technical mistakes such as missed symmetry, forgotten H atoms, an incorrect chemical formula, *etc.*, by comparing the various items in the CIF. They can also highlight features of a structure that appear to be unusual compared to expectations and therefore warrant checking, such as unusual bond lengths or H atoms that do not form an expected hydrogen bond. Such warnings can be a reminder to check aspects of the refinement or structure that may have gone unnoticed, for example where it may be advisable to introduce restraints or where anisotropic ADPs have become excessively distorted. Finally, the systems attempt to highlight experimental deficiencies, such as unsatisfactory resolution, completeness, data-to-parameter ratio, *etc.*

Alerts and warnings are usually categorised according to their perceived severity. For example, we have seen that measuring data to only 1.20 Å resolution will have quite significant consequences for the results, so this might be highlighted as a "category A" warning. Failing to report the colour of the crystal, on the other hand, would warrant only a gentle reminder that this might be helpful. This approach to categorization is helpful to put each alert in context, but unfortunately it provides an easy route to "threshold" judgements. For example, *"if a structure generates no category A alerts, it must be correct"*. This should be viewed rather like the discussion of *R*-factors in Chapter 12. Crystallographic results must be judged holistically against numerous criteria, and less-than-optimal results can still be useful if considered with appropriate understanding.

Validation warnings act as a reminder that results should be treated with appropriate caution, but they should never come as a surprise to a well-informed user who has carried out a careful and thoughtful analysis.

14.4.3 Crystallographic Databases

The most relevant resource for small-molecule pharmaceutical crystal structures is the *Cambridge Structural Database* (CSD),[3] which records and distributes almost all published crystallographic information. Other public databases exist, but the CSD benefits from being long-established, carefully edited and essentially complete. It also provides a comprehensive suite of software tools to extract information and derive knowledge from the core collection of crystal structures. At the time of writing, the CSD is approaching one million database entries, although this includes many crystal structures that have been determined more than once. It is probable that private databases within academic and industrial facilities contain at least as many crystal structures as the CSD, which may never reach the public domain. This is particularly relevant in the pharmaceutical industry where solid-state information may be commercially sensitive. For this reason, the content of the CSD is not necessarily representative of *all* knowledge that exists for pharmaceutical crystals.

There can be many reasons to consult a crystallographic database. Any project involving pharmaceutical solids will invariably begin by considering what crystallographic information exists in the literature. During a single-crystal analysis, it is often helpful to search a database to check whether established unit-cell parameters correspond to a known crystalline phase. If it becomes necessary to apply restraints in a refinement, these should be based on expectations established from similar crystal structures. Likewise, expectations for acceptable bond distances or angles in validation systems are based on distributions established from a large sample of known structures.

When searching a database it is instructive to recall how the various elements of the crystal structure are obtained, and the extent to which these might vary. We have seen numerous times that the same crystal structure could be reported with different unit-cell parameters or space-group settings, so a search on unit-cell parameters must use the reduced cell (Section 2.2). The cell parameters are likely to vary significantly with temperature, and crystal structures tend to be reported at various temperatures, so a suitable tolerance should also be applied. Probably the most common way to search a database is to define a target molecule. As discussed in Section 14.3.6, chemical bonding is inferred from interatomic distances. For most element types in typical pharmaceutical solids, it is usually clear whether atoms are bonded or not because the bond distances are well-defined and the difference between a bonded and non-bonded distance is substantial. For example, a pair of bonded C atoms will usually have an interatomic distance in the approximate range 1.20–1.54 Å, while a non-bonded $C \cdots C$ distance will usually be greater than 3 Å. The bond type, however, can sometimes be harder to specify, particularly where delocalised bonding or aromaticity could be possible, so specific searches for particular bond types may miss relevant structures. To handle this scenario, the CSD allows searches for "any" bond type, which should generally be used as a default option. Similarly, the inevitable variation in treatment of H atoms means that it is advisable to omit them from any molecular search query.

14.5   Summary of Key Points

- For a single-crystal X-ray analysis, parameters such as atomic coordinates are not measured directly, but there are many indications of each refined parameter within the collection of structure factor equations. These can be unravelled in the least-squares calculations to yield estimates for the experimental uncertainties associated with each parameter.
- The precision of each parameter is inversely related to its standard uncertainty, and the overall "precision of a crystal structure" refers qualitatively to the magnitudes of the uncertainties.
- A parameter should be reproducible within the specified uncertainty under the same measurement conditions, but may still be inaccurate due to systematic errors. A parameter that has been determined accurately should be reproducible within the specified uncertainty under any measurement conditions.
- The uncertainties on primary parameters such as unit-cell parameters and atomic coordinates carry over into derived parameters such as bond distances and angles. Parameters must be compared relative to their associated uncertainties.
- The precision of a crystal structure generally increases as the data resolution increases. A low-resolution structure may be suitable to reveal molecular structure and packing patterns, but may not be sufficient to enable detailed comparisons of bond distances and angles due to the associated higher uncertainties.

- Conclusions involving H atoms in X-ray crystal structures generally require careful consideration of the extent to which the information has been extracted from the data rather than imposed by the analyst.
- Absolute-structure determination involves assessment of the Flack parameter relative to its associated uncertainty. However, least-squares refinement generally overestimates the uncertainty on the Flack parameter. Alternative methods can be implemented post-refinement to produce more precise indications of the Flack parameter.
- The standard electronic format for reporting small-molecule crystal structures is the Crystallographic Information File (CIF). The CIF contains the entire record of the crystallographic analysis after integration of the primary diffraction images. CIFs can be checked by automated systems which aim to identify problems or inconsistencies in the results and to advise when action might need to be taken.
- Published crystal structures are collected and distributed in crystallographic databases. The most relevant resource for small-molecule pharmaceutical crystal structures is the Cambridge Structural Database (CSD).

14.6 Case Study: Perindoprilat Monohydrate

Perindopril is used to treat high blood pressure and heart failure. The prodrug undergoes hydrolysis after ingestion to produce the active dicarboxylic acid perindoprilat (Figure 14.10). This case study considers the crystal structure of perindoprilat monohydrate,[4] with particular focus on determination of absolute structure. The molecule contains five chiral centres (indicated with an asterisk in

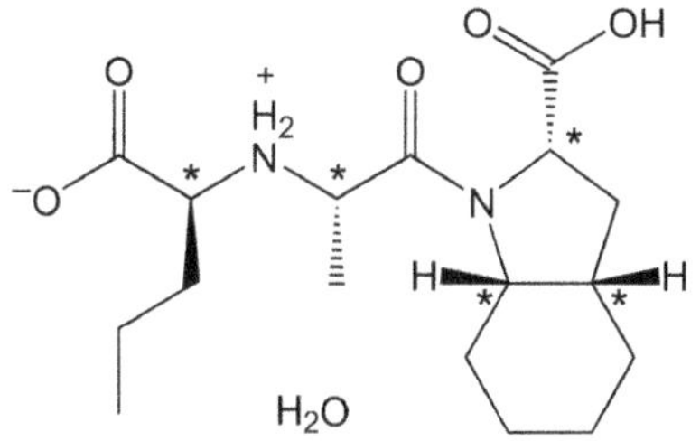

Figure 14.10 Molecular structure of perindoprilat monohydrate, $C_{17}H_{28}N_2O_5 \cdot H_2O$, which is zwitterionic in the crystal. All chiral centres (marked *) are known to have the *S* configuration.

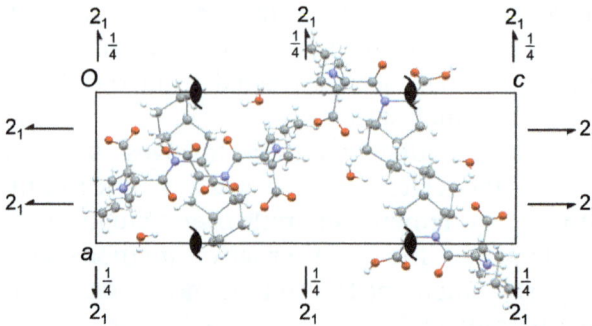

Figure 14.11   Crystal structure of perindoprilat monohydrate projected along the *b* axis, showing the symmetry elements in space group $P2_12_12_1$.

Figure 14.10), all of which have the *S* configuration. The absolute configuration is *known* for the molecule in this case, but we are interested to see how it is revealed by the X-ray analysis. The crystal structure adopts space group $P2_12_12_1$, which is orthorhombic with non-intersecting 2_1 screw axes running parallel to each of the unit-cell axes (Figure 14.11). This is one of the 65 chiral (Sohncke) space groups, which does not contain any improper symmetry operations. The following discussion is based on data published in ref. 4, which were measured at 100 K using Cu radiation. A total of 3321 unique reflections were measured, with 1396 Friedel pairs among this set.

Structure determination is straightforward and refinement with the correct absolute structure (corresponding to the *S* absolute configuration for all chiral centres in the molecule) produces $R1 = 0.027$ and $wR2 = 0.069$. Refinement of the Flack parameter (x) within the least-squares process produces a value of 0.04(18). This is obviously close to the expected value of zero, and therefore consistent with what we know about the absolute structure. If the structure is inverted so that all chiral centres in the molecule have the incorrect *R* configuration, the Flack parameter refines to 0.96(18). Hence, this data set produces a clear indication of the correct absolute structure *via* the refined value of *x*. However, the standard uncertainty of 0.18 is larger than the suggested guideline of 0.08 discussed in Section 14.3.4. At the 3σ level, 0.04(18) indicates that the true value of *x* could conceivably be as large as 0.54. Since we know that the crystal is enantiopure, the true value of *x* cannot lie within 3σ of 0.5, which a value of 0.54 obviously does. So this is an example where the uncertainty associated with the determined value of *x* is too large to make a valid conclusion. The value does indicate the absolute structure that we know to be correct, but if we didn't know the answer we couldn't use this result to prove conclusively the existence of the *S* enantiomer.

Given that we know the *S* configuration is correct, we are left with a feeling that the data contain the information to determine the absolute structure but that traditional least-squares refinement of the Flack parameter is not able to show it conclusively. The alternative quotient method described in Section 14.3.4 produces a similar value of $x = 0.06$, but crucially now with an uncertainty of 0.04. Recall that this comes from a least-squares fit of a plot of $Q(hkl)_{obs}$ *vs.* $Q(hkl)_{calc}$ (as in Figure 14.6). Bayesian statistical methods (with the assumption of enantiopurity) suggest a probability of 1.000 for the absolute structure being correct. Clearly, these two measures enable confident determination of the absolute structure for this light-atom API.

References

1. (a) P. Narula, M. Haridas and T. P. Singh, *Indian J. Phys. A*, 1987, **61**, 132; (b) R. E. Marsh, *Acta Crystallogr., Sect. B: Struct. Sci., Cryst. Eng. Mater.*, 1995, **B51**, 897, [CSD: SFDMOX02/03].
2. www.checkcif.iucr.org (accessed February 2019).
3. (a) Cambridge Structural Database, Cambridge Crystallographic Data Centre, UK; (b) C. R. Groom, I. J. Bruno, M. P. Lightfoot and S. C. Ward, *Acta Crystallogr., Sect. B: Struct. Sci., Cryst. Eng. Mater.*, 2016, **B72**, 171.
4. J. Bojarska, W. Maniukiewicz, L. Sieroń, P. Kopczacki, K. Walczyński and M. Remko, *Acta Crystallogr., Sect. C: Struct. Chem.*, 2012, **C68**, o443, [CSD: FEFKEI].

Subject Index

Page numbers in *italics* indicate the subject is only in a figure on that page. Other page references may include figures. Page numbers with a suffix T indicate a table. PXRD is used as an abbreviation for powder X-ray diffraction.

absolute structure 63
  determination of 118–21, 230–31, 251, 271–74
  perindoprilat monohydrate (case study) 283–85
  PXRD and 171
absorption correction *see* multi-scan correction
accuracy 263–64
acetaminophen *see* paracetamol
ADPs *see* atomic displacement parameters
amorphous solids 3
  co-amorphous solids 9–10
  polyamorphism 6–8
  X-ray diffraction of 83
amplitude (wave) 83–84
angle of diffraction *see* diffraction angles
angstrom units 24
angular divergence 180–83
angular (spatial) resolution 130, 137–38, 183, 187
angular width of diffracted beam 92–93, 154
anhydrates 11
anisotropic absorption 132, 159, 269
anomalous scattering 118–22, 127, 199–200, 271–72
anti-scatter slits 182–83
Argand diagrams 96, 99
  centrosymmetric structures 117
  Friedel's law expressed on *111*
  non-centrosymmetric structures 119, 120–21
  structure factor equation *100*
L-aspartic acid (twinning case study) *244*, 255–59
aspirin, form I, single crystal X-ray analysis (case study) 145–48, 165–67
asymmetric unit 53–54
atomic displacement parameters (ADPs) 221–25, 227, 229, 267–69
atomic scattering factors 97, 99–101, 119–20, 221–22, 224–25
automated validation systems 275, 280–81
axial divergence 180

basis vectors 19–20
BFDH *see* Bravais–Friedel–Donnay–Harker method
Bijvoet pairs *see* Friedel pairs
bonding 267, *268*, 276–77, 282 *see also* hydrogen bonding
Bragg equation 88–90, 93
Bragg planes 88–91, 93, 105–6, 127–28, *172*
Bragg–Brentano (reflection) geometry 176–78, 181–83, 184
Bragg's description of diffraction geometry 87–90, 104–5
Bravais lattices 58–59, 140–41, 151, 159, 163
Bravais–Friedel–Donnay–Harker (BFDH) method 77, 78–81

Cambridge Structural Database (CSD) 281–82
carbamazepine *2*
carbamazepine-saccharin co-crystals *10*
celecoxib (PXRD case study) 190–93
centred unit cells 56–61, 113–14, 114T
centring
  crystals 134–35
  powder samples 184
centroids 29
centrosymmetric crystal structures 116–18, 121–22, 150, 151–52
centrosymmetric space groups 62–63, 111, 116
chiral molecules 30, 32, 34, 46 *see also* absolute structure
chiral space groups *62*, 63
CIFs (Crystallographic Information Files) 278–80
classification of pharmaceutical solids 1–15
  summary 1–4, 15
  chemical 8–13
  inconsistencies (literature *vs.* regulatory guidelines) 14
  legal/regulatory implications 11–13
  structural 4–8

co-amorphous solids 9–10
co-crystal formers (co-formers) 9, 14
co-crystals 9–10, 12–13, 14
completeness 153–54
complex numbers 95–97
constraints and restraints 225–27, 239–40, 266, 270
crystal habit 75–78, 143–44
crystal morphology 68–81
   summary 68–69, 78
   case study (BFDH morphology of famotidine form A) 78–81
   crystal habit and 75–78, 143–44
   faces 74–78
   measuring 143–44
   planes 69–74
crystal structure 16–27, 82–106
   summary 16–17, 24–25, 82, 102–3
   case studies
      mefenamic acid 25–27
      ropivacaine hydrochloride form 2 103–6
   fractional coordinates 23–24, 30–32
   geometry of diffraction and 83–93, 101–2, 103–6
   intensities of diffraction and 93–101, 102
   lattices 17–20
   metric properties 24, 55–56
   structure comparison (same or different?) 277–78
   unit cells (*see* unit cells *main entry*)
   see also Bravais lattices
crystal systems 54–56, 110T
crystalline solids 3, 4–6
   solid form identification (PXRD "fingerprint") 186, 278
crystallographic databases 281–82
Crystallographic Information Files (CIFs) 278–80
crystallographic results 260–85
   summary 260–61, 282–83
   bonding and molecular geometry 267, *268*, 276–77, 282
   case study (perindoprilat monohydrate) 283–85
   interpreting 266–78
   missed symmetry 274–76
   reporting 182, 243–44, 262, 278–82
   structure comparison (same or different?) 277–78
   uncertainties 261–66, 270–74
crystallographically distinct (independent) molecules 54

CSD (Cambridge Structural Database) 281–82
Cu radiation 127–28, 153, 155, 178–80, *181*
cubic crystal systems 55T, 56, 110T

d-spacing 71, 72–73, 77
   of Bragg planes 88–89, 127–28
   resolution and 89, 152–53
databases, crystallographic 281–82
Debye–Scherrer (transmission) geometry 175–76, 180–81, 184
dehydrated hydrates *see* anhydrates
desolvated solvates 13
detectors
   powder X-ray diffraction (PXRD) 183
   single crystal X-ray diffraction 128–31
difference curve 188
difference density 208–9, 219–21
diffracting position 91–92
diffraction angles 89–90, 91
   angular width 92–93, 154
   monochromated radiation in PXRD 179–80
   single crystal experiments 129–30, 136, 152–53, 155–57
diffraction patterns 82–106, 107–24
   summary 82, 102–3, 107, 122
   anomalous scattering 118–22, 127
   case study (ropivacaine hydro-chloride form 2) 123–24
   centrosymmetric structures 116–18, 121–22
   Friedel's law 111, 116, 118, 120
   geometry of
      crystal structure and 83–93, 101–2, 103–6
      measuring 135–43, 158–59
   indexing 138–40, 141–43, 145–46, 186–87
   intensities of
      crystal structure and 93–101, 102
      measuring 150–60
      PXRD patterns 170–71, 173, 187–89
      R_{int} (merging or internal R-factor) 162–63
      symmetry of 109–11, 123–24
intensity statistics 116
Laue groups 110T, 111, 116, 122, 163, 207

diffraction patterns (*continued*)
    non-centrosymmetric crystal
        structures 116–17, 118–21
    point groups of 109–10, 110T, 111,
        150–51, 159–60
    predicted and observed positions
        compared 142–43
    space group determination from
        115, 163
    symmetry in 109–16
    systematic absences 111–14, 112T,
        114T, 115, 124
diffractometers
    powder X-ray diffraction (PXRD)
        175–84
    single crystal X-ray 126–31
direct methods (phase problem) 202–4,
    206, 207
disorder 236–44
    summary 236–37, 252
    case study (griseofulvin–nitroethane
        (1 : 1)) 252–55, *256*
    solvent masking 240–44
    see also twinning
displacement ellipsoids 221–25, 226, 227,
    267–69, 280
divergence slits 181–82
domain size, crystallite 172–73
dual-space methods (phase problem)
    204–6, 207
dynamic range of detectors 130–31

elastic scattering 118
electron density 94, 99–100
    determination from structure
        factors (Fourier summation)
        195–202, 208–9, 211–13
    resolution and 152–53
    solvent masking and 241–42
    structure refinement and 219–20,
        228
ellipsoids *see* displacement ellipsoids
EMA (European Medicines Agency) 14
enantiomers *see* absolute structure;
    chiral molecules
equivalent positions 30–31, 52–53,
    59–60, 226
errors
    ADPs as "dustbin" for 269
    series termination 200, 206
    systematic 159–60, 263–64
    uncertainties 261–66, 270–74
Euler's notation 96, 99
European Medicines Agency (EMA) 14

face indexing 143–44
faces, crystal 74–78
famotidine, BFDH morphology of form A
    (case study) 78–81
fast scans 156
FDA (United States Food and Drug
    Agency) 14
figure-of-merit (FOM) 146–47
"fingerprint", PXRD pattern as 186, 278
Flack parameter 230–31, 251, 271–74,
    284–85
FOM (figure-of-merit) 146–47
formoterol fumarate dihydrate 44–47, 53
Fourier maps 197, 211–12
Fourier summation (synthesis) 195–202,
    208–9, 211–13
fractional coordinates 23–24, 30–32
Friedel pairs 111, 171, 230–31
Friedel's law 111, 116, 118, 120, 199–200

general equivalent positions (GEPs)
    52–53, 59–60
geometry of diffraction
    crystal structure and 83–93, 101–2,
        103–6
    measuring 135–43, 158–59
glide operations 42–43, 53, 60–61
glide planes 42–43, 111–13, 112T, 115
goniometers 128, 133–34, 176–77
goodness-of-fit (GooF) 219
griseofulvin–nitroethane (1 : 1) (disorder
    case study) 252–55, *256*

H atoms 227–30, 233–35, 266, 269–71,
    276, 282
habit, crystal 75–78, 143–44
harvesting diffraction images 138–39
Hermann–Mauguin (H–M) notation 36,
    41, 45, 61–62
hexagonal crystal systems 55T, 56, 110T
Hirshfeld rigid bond test 269
hydrates 10–11, 13
hydrogen atoms *see* H atoms
hydrogen bonding 12, 229, 230, 270–71

identity operation 37–38
image width 154–55
improper rotation operations (rotoinver-
    sions) 35–37, 39
indexing diffraction patterns 138–40,
    141–43, 145–46, 186–87
indexing, face 143–44
instrument profile function 172–73

integration 157–59
intensities of diffraction
 crystal structure and 93–101, 102
 measuring 150–60
 PXRD patterns 170–71, 173, 187–89
 R_{int} (merging or internal R-factor)
  162–63
 symmetry of 109–11, 123–24
intensity (wave) 83–84
intensity statistics 116
intensity-weighted reciprocal lattice
 109, 115
interference of waves 84, *85*, 86–87, 94–97
inversion-distinguishing power 271–72
inversion operations 34–35, *36*
inversion twinning 250–51, 273

lattice vectors 18, 19–20, 23
lattices 17–20 *see also* Bravais lattices;
 reciprocal lattices
Laue groups 110T, 111, 116, 122,
 163, 207
least-squares fitting 215–18

matrix strategy 138
mefenamic acid 25–27
merohedral twinning 245–47, 250–51
Miller indices 69–71, 72, 74, *75*
 Bragg planes described by 89
mirror operations 30–32
missed symmetry 274–76
Mo radiation 127–28, 153
molecular crystal structure *see* crystal
 structure
monochromators 179–80, *181*
monoclinic crystal systems 55T, 56,
 58–59, 110T, 114T
 paracetamol (acetaminophen)
  (case study) 64–67
morphology *see* crystal morphology
mosaicity 92–93, 136, 143
mounting crystals 133–34
multi-component solids 2, *3*, 8–11
multi-scan correction 154, 159–60
multiplicity (redundancy) 154, 157,
 159, 171

non-centrosymmetric crystal structures
 116–17, 118–21, 150, 230–31, 250–51,
 271–74
non-centrosymmetric space groups
 62–63
non-merohedral twinning 245, 247–50

non-positive definite (n.p.d.) matrices
 268–69
normal vectors 71–72

orientation matrix 138, 139–40
orthorhombic crystal systems 55T,
 110T, 114T
overloading (topping) of detector
 131, 156

pair relationships 203–4
paracetamol (acetaminophen) *2*
 difference density *209, 220*
 hydrogen-bonded layers *6*
 monoclinic form 64–67, *181, 277*
 polymorphs 6, *7*
 simulated PXRD pattern *181*
 systematic absences in diffraction
  pattern *113*
Pawley refinement 185, 187–88
peak profiles (PXRD) 170–73, 187–89
perindoprilat monohydrate (case study)
 283–85
pharmaceutical solids *see* classification
 of pharmaceutical solids
phase (wave) 84
phase difference 98–99
phase problem 202–6, 207
planes
 summary 68–69, 78
 Bragg 88–91, 93, 105–6, 127–28, *172*
 crystal morphology 69–74
 glide 42–43, 111–13, 112T, 115
point groups 29, 109–10, 110T, 111,
 150–51, 159–60
point symmetry operations 29–39
 identity operation 37–38
 improper rotations (rotoinversions)
  35–37, 39
 inversions 34–35, *36*
 limitations in crystals 38–39
 mirrors 30–32
 rotations 32–34, 39
polyamorphism 6–8
polymorphs 3, 14
 celecoxib (PXRD case study)
  190–93
 sulfathiazole
  structure refinement case
   study 232–35
  structure solution case study
   211–13, 267, *268*, 276–77
position vectors 23
powder rings 169–70, 173

powder X-ray diffraction (PXRD) 168–93
   summary 168–69, 189–90
   angular divergence 180–83
   axial divergence 180
   case study (celecoxib) 190–93
   centring samples 184
   detectors 183
   diffractometers 175–84
   as "fingerprint" (solid form
      identification) 186, 278
   incident beam path 180–83
   instrument profile function 172–73
   interpretation and uses 185–89,
      278
   Pawley refinement 185, 187–88
   peak profiles 170–73, 187–89
   preferred orientation 173, 175, 185
   reflection (Bragg–Brentano)
      geometry 176–78, 181–83, 184
   reporting PXRD patterns 182
   Rietveld refinement 189
   samples for 173, 176, 177–78,
      184–85
   simulating PXRD patterns 173–75,
      278
   transmission (Debye–Scherrer)
      geometry 175–76, 180–81, 184
   wavelength of X-rays 93, 170,
      178–80
   X-ray source 178–80, *181*
precision 262, 263–64, 267
preferred orientation 173, 175, 185
primitive unit cells 21
profile of diffracted beam 154, 158
profile of PXRD peaks 170–73, 187–89
pseudopolymorphs 11
PXRD *see* powder X-ray diffraction

Q peaks 219–20, 233–34
Q plots 231, 273, 285
quartet relationships 204

R-factors 162–63, 218–19, 220, 227
reciprocal lattices 71–74
   Bragg planes and 90–91, 105–6
   indexing diffraction patterns
      and 138–40, 145–46
   intensity-weighted 109, 115
reduced basis 20
reduced unit cells 20, 27, 281
redundancy (multiplicity) 154, 157,
   159, 171
refining X-ray crystal structures 214–35
   summary 214–15, 231–32

absolute structure 230–31, 251
atomic displacement parameters
   (ADPs) 221–25, 227, 229, 267–69
   case study (sulfathiazole) 232–35
   H atoms 227–30, 233–35, 266,
      269–71, 276
   restraints and constraints 225–27,
      239–40, 266, 270
   theory 215–20
reflection (Bragg–Brentano) geometry
   176–78, 181–83, 184
reporting results 182, 243–44, 262, 278–82
resolution
   angular (spatial) 130, 137–38,
      183, 187
   d-spacing and 89, 127–28
   diffraction angle and 152–53, 200
   lower *vs.* higher 200, 206, 211–13,
      266–67
   time available and 157
restraints and constraints 225–27, 239–40,
   266, 270
results *see* crystallographic results;
   powder X-ray diffraction (PXRD),
   interpretation and uses
Rietveld refinement 189
right-handed axes 19–20
R_{int} (merging or internal *R*-factor) 162–63
ropivacaine hydrochloride form 2 (case
   study) 103–6, 123–24
rotation operations 32–34, 39
rotoinversions *see* improper rotation
   operations

salts 8–9, 12, 14 *see also* H atoms
Schoenflies notation 45
screw axes 40–42, 111–13, 112T, 115, 124
screw operations 40–42, 53
series termination errors 200, 206
shutterless mechanism 136, 156
signal-to-noise ratio 155
sildenafil citrate *(Viagra)* 9
simulating PXRD patterns 173–75, 278
single-component solids 2, *3*, 11
single crystal X-ray diffraction 83–103,
   125–48, 149–67
   summary 102–3, 125–26, 144–45,
      149–50, 163–64
   angular width of diffracted beam
      92–93, 154
   atomic scattering factors 99–101
   Bragg's description 87–90, 104–5
   case study (aspirin form I) 145–48,
      165–67

crystal centring 134–35
crystal choice 131–33
crystal morphology, measuring
143–44
crystal mounting 133–34
detectors 128–31
diffracting position 91–92
diffractometers 126–31
fast scans 156
geometry of diffraction 83–93,
101–2, 103–6
measuring 135–43, 158–59
image measurement time
155–57
image width 154–55
integration 157–59
intensities of pattern 93–101, 102
measuring 150–60
"measure first, analyse later"
strategy 160–61
multi-scan correction 159–60
overloading (topping) of
detector 131, 156
reciprocal lattices and 90–91,
105–6, 138–40, 145–46
reporting results 243–44, 262,
278–82
R_{int} (merging or internal *R*-factor)
162–63
structure factors 97–99, *100*, 101,
102, 117, 120–21, 122
summing X-rays 94–101
troubleshooting 140, 162
wavelength of X-rays 93,
127–28
see also crystallographic results;
refining X-ray crystal structures;
solving X-ray crystal structures
site occupancy factors (SOFs) 238–39
Sohncke groups 63
solid form identification (PXRD
"fingerprint") 186, 278
solid-form landscape 4
solids, classification *see* classification of
pharmaceutical solids
Soller slits 180, *182*
solvates 10–11
disorder in 240–44, 252–55
distinction between co-crystals
and 12–13, 14
solvatomorphs 11
solvent-accessible volume 242, *243*
solvent masking (*SQUEEZE* procedure)
240–44, 255, *256*

solving X-ray crystal structures 194–213
summary 194–95, 209–10
atomic position determination
200–201
case study (sulfathiazole) 211–13,
267, *268*, 276–77
developing the structure model
207–9
electron density determination
(Fourier summation) 195–202,
208–9, 211–13
Friedel's law and 199–200
phase problem 202–6, 207
practical considerations 206–7
space group choice 207
troubleshooting 206, 207
space groups 48–67
summary 48–49, 63–64
asymmetric unit 53–54
case study (paracetamol) 64–67
centred unit cells 56–61
centrosymmetric 62–63, 111, 116
chiral *62*, 63
combining symmetry operations
49–52
crystal systems 54–56
determination from diffraction
patterns 115, 163
general equivalent positions
(GEPs) 52–53, 59–60
non-centrosymmetric 62–63
settings 49, 167, 211
special equivalent positions (SEPs)
53, 226
structure solution and 207
symbols 61–62
space symmetry elements 40–43, 111–13,
112T, 115, 124
space symmetry operations 40–43
glides 42–43, 53, 60–61
screws 40–42, 53
translation 17–18, 40, 49
spatial (angular) resolution 130, 137–38,
183, 187
special equivalent positions (SEPs)
53, 226
SQUEEZE procedure (solvent masking)
240–44, 255, *256*
standard deviation 261–62
standard uncertainty 262, 263
statistical variance 264
stoichiometry 12–13
strain, crystallite 172–73
structure, crystal *see* crystal structure

structure factors 97–99, *100*, 101, 102, 117, 120–21, 122 *see also* refining X-ray crystal structures; solving X-ray crystal structures
structure refinement *see* refining X-ray crystal structures
structure solution *see* solving X-ray crystal structures
sulfathiazole
structure refinement case study 232–35
structure solution case study 211–13, 267, *268*, 276–77
sulfodimethoxine 274–75
symmetry
in crystals (*see* symmetry operations)
in diffraction patterns 109–16, 123–24
equivalent positions 30–31, 52–53, 59–60, 226
missed 274–76
translational 17–18, 40, 49
see also symmetry elements; symmetry operations
symmetry elements 29, 40–43, 111–13, 112T, 115, 124 *see also* symmetry operations
symmetry operations 28–47
summary 28–30, 43–44
case study (formoterol fumarate dihydrate) 44–47
combining 49–52
point symmetry operations 29–39
identity operation 37–38
improper rotations (rotoinversions) 35–37, 39
inversions 34–35, *36*
limitations in crystals 38–39
mirrors 30–32
rotations 32–34, 39
space symmetry operations 40–43
glides 42–43, 53, 60–61
screws 40–42, 53
translation 17–18, 40, 49
see also point groups; space groups; symmetry elements
systematic absences 111–14, 112T, 114T, 115, 124
systematic errors 159–60, 263–64

tetragonal crystal systems 55T, 56, 110T
thermal motion *see* atomic displacement parameters (ADPs)

thick-slice and thin-slice images 154–55
topping (overloading) of detector 131, 156
translational symmetry 17, 17–18, 40, 49
transmission (Debye–Scherrer) geometry 175–76, 180–81, 184
triclinic crystal systems 55T, 56, 110T
trigonal crystal systems 55T, 56, 110T
triplet relationships 204
twin laws 247
twinning 236–37, 244–52, 269
case study (L-aspartic acid) *244*, 255–59

U_{ij} *see* atomic displacement parameters (ADPs)
U_{iso} 221–22
uncertainties 261–66, 270–74
unit cells 20–22, 49
centred 56–61, 113–14, 114T
crystal faces and 74–75
in Crystallographic Information Files (CIFs) 279
in database searches 281–82
dimensions for crystal systems 55T
from PXRD patterns 186–87
reduced 20, 27, 281
United States Food and Drug Agency (FDA) 14

validation systems 275, 280–81
variable divergence slits 182
Viagra (sildenafil citrate) *9*

wave characteristics 83–84
wave interference 84, *85*, 86–87, 94–97
wavelengths, X-ray 93, 127–28, 170, 178–80
weighted least-squares fitting 217–18

X-ray diffraction 83 *see also* diffraction patterns; powder X-ray diffraction (PXRD); single crystal X-ray diffraction
X-ray sources 127–28, 153, 155, 178–80, *181*
X-ray wavelengths 93, 127–28, 170, 178–80
XRPD *see* powder X-ray diffraction